T0189981

Intelligent Systems Reference Library

Volume 71

Series editors

Janusz Kacprzyk, Polish Academy of Sciences, Warsaw, Poland
e-mail: kacprzyk@ibspan.waw.pl

Lakhmi C. Jain, University of Canberra, Canberra, Australia
e-mail: Lakhmi.Jain@unisa.edu.au

About this Series

The aim of this series is to publish a Reference Library, including novel advances and developments in all aspects of Intelligent Systems in an easily accessible and well structured form. The series includes reference works, handbooks, compendia, textbooks, well-structured monographs, dictionaries, and encyclopedias. It contains well integrated knowledge and current information in the field of Intelligent Systems. The series covers the theory, applications, and design methods of Intelligent Systems. Virtually all disciplines such as engineering, computer science, avionics, business, e-commerce, environment, healthcare, physics and life science are included.

More information about this series at http://www.springer.com/series/8578

Andrzej Piotr Wierzbicki

Techne$_n$: Elements of Recent History of Information Technologies with Epistemological Conclusions

Springer

Andrzej Piotr Wierzbicki
National Institute of Telecommunications
Warsaw
Poland

ISSN 1868-4394
ISBN 978-3-319-38483-2
DOI 10.1007/978-3-319-09033-7

ISSN 1868-4408 (electronic)
ISBN 978-3-319-09033-7 (eBook)

Springer Cham Heidelberg New York Dordrecht London

Printed on acid-free paper

Springer is part of Springer Science+Business Media (www.springer.com)

Acknowledgments

This is a considerably revised and enlarged translation of the Polish book $Techne_n$; the translation and enlargement were supported by a statutory research of the National Institute of Telecommunications in Poland. I would like to express my thanks to an internal reviewer (my wife, Dr. Maria Wierzbicka) and other reviewers, Professors: Jerzy Kleer, Józef Lubacz and Kornel Wydro for their numerous and incisive comments that helped to clarify and enrich the material of this book.

Contents

Chapter 1
Introduction

1.1 Message

This book expresses the conviction that *the art of constructing tools,* Greek *techne*, changes its character along with changes of civilization eras and co-determines such changes. This does not mean that tools typical of a civilization era fully determine it, but they change our way of perceiving and interpreting the world, and thus co-determine this era. There might have been many such civilization eras (the era of cut stone, of bronze tools, of iron and steel, thus much more than *three waves* of rural, industrial and informational civilization, see Toffler and Toffler 1980). Therefore, *Techne_n* in the main title of this book means art of constructing tools in the *n*-th, subsequent civilization era.

A huge civilization change occurred during the last 50 years, both in the world and in Poland. As a technician,[1] I took part in creating this change; I am convinced that around 1980 an *informational revolution* has started, related to gradual dissemination (socio-economic penetration) of personal computers and computer networks. I consider this as a phenomenon of similar importance as the invention of James Watt that around 1760 started, according to Braudel (1979), a new *long duration historical era* commonly called the era of industrial civilization. Thus, we can reckon that around 1980 a new, subsequent civilization era has started, an era which is diversely called and interpreted (post-industrial society, third wave, information society, post-capitalist society, knowledge economy or society). I will nevertheless use the concept of *informational revolution leading to knowledge civilization.*

---

[1] The word *technician* is interpreted broadly in this book—as a creator of tools and technical artefacts, and in my case also a representative of technical sciences. For many reasons I prefer this word to the denotation *technologist* (the word *technology* is used ambivalently, sometimes even with pejorative connotations, by philosophy of technology, see further chapters; in Polish, it is also often used incorrectly, because its precise meaning in Polish corresponds to a *technical process*). The word *technician* has also a more specific meaning as a technical working post, lower than the post of an *engineer*; but also in this case a technician creates tools.

© Springer International Publishing Switzerland 2015
A.P. Wierzbicki, *Techne_n: Elements of Recent History of Information Technologies with Epistemological Conclusions*, Intelligent Systems Reference Library 71,
DOI 10.1007/978-3-319-09033-7_1

Some representatives of social sciences and humanities, see, e.g., (Dusek 2006), express the opinion that any considerations concerning information society, informational revolution etc. are symptoms of *technocracy* and *technological determinism*. Such opinions are similar, however, to burying one's head in the sand, because even earlier many publications, see. e.g., (Bard and Söderquist 2002), indicated how large and important in developed countries are today social and economic changes resulting from informational revolution. Therefore, opinions negating this revolution stem from the reluctance to abandon old paradigms, in particular, paradigms of philosophy of technology. If the informational revolution actually happens, then the classical philosophy of technology, concentrated on the critique of industrial society and avoiding closer contacts with *technology proper* (or *techne*),[2] has no chances: it must address new problems and, in order to understand them, must develop closer contacts to technology proper. Somewhat different are the attitudes of cultural anthropology or contemporary sociology; they try holistically, preserving the dominant role of culture or sociology in understanding of social developments, comment on the changes of the current world, but they do it usually ex post, while questioning (with diverse justification, often weighty but not fully convincing) the role of anticipation and any analysis ex ante. Further on I will show that such attitude is not sufficient to understand the informational revolution because of the important role of diverse delays characteristic of technical developments.

There are also opinions that the world (and, especially, Poland) is yet far away from any *knowledge civilization*. Obviously, the penetration of informational revolution and knowledge civilization is not large today, both in Poland and in the world; world-wide, aspects of knowledge civilization coexist with much larger aspects of industrial and rural civilization. Even if we assume that the pace of social and civilization development doubled since the time of James Watt, we should compare current advancement of information society and knowledge civilization with the advancement of industrial civilization in the years 1820–1840, when it was already clear that industrial civilization had started, but it was yet far away from any matured form. Even today, industrial civilization is not dominant for all countries in the world, and there are opinions that this is good, because the newly developing countries can (even if it is not certain that they will) avoid errors in developing industrial civilization with its negative impacts on and deterioration of natural environment.

It is impossible, however, to stop civilization evolution, driven by the needs of people, inventions of technicians, and profits of capitalists resulting from socio-economic dissemination of these inventions to satisfy these needs (together with advertising and creating demand for acceleration of such penetration). We can,

---

[2] We shall see in further chapters that *technology proper*—that is, technology or technique as understood by technicians, people who perform it—most closely corresponds to the meaning of *techne*.

however, try to predict directions of this evolution and counteract negative aspects of such development. There is no doubt that informational revolution will bring further development of mobile smart telephony, of mobile personal computing devices using a complex structure of local and global computer networks (new generations of Internet and WWW), of automating production with dematerialization of work and widespread use of robots, etc. For example, we cannot stop the advance of robotics because recent accidents in mines in many countries indicate how important it is to replace human work in dangerous posts with robots. The big related questions, however, are what types of robots to construct, how versatile and autonomous they should be, and how not only to ensure their control and oversight by human users, but also to prevent their use for criminal goals.

A related, more essential question is: *whether the civilization development of humanity,* based on an avalanche growth resulting from a reciprocal, positive feedback between science and technology on one side and socio-economic applications on the other, *is not dangerous in itself, does not lead to self-destruction?* In the context of this question, an important position is the book of Paul Davies *The Eerie Silence: Renewing Our Search for Alien Intelligence* (Davies 2010). It considers possible interpretations of the fact that for 50 years during which we have been sending radio signals into space, evidencing the presence of intelligence on Earth and listening for possible answers, we have not yet obtained any answer, in other words, the answer until now is silence.

Obviously, diverse interpretations of this fact are possible, for example, the level of civilization advancement based on radio communication could be considered too primitive and environment-polluting by a truly developed intelligence (today even we know that truly broadband access to information cannot be based on wireless radio communication alone, but must be supported by fibre optic networks). However, the most distressing interpretation might be that a fast growth of technical civilization combined with market economy dominating on Earth today could be internally unstable, lead to avalanche growth and disaster, and that civilizations such as ours are rare and temporary ephemerids in the universe. The mere possibility of such interpretation should induce us to a deeper and more serious reflection on the threats of the future.

Questions of this type demand participation of philosophy of technology, but it is not sure whether it is well prepared to ponder them. The philosophy of technology was formed during industrial age and its main object was to criticise the industrial civilization, often erroneously equated with technology, because philosophy of technology, as I discuss more thoroughly in further chapters, does not define the meaning of the concept *technology* with a sufficient degree of precision and uses it in diverse senses, often convenient for proving a thesis currently being proved. But if we assume that industrial civilization is undergoing radical changes, we also need to radically change the philosophy of technology. In order to notice the necessary directions of such a change, it seems indispensable to analyse internal, recent history of information technology, and answer the question on how the informational revolution came about and what conceptual surprises, new opportunities and threats can it yet bring about?

This is precisely the message of this book: the conviction that because of the informational revolution it is necessary to review the recent history of information technology, the consequent conceptual changes and epistemological conclusions related to them, both from the point of view of philosophy of science as well as philosophy of technology, however formulated from a technological perspective.

1.2 Some Important Concepts for Further Investigations

I would like to stress once more that I consider myself a technician (see Annex), with enough time in retirement to read diverse texts in philosophy and to work on models of knowledge creation, but certainly not a philosopher. What differs me from philosophers is the *episteme*, understood in Foucauldian sense (1972) as the way of creating and substantiating knowledge characteristic of a given cultural era, but I extend this concept with the conclusion that difference of episteme can result from different cultural spheres (such as cultural spheres of exact sciences, technology, and social sciences with humanities). From personal experience and many scientific discussions I know that today technology has a different episteme than philosophy or social sciences and humanities, different also from exact and natural sciences.

In further investigations I shall use several concepts of particular importance that I will only mention here. They will be discussed and developed further in subsequent chapters.

The first one is the already mentioned concept of *delay*. I use it in the technical sense: in dynamic systems, the consequences of our actions accumulate and grow in delay when compared to these actions. There are diverse types of delay, and between them the so-called *pure delay* (the time span between an action and the moment the first consequences of this action occur) and *inertial delay* (related to an accretion of consequences of an action in time). The most important fact for the development of technology is that new, fundamental inventions are applied in broad socio-economic context with substantial delays, both pure and inertial ones (in the case of computers or cellular telephony the pure delay exceeded 40 years, the inertial delay lasts until today).

The next one is the concept of *episteme*, also already mentioned. I do not use episteme in the classical, Greek sense (the whole of rationally substantiated knowledge), but rather in the (post)modern, contemporary sense as suggested by Foucault (1972): the way of creating and justifying knowledge characteristic of a given cultural or civilization era (however, I broaden this concept to cover also a given, individual cultural sphere). Further chapters present substantiation of the conclusion that an old episteme might disintegrate at the end of a long duration era, for example: at the end of industrial civilization era the old episteme developed differently for different cultural spheres (e.g., the sphere of hard and natural sciences, social sciences and humanities, technical sciences) and we may need a new synthesis, a new episteme to be constructed for the needs of a new era. As mentioned above, the concept of episteme is fundamentally related to the concept of *long duration historical structure* of Braudel (1979), the permanence of basic concepts about the world during a given civilization era.

The third fundamental concept is the *fallibility and power of intuition* and its role in technical creativity. From Plato to Kant, intuition was considered to be an internal source of certainty about infallible knowledge. Further development, *inter alia* in mathematics, has undermined such conviction, but the understanding of intuition retained its super-natural and transcendental aspects. Nevertheless, from my personal experience I know that intuition is a basic source of technical creativity, and I felt the need to explain rationally and technically, and, as it turned out, also naturalistically and evolutionary, both the power and the fallibility of intuition. According to such technical and evolutionary theory of intuition (Wierzbicki 1997), presented in more detail in one of next chapters, intuitive reasoning is many thousand times more powerful than rational reasoning (just as a picture is worth many thousands of words), but it does not imply that intuitive reasoning is infallible.

The fourth concept is that of *logical pluralism* and the necessity of choosing the type of logics adequate for a given class of problems. Since the work of Łukasiewicz (1911), many types of non-classical logics and their diverse applications have been developed for entire century, but they often had to be re-discovered, and the relevance of a given type of logics for a given application area had to be substantiated, because of the very theoretical approach of mathematical logicians. All this occurred with a substantial contribution of Polish researchers, e.g., Pawlak (1991) justified anew the practical necessity of using three-valued logics for large computerized data sets and developed *rough set theory* with a broad set boundary and without excluding middle (hence, *a third way* always exists) as well with data mining applications. In technical applications, we use many types of logic, rough sets, fuzzy sets, temporal and modal logics, as well as the logic of feedback which might be interpreted as a specific temporal logic and is addressed in more detail in further chapters. However, such logical pluralism has profound consequences. If we are to select a type of logic that is adequate for a given class of problems, then how do we know that the classical binary logic is always adequate for diverse epistemological or philosophic problems? I will show further that some classic philosophic conclusions are based on inadequate classical logic applied to problems that require a different one.

The fifth concept is the concept of *hermeneutic horizon* and its relation to a *paradigm, research tradition and episteme*. Diverse meanings of this concept are know from the works of Husserl, Heidegger (1927) and Gadamer (1960), but a more precise meaning was given to it recently by Król (2006, 2007). According to Król, hermeneutic horizon is an intuitive conviction about the correctness and interpretation of basic axioms of a given field of science; it is objective in the sense that it is independent from individual, subjective convictions, but it is an element of a Braudelian historical long-duration structure. Hermeneutic horizon is the foundation of scientific paradigms. Later on, I will propose a somewhat extended understanding of hermeneutic horizon treated as an hermeneutic perspective, characteristic of a given scientific school (even in one scientific discipline, such as economics, we can have several scientific schools, such as neoliberal school[3] and post-Keynesian school).

The sixth is the concept of *intellectual heritage of humanity,* with its subdivision into rational, intuitive and emotional parts. It is related to a critique of phenomenological reduction: in our interpretation of reality we never extract from brackets what we know about this intellectual heritage, we always remain under influence at least of what Popper (1972) called *third world* or rather *world 3,* postulating its objective existence. However, Popper thought rather about the rational part of our intellectual heritage, while (according to the evolutionary theory of intuition discussed in this book) its intuitive part (to which we can count e.g. the Kantian synthetic a priori judgments), or its emotional part (myths, films etc.) may be much more powerful.

The seventh concept, that of *micro-theories* or rather *micro-models of knowledge* (and technology) *creation* and their relation to the philosophy of science and technology, is a somewhat related concept. In the last decade of the 20th century a new demand emerged, in relation to a concurrent development trend towards knowledge based economy, to propose models of knowledge creation needed right here and now, for today or tomorrow. As opposed to theories known in philosophy and epistemology, addressing the creation of knowledge in a long-term historical perspective, which we might call *macro-theories of knowledge creation,* such micro-theories or micro-models should describe how knowledge is usually created on a daily basis, or how knowledge should be created in a micro-scale. Such micro-models appeared not in philosophy, but within management theory or technical systems science, etc. One of such micro-models is the model of *brainstorming* (Osborne 1957), but in the last decade of the 20th and the first decade of the 21st century, in response to the demand mentioned above, many of such micro-models emerged; they will be described in a separate chapter further in this book. Some philosophers of science maintain that only macro-theories of knowledge creation are justified, but such belief is burying one's head in the sand, since micro-models respond to different needs and are not necessarily consistent with the paradigm of philosophy of science.

---

[3] By *neoliberalism* I understand not a variant of *liberalism* (the noble conviction that the respect for individual freedom should constitute the foundation of a contemporary society), but its extreme economic interpretation, a belief that it is sufficient to leave the freedom of market to solve all the problems. I appreciate free market as a robust tool of economic equilibration, but I recognize also the fact not noted by neoliberal economists: due to changes brought by informational revolution, the free market left to itself can be subject to corruption (each tool can be broken or corrupted, and free market is based on the desire to gain which in new conditions can lead to actions not foreseen by the classical market theory, such as it happened during the crisis of 2008–2010, see Chap. 14) and so it requires a repair or at least new regulation.

1.3 Examples of Non-paradigmatic Episteme of Technology

I would like to return to my conviction that the episteme of technology, or technical sciences, differs essentially from the episteme of exact and natural science, and the latter also differs essentially, as it will be shown later, from the episteme of social sciences and humanities. I shall illustrate this conviction with examples (anecdotes) from my own experiences.

In the end of 1960s, I co-organized a small symposium of the Polish Society of Theoretical and Applied Electrotechnics in Mielno near Kołobrzeg on the Baltic shore. A specialist in graph theory and its applications, professor Stanisław Bellert, presented there a paper *on the detection of electromagnetic waves from large distances*. It turned out that the paper presented an alternative cosmogonic theory explaining the red shift of light waves when coming from great distances. While the classical theory explains red shift by cosmic explosion, escape of distant galaxies, professor Bellert assumed a stationary character of the universe and explained the red shift by a different curvature of space than assumed in the cosmogonic theories of physics. Already at this symposium a heated dispute sparked off; the dividing line was between physicist's position with the representative opinion "you cannot make assumptions that contradict the accepted relativity theory" and the technical position "interesting, but how should we construct a critical experiment that will either confirm or invalidate the theory proposed by Bellert?" Today I would say that this was a debate between a paradigmatic position (Kuhn 1962) represented by physicists and a falsification attempt suggested by technicians (in a later, more ripe understanding of falsificationism suggested by Popper (1972), even if Popper misunderstood technology and suggested that technicians do not use falsification). On this ground I maintain that technology and technical science is more prone to falsification than exact and natural science: technician prefers to have several tools, hence also several theories, if they can be experimentally compared. To finish the story I would only add that, obviously, the paradigm has won: since the theory of Bellert concerned physics, physicists attacked Bellert very strongly and he practically abandoned any further work on his interesting theory.

From that time I was against paradigmatic approaches to theories, particularly in technical science.[4] I have several other examples but I shall quote only one, concerning the already mentioned theory of *rough sets* developed by Pawlak, published internationally in (Pawlak 1991); today, this theory is considered to be one of the greatest achievements of the Polish school of decision support and

---

[4] Later I realised that this made me a follower of the concept of *open society* (Popper 1962), even if differently interpreted, differently also from (Soros 2006): not only the openness of political system, but also social tolerance for the diversity of opinions and beliefs, historically developed e.g. in Netherlands, but still (despite the change of political system to a more open one) unavailable in Poland.

information sciences. In the second part of the decade 1980–1990, still in the times of the so-called real socialism in Poland, the officers of the then Ministry of Science and Higher Education asked me to prepare a super-review of a troublesome case. They knew that in 1985 I had returned to Poland from abroad after 7 years of absence and that I was known not only as an expert in the decisions theory, but also because of the highly independent way in which I expressed my beliefs and prepared reviews. They funded the research of professor Pawlak, but some critics attacked his concepts saying either that they are not original (after all, three-valued logics with values "yes", "no" and "maybe" was proposed in 1911 by Łukasiewicz) or that they are politically incorrect (because a political truth should be unique, not rough!).

I realised that this criticism was not only internally inconsistent, but also represented diverse paradigmatic positions, this time not exclusively connected to exact sciences. Therefore, I argued in my review that the theory of Pawlak was an attempt to give an experimental (and not axiomatic) basis to three-valued logics: Pawlak noted that in each large data set, with a logical rule given to characterise the elements of this set, there are many elements confirming the rule, also many elements negating the rule, but there might be a large number of elements neither confirming nor negating the rule (the latter form a rough, broad boundary of the set of elements confirming the rule). I argued also that multi-valued logics is a Polish specialty (since Łukasiewicz 1911) and it is high time to develop nontrivial applications of this logics. Thus, Pawlak research continued to be funded, and an English publication of his results occurred in the form of a book (Pawlak 1991)[5]; from this time until his recent death we were on friendly terms.

This two examples illustrate my conviction that technology, even in its theoretical aspects and technical science, should not be paradigmatic. In technical practice we use falsification, because technical products, tools and artefacts, must be comprehensively tested, including destructive tests (such as automobiles are subjected to crash tests) in order to determine the limits of their safe operation.

All this results, when I read books in the field of philosophy of technology, in my concern that this philosophy does not fully understand technology and keeps straying around when trying to help people lost in contemporary world of high technology. And such help from philosophy of technology is necessary, because technical products are not fully neutral[6]; they have positive and negative aspects, but without a good understanding of technology we can face a therapy based on a wrong diagnosis. A wrong diagnosis of threats related to the use of technology might probably lead to focusing on wrong conclusions concerning corrective

---

[5] Accidentally, the publication of the book by Pawlak coincided in time with the slogan "there is no third way" used by L. Balcerowicz for political purposes; this slogan was, of course, clearly incorrect according to the theory of rough sets—and also according to later practical economic developments, e.g. in China or during the financial crisis of 2008.

[6] The opinion that technical products are fully neutral (because the use of a hammer depends on the user, a human wielding the hammer) is very simplistic; even we, technicians, know that tools can fascinate people and provoke their actions.

measures. At first, my fears was of intuitive nature and only recently I succeeded in rationalising them and ordering arguments concerning the issue where we, technicians, need a help of philosophy of technology and what are the reasons for changing the traditional attitude of philosophy of technology.

1.4 Concern of a Technician When Reading Texts on Philosophy of Technology

Against this background I can explain the reasons for writing this book. As I already mentioned, the main reason is the concern, the Heideggerian *Sorge,* resulting from the perception of inadequate understanding of technology by philosophy of science and technology. Perhaps we, technicians, are responsible for that, but certainly not for the vision of technology as a dark force, in an inexplicable but noticeable way emerging from the socio-economic system of applications of technology and causing an enslavement of people participating in this system (see, e.g., Postman 1992), but rather for the lack of opposition to this vision, disregarding it as unjust and unfounded, hence not requiring any response. Admittedly, technology has both positive and negative aspects that can be diversely, not always positively exploited by excessively ambitious politicians or technology brokers (that is, entrepreneurs using technology) greedy for profit. But in order to counteract this, a better, deeper understanding of technology is needed; a wrong diagnosis of the dangers accompanying applications of technology can lead to a focus on wrong conclusions on how to act.

I was searching for such a sufficiently deep understanding of technology in many books, often excellent ones, but unfortunately, usually written almost entirely from the position of an episteme characteristic for social sciences and humanities, different from the episteme of technology. Since it seems that it is difficult for a philosopher of technology to understand technology without understanding this difference of episteme, I shall stress it and discuss in more detail in further parts of this book. This epistemic difference developed mostly during last 50 years, in relation to postmodernism on the one side and the development of information technology on the other side. To this difference is also related another, paradoxical aspect of philosophy of technology: it's frequent anti-technological attitude. This aspect, perhaps, results from a small number of approaches that are positive toward technology, but it deepens the feeling of a lack of understanding of technology by a majority of philosophers of technology.

Since the dangers related to this lack of understanding are serious, one of the goals of this book is at least a partial filling of this gap, a partial correction of the conclusions of philosophy of technology, and also philosophy of science, strongly used by philosophy of technology. These are, naturally, only epistemological reflections of a technician, but I think they should be presented. As a method, I try to compare typical theses of philosophy of technology and science with some conclusions that can be derived from recent history of information techniques. Another part of the method is a comparison of epistemic assumptions, combined with a critique of hermeneutic horizons or perspectives. This defines the subject of this book: elements of recent history of information techniques and epistemological conclusions.

Although the entire book is a specific critique of contemporary philosophy of technology, it does not focus on critique but on showing how *tools typical for a given civilization epoch, or the art of their creation, techne, co-determine new concepts and new image of the world characteristic for this epoch.* This substantiates the main title of this book, *Techne$_n$*, that expresses the conviction that we are living at the beginning of a subsequent (n-th, at least fifth) civilization epoch in the punctuated evolution of human civilization, in which new tools co-determine new concepts, change significantly not only the way of perceiving the world, but also the way of reasoning accepted as correct, see, e.g., Chap. 6. Therefore, I concentrated more on a positive description of changes contributed by technology to our understanding of the world than on the critique of classical philosophy of technology; I only devote one chapter at the end of the book to this critique.

1.5 A Review of Contents of the Book

As already mentioned, the main title of the book, *Techne$_n$*, expresses my conviction about punctuated[7] evolution of technology. This title refers also to Heidegger (1954), which I discuss in more detail in the next chapter, Preface to Part I of this book. The book consists of four different parts: the first one presents general epistemological observations, the second one discusses selected elements of recent history of information techniques, and the third one provides specific epistemological and other conclusions the fourth presents general conclusions.

Part I, after a general preface, addresses the question *what is technology,* followed by a discussion of the issue of *delays in technology development,* turns to a technical but evolutionary *theory of intuition,* then discusses the issue of *truth and objectivity* and indicates the emergent, new episteme on the basis of fundamental naturalism.

Part II starts with an outline of the history of telecommunications, based on a selection of its important elements, but stressing their social and conceptual importance. The history of automatic control and analog computers is treated similarly, then the history of digital computers and transistor circuits, the history of systems theory and technology, and finally, the history of information revolution.

---

[7] Since the works of Lorentz (1965) it is known that biological evolution does not have a continuous, incremental character, but is *punctuated:* it is marked by periods of fast, in a sense revolutionary changes, after which slow evolution periods return. I believe that the same applies to the evolution of technology and civilization.

Part III presents more detailed issues: micro-models of knowledge creation, the relation of the history and the philosophy of technology, and some general conclusions: the issue of how to predict development of technology, and also challenges and threats brought by the new era. Part IV presents final conclusions.

Because of the diversity of problems and approaches in these three parts, some arguments, especially those I believe to be fundamental, I simply repeat sometimes and I ask the Reader for understanding and indulgence.

References

Bard, A., Söderqvist, J.: Netocracy—The New Power Elite and Life After Capitalism. Reuters/ Pearsall, UK. Polish translation (2006) Netokracja. Nowa elita władzy i życie po kapitalizmie. Wydawnictwa Akademickie i Profesjonalne, Warsaw (2002)

Braudel, F.: Civilisation Matérielle, Economie et Capitalisme, XV-XVIII siècle. Armand Colin, Paris (1979)

Davies, P.: The Eerie Silence: Renewing Our Search for Alien Intelligence. Harcourt, Houghton Mifflin (2010)

Dusek, V.: Philosophy of Technology—An Introduction. Blackwell Publishing, Oxford (2006)

Foucault, M.: The Order of Things: An Archeology of Human Sciences. Routledge, New York (1972)

Gadamer, H.-G.: Warheit Und Methode. Grundzüge einer philosophishen Hermeneutik. J.B.C Mohr (Siebeck), Tübingen (1960)

Heidegger, M.: Sein Und Zeit. Niemayer, Halle (1927)

Heidegger, M.: Die Technik und die Kehre. In: Heidegger, M. (eds.) Vorträge und Aufsätze, Günther Neske Verlag, Pfullingen (1954)

Król, Z.: Matematyczny Platonizm I Hermeneutyka (Mathematical Platonism and Hermeneutics). Wydawnictwo IFiS PAN, Warsaw (2006)

Król, Z.: (2007) The Emergence of New Concepts in Science. In: Wierzbicki, AP., Nakamori, Y. (eds.) Creative Environments, op.cit

Kuhn, T.S.: The Structure of Scientific Revolutions, 2nd edn. Chicago University Press, Chicago (1962). (2nd ed., 1970)

Lorentz, K.: Evolution and Modification of Behavior: A Critical Examination of the Concepts of the "Learned" and the "Innate" Elements of Behavior. The University of Chicago Press, Chicago (1965)

Łukasiewicz, J.: O wartościach logicznych (On Logical Values). Ruch Filozoficzny I, 50–59 (1911)

Osborn, A.F.: Applied Imagination. Scribner, New York (1957)

Pawlak, Z.: Rough Sets—Theoretical Aspects of Reasoning About Data. Kluwer, Dordrecht (1991)

Popper, K.R.: Objective Knowledge. Oxford University Press, Oxford (1972)

Popper, K.R.: Open Society and Its Enemies. Tłumaczenie polskie (1987) Społeczeństwo otwarte i jego wrogowie. Wyd. Krytyka, Warsaw (1962)

Postman, N.: Technopoly. The surrender of culture to technology, Knopf, New York Polish translation (1995) Technopol. Triumf techniki nad kulturą. Państwowy Instytut Wydawniczy, Warsaw (1992)

Soros, G.: The Age of Fallibility: Consequences of the War on Terror. PublicAffairs, New York. Polish translation (2006) Nowy, okropny świat: era omylności, Świat Książki, Warsaw (2006)

Toffler, A., Toffler, H.: The Third Wave. William Morrow, New York (1980)

Wierzbicki, A.P.: On the role of intuition in decision making and some ways of multicriteria aid of intuition. Multiple Criteria Decis Making 6, 65–78 (1997)

Part I
Epistemological Observations

Chapter 2
Preface: New Epoch, It's Conceptual Platform and Episteme

2.1 Preface

The main goal of this book is to indicate how much our understanding of contemporary world has changed together with the development of information techniques (see also Lubacz 2008) and related fields, such as mathematical computational techniques. Therefore, I start with general epistemological observations related to this change, and only later I document this change in more detail through the analysis of selected elements of recent history of information technology. Through the latter I understand the history, with the epoch of industrial civilization included, from around 1760, although clearly light signalling was known already in ancient times. As I could not present a comprehensive history (it would require much more space and time), only "elements" are presented, and obviously treated selectively. The method of selection concentrates not on technical or instrumental importance of various inventions that contribute to this history, but on their social or even conceptual importance.

In the above sense this book differs essentially from a number of similar works which are certainly worth reading but pursue different goals (see, e.g., Okin 2005). I understand the concept of information technology or techniques differently; most texts limit this concept to computers and digital techniques or their applications, while personally I believe it is important to stress the role of several other techniques of information processing, especially analog processing, including also the specific analog-digital processing of information in neurons of our brains. Between several contemporary works on the history of information technology, the excellent, deeply technical monograph *From Gutenberg to Internet* (Norman 2005) is so dominated by the horizon or hermeneutic perspective of digital techniques that while Vannevar Bush, the actual inventor of the first electronic (although analog) computer, is mentioned in various contexts in several places, it is not stressed that it was he who constructed the first and rather broadly used type of electronic computer. Sociological approach to this history, e.g. (Mattelart 2001), (Bard and Söderqvist

© Springer International Publishing Switzerland 2015
A.P. Wierzbicki, *Technen: Elements of Recent History of Information Technologies with Epistemological Conclusions*, Intelligent Systems Reference Library 71,
DOI 10.1007/978-3-319-09033-7_2

2002) is also influenced by the perspective of "digital world", even if it underlines the social importance of information techniques, and sometimes even their impact on the way of perceiving the world, in a stronger and better way. Purely historical approach to the history of technology can be excellent, see, e.g., (Kopczyński 2003), when it comes to the details of development of diverse techniques, but it focuses on other techniques than informational and does not notice the impact of technology on the way the world is perceived.[1] There are also many fundamental works highlighting the impact of changes in contemporary technology on diverse aspects of civilization development, such as the emergence of media society (McLuhan 1964), end of industrial civilization (Bell 1973), wave-like character of civilization development (Toffler and Toffler 1980), the importance of knowledge in information society (Stehr 1994), or finally, the networked character of contemporary society (Castells 2000). The closest to my goals is the collective work (Lucertini et al. 2004), stressing the impact of such changes on the set of concepts that determine our perception of the world and close to the ideas presented by me originally in Wierzbicki (1988).

In these diverse works we can notice a general agreement to the diagnosis that *we live in a period of information revolution*, that this revolution will essentially change not only technology but also socio-economic system, culture and civilization in general, and lead to a new civilization epoch. There are notable exceptions from this general agreement, mostly between humanists working on history or philosophy of technology. E.g., (Dusek 2006, p. 49) maintains that "the theory of post-industrial society was advanced by a number of technocratic thinkers"; in his fight against "technological ", Dusek uses the term "technocratic" as a strongly negative epithet. Similarly, (Kopczyński 2003) does not notice at all the informational revolution and concentrates on the thesis that industrial revolution was the greatest achievement of humanity; Kopczyński also uses the term "technocratic" as an epithet. But the very concept of *technocracy* is used imprecisely by the philosophy and history of technology, because it emerged from transferring Henry Ford's ideas of manufacturing organization to other fields, so it has more to do with management science and *technology brokers* (i.e. entrepreneurs that do not create technology but use it for profit) than with technical creativity. On the other hand, *the evidence of great social and economic changes resulting from applications of computers and network techniques is obvious today*, see, e.g., (Castells 2000), (Bard and Söderqvist 2002). Therefore, the opinions of authors denying these changes can be interpreted as either a *cognitive gap* or a defense of an old paradigm; to repeat: *if the thesis about a beginning of a new epoch is true, then philosophy of technology must take up quite new problems and ask technicians about their opinions.*

---

[1] For example, the rotary speed controller of James Watt is commented in Kopczyński (2003) with the sentence "The possibility of control of rotary speed provided an ingenious centrifugal controller" without noting its historical conceptual importance and a large number of continuators of this concept that will be discussed in Chap. 8.

2.2 New Epoch with Its Conceptual Platform and Episteme

As already noted, there are many names used to describe the new civilization epoch coming after the informational revolution: *media society, service society, post-industrial, post-Fordist, post-capitalist, information* or *informational society, Third Wave, networked society, knowledge based economy* (each economy was based on knowledge, but it is a matter of degree: after the informational revolution, knowledge becomes the essential productive factor). Personally, I prefer the term *knowledge society* (Stehr 1994) or even *knowledge civilization* (Wierzbicki and Nakamori 2006). This follows from the conviction that one of the main megatrends of the informational revolution, the megatrend (discussed in detail later) of *dematerialization of work*, the usage of informational techniques (computers, robots etc.) to replace people in activities of mechanical, scarcely creative character, leads inevitably to an increased role of knowledge not only in economy, but in social life as a whole. Knowledge is understood here as more than just information, even if information is an essential constituent of knowledge, e.g., one of many possible definitions of knowledge is *information organized for a defined purpose*. Knowledge is also more than just rationally substantiated and organized *explicit knowledge*: the concept of knowledge encompasses also *tacit knowledge* of *preverbal* character, including *tacit knowing* but also *intuition, emotions* and *instincts*.

The critics of the concepts of *knowledge civilization* or *knowledge based economy* maintain that we are yet far away in the global scale from the situation when knowledge will be a decisive production factor. This critique is, however, a-historical: even 60 years after the inventions of J. Watt, who in the years 1760–1781 improved the steam engine (which was known many decades earlier, but was too dangerous to be broadly used before Watt made it more secure), we did not have an industrial society in Poland and only in the years 1820–1830, great efforts of Stanisław Staszic and other adherents to industrial civilization resulted in its beginnings in Poland. Nevertheless, industrial civilization slowly developed world-wide. The great historian Fernand Braudel, who accepted the approximate date 1760 of Watt's inventions as the beginning of the epoch of industrial civilization, knew very well that in the year 1790, 30 years after these inventions, even in England not much was changed.

Following the example of Fernand Braudel, *I accept the date of 1980 as the beginning of informational revolution and thus of the epoch of knowledge civilization:* even if computers were known for circa 50 years earlier (see Chaps. 8 and 9), their broad social application started around 1980 when the competition of Apple and IBM resulted in market supply of personal computers. Just after 1980 the techniques of computer networks were sufficiently developed and declassified for their broader social use (before that date, they were used in military and in some specific applications, such as airplane ticket distribution). Thus, similarly as Braudel, I am aware that today, approximately 30 years after 1980, we cannot expect that the knowledge based economy will be strongly developed. Yet the speed of socio-economic change today is greater than it was 200 years ago and thus informational techniques are much more broadly widespread, their socio-economic penetration is much greater today than that of industrial techniques in 1790; if we assume that the speed of change doubled or even tripled since that time, a reasonable comparison should concern the spread of industrial civilization in 1820 or even 1850.

I am also aware that historical caesuras are stipulated and some philosophers of technology would criticise even Fernand Braudel for his "technocratic determinism", expressed by the fact that he accepted the date of the invention of Johann Gutenberg, circa 1440, as the beginning of the pre-industrial era of banking, geographical discoveries and formation of capitalism, and the date of inventions of James Watt, circa 1760, as the end of that epoch and the beginning of the industrial era. However, some caesuras are necessary and they can be based on the conviction that *it is the nature of broadly socially used tools, characteristic of a given civilization epoch, which co-determines (even if not fully determines) many aspects of social life in that epoch.*[2] This does not mean full technological determinism, since technicians are also people, members of a given society, and they develop tools according to their perception of the needs of that society in a given epoch.

The life in Europe in the years 1000–1440, in the epoch of late middle ages, the civilization of monasteries and gothic cathedrals, was also co-determined by tools typical of this epoch; and the life before the year 1000 was also different, but I know not enough about the tools used then and I think that they were not much different from the tools used in ancient Greece and Rome. If we count in this way, we can distinguish at least four subsequent civilization epochs with their specific *techne* (different variants of the art of constructing tools) that co-determines them. Therefore, I use the term *techne$_n$* to describe the art of constructing tools specific for informational revolution and knowledge civilization. The number n is stipulated, and it actually means *subsequent* or *many*, since to account for former civilization epochs we should speak about *techne$_5$* or even *techne$_9$* (depending on the way in which we treat eras of hunting and gathering with their stone tools, eras of bronze and iron tolls, tools related to the invention of a wheel, etc.) Thus, I believe that the thesis of Tofflers about *the third wave* is oversimplified.

Irrespectively of the numbering of civilization epochs, we are perhaps correct in the judgment that the inventions of Gutenberg and Watt enabled (not caused,

---

[2] The concept of *historical materialism* of Karl Marx, according to which socio-economic changes are determined by the development of *productive forces*, must be modified after the informational revolution: equally or more important are *tools* commonly used by people in a given civilization epoch, and *techne*, the art of constructing such tools. For example, it is computers together with robots and automatic washing machines that enable nowadays the realization of the goal of full equality of women (socio-cultural changes are also needed to reach this goal, but it would be not attainable without these tools). Therefore, we should rather interpret it as a *techno-cultural co-evolution*. See also Sadowski (2009).

because there were many other causes) the development of socio-economic processes that resulted in the civilization of print and industrial civilization.

Significantly, both these inventions were not entirely new, they were considerable improvements of older inventions, crucial for a broad social application and penetration. Print was known earlier in China, but did not lead to an era of print for many reasons, technically because Chinese printing matrix was very difficult and laborious to engrave. The invention of Johann Gutenberg was to use separate type letters to set up a printing matrix: this enabled broad social penetration of typed books, at first naturally expensive and accessible only for rich families, but later gradually less and less expensive. Similarly, steam engine was known before James Watt (since around 1698; Thomas Newcomen improved it around 1710, 50 years before Watt inventions) but was inefficient in terms of energy use and dangerous, and it had a tendency to explode because of instability of its rotary speed. James Watt constructed an efficient and safe steam engine by adding some important elements: additional steam chamber improving efficiency, etc., but the most essential element was a centrifugal rotary speed controller, in a sense a prototype of the concept of feedback, discussed in detail in further chapters. Similarly, computers were known before 1980, but before personal computers they were reserved for a small group of "computer priests", and computer networks were reserved for military use.

As already mentioned, instead of more precise accounting of historical periods, some authors (Toffler and Toffler 1980) discuss three civilization waves: *agricultural, industrial and information civilization*. A book by Tofflers, entitled *The Third Wave*, had a very important impact, also in preparing the democratization of Poland[3]; however, their approach is too simplified, because agricultural civilization had many phases. Therefore, I prefer to discuss *three recent periods of long-duration or civilization epochs*:

- *pre-industrial or print civilization (formation of capitalism)* 1440–1760;
- *industrial civilization* 1760–1980;
- *knowledge civilization* 1980–2100(?).

The date 2100(?) is not only a simple prediction based on the shortening time spans of these periods (320, 220, at least 120?) but also can be substantiated differently, see (Kameoka and Wierzbicki 2005) and Chap. 8. We observe, as will

---

[3] In fact, Tofflers predicted the fall of the so-called communist system: in *The Third Wave* they state that robotization and automation will lead to the destruction of the class of manufacturing proletariat and that information society will develop, but it can develop only in democratic and market economy states (thus, Nasim Taleb in his book *The Black Swan*, 2007, is incorrect when saying that nobody predicted this fall). Ronald Reagan knew these opinions and acted by promoting high tech space weapons, thus putting a pressure on the communist system. On the other hand, from personal experience (I have promoted the idea of information society in Poland since around 1985 and obtained even the T. Hoffmokl Award for these efforts) I know that the leaders of Polish government at that time, Wojciech Jaruzelski and Mieczysław Rakowski, read the book of Tofflers, which might have helped to convince them about the need of democratization of Poland and negotiation with Solidarity movement.

be discussed later in detail, a shortening of civilization delays or delay times, defined as the time elapsed between an essentially new idea and its broad socio-economic application. If such delays would result in a wave-like civilization development (which is a fully justified conclusion in the theory of feedback systems), then it can be shown that the period of such a wave or cycle corresponds to approximately four times the delay. If we estimate civilization delay times today as 30–40 years, then the corresponding period of a civilization wave, the time of knowledge civilization epoch, can be estimated as 120–160 years.

Perhaps more interesting are answers not to the question how long the new civilization epoch will last, but to the question what changes it will bring. These changes will be discussed in detail in the last chapters of this book; but generally they are related to three main megatrends[4] of the informational revolution, discussed in Wierzbicki (2000, 2008). Below I recall them shortly, together with short discussion:

    I. Technical megatrend of *digital integration*;
    II. Socio-economic megatrend of *dematerialization of work and change of professions;*
    III. Intellectual megatrend of *change of the way of perceiving the world.*

The technical megatrend of *digital integration* is sometimes also called *convergence megatrend*. All signals, results of measurements, data etc. could be transformed and transmitted in an uniform digital standard form, but this requires time and adaptation. From a technical perspective, *digital integration could be much more advanced,* if its speed were not limited by economic, social and political aspects.

Telecommunication networks, including computer networks, are being integrated, but this process is slow because *a full integration of standards would allow new and small enterprises to offer various services* on this profitable and fast growing market. If standards are not integrated, it is easy to defend monopolistic or oligopolistic positions on this market, e.g. by formulating sufficiently complex requirements concerning connection of networks, so-called *interconnect* conditions.[5]

Diverse aspects of the intelligence of networks, computers, decision support, even generally of our ambient habitat, are subject of integration. The miniaturization of microprocessors and diverse sensors enables the development of *ambient intelligence*, sometimes called also *Internet of things,* in the form of an intelligent office, room, shop, car, highway etc.

Diverse media of communication, newspapers, books, radio, television, are subject to integration. A slow transition from paper to electronic form of recording

---

[4] I understand the concept of *megatrends* slightly differently than their original definition by John Naisbit (1982) who required that megatrends should be new directions, while for me megatrends must be important social phenomena of long duration, lasting at least several decades.

[5] I once checked personally how voluminous are interconnect conditions of NTT (the former telecommunications monopolist in Japan): about two thousand pages.

takes place, although a change of human customs requires long time. Economic and political consequences of media integration are well understood and we already observe fierce *struggles to control the integrated media.* A related issue is the struggle to control knowledge by its privatization; but academic environments respond to that: universities and research institutes postulate today an open publication of all research results financed from public funds, using Internet portals and the principles of *Open Access.*

The socio-economic megatrend of dematerialization of work and change of professions is even stronger than the megatrend of digital integration. The ideal of technology resulting in less heavy and monotone work drove the technical advancement of the epoch of industrial civilization, at least it was so perceived by technicians, but not necessarily entrepreneurs or philosophers of technology. The industrial civilization epoch ended when this ideal started to materialize, when robots and computers started to replace heavy work of people. In fact, we can concur with the diagnosis of (Toffler and Toffler 1980) that robotization and automation lead to the destruction of the classical proletariat class and that the informational revolution will cause (has now already caused) also the fall of the so-called communist system. However, related fast changes of technology have caused also fast changes in professions and a resulting phenomenon of the so-called *structural unemployment.* This term is actually erroneous, as it stems from a static way of thinking: it indicates that the structure of economy has changed and the resulting unemployment will be observed as long as the structure of labour will adapt to the structure of economy. But after the informational revolution, we actually observe a continuous change of the structure of economy and the speed of this change is limited precisely by the speed of adaptation of the labour force. *Already today we can build fully automated and robotized factories, but what to do with people working in existing ones?* The Economist (issue April 21–27, 2012) promoted the construction of such factories under the name of *third industrial revolution,* but completely disregarded its socio-political aspects. A sudden third industrial revolution would mean sudden 50 % of unemployment, which is politically unacceptable. If old professions disappear, we must *devise new professions* that will replace the old ones. We must also *provide relatively permanent working places* for young people starting their professional life. The neoliberal slogan (popular, e.g., in Poland) that we must not interfere with labour market because it will result in difficulties for entrepreneurs and will drive them away, is not acceptable. Intervention on the labour market should take the form of a *new industrial policy* that would promote the formation of new, relatively permanent working places, naturally adapted to the conditions after the informational revolution. It will be discussed in more detail in Chap. 14.

Dematerialization of work has certain undoubted advantages. It *enables actual satisfaction of the women's rights slogan.* Women liberation movements remained utopian in the era of industrial civilization. It was only the computer and the robot (considering contemporary washing machine as a kind of robot) that actually enabled equal rights of women, even if for the full realization of those rights a corresponding change of social customs and mores is also necessary. Paradoxically, *feminist activists* often do not notice these elementary conditions.

Dematerialization of work results in further great dangers beside unemployment and precariat formation. Not all people are sufficiently elastic and capable of changing profession several times in life. This results in generational divide, between young people, more easily learning how to use new tools, and older people. Generational divide is a part of digital exclusion or digital divide[6]—between those who gain by using digital techniques and those who cannot use them, whatever the reasons. The digital divide is a phenomenon of long social duration; exposed to market forces, it would finally disappear, but not earlier than after many decades. An obvious method of counteracting it is a reform and intensification of education, but such a reform must address the needs of the new epoch and include life-long education.

A fundamental reform of education is also necessary because of the third *megatrend of intellectual challenges including the change of the way of perceiving the world and a conceptual revolution*, the most difficult one to cope with. This megatrend brings the greatest challenges and we shall concentrate on it, starting with the concept of *episteme*.

Episteme, in its postmodern meaning as the way of creating and justifying knowledge characteristic of a given epoch, develops, according to Foucault (1972), in the beginning period of that epoch. Michel Foucaultdescribes the formation of *modern episteme* (characteristic of the period of industrial civilization) at the end of eighteenth and the beginning of nineteenth century, while the beginning of industrial civilization is typically ascribed to the year 1760. However, even before Watt many new concepts emerged, starting with Cartesians, Newton, French encyclopaedists, that formed a new *conceptual platform,* see Wierzbicki (1988). Thus, a reciprocal and preceding concept in relation to episteme is the concept of a conceptual platform, a set of new concepts that emerge towards the end of a civilization epoch and prepare the formation of a new episteme. I shall discuss in next chapters in more detail the concepts of conceptual platform and episteme, their prognostic use (Foucault used the concept of episteme only in a historical context), the process of destruction of an old episteme, etc.

Here I anticipate the results of future discussions by stating that in the second half of twentieth century, the process of destruction of old episteme resulted in a *divergent development of differing epistemai of three cultural spheres:*

- *Strict and natural ("hard") sciences;*
- *Social sciences and humanities ("soft sciences");*
- *Technical sciences.*

---

[6] The European Union gives a priority to counteracting the phenomenon of digital divide and teaching the use of digital techniques. In Poland, however, the dangers of digital divide are not fully perceived, see also final chapters of this book.

Thus, we should speak not about *two cultures* (Snow 1960), but about *three separate epistemai* characteristic of different cultural spheres (Wierzbicki 2005). These cultural spheres adhere to different values today, use different concepts and languages, different paradigms or hermeneutical horizons on which these paradigms are based; *such differences gradually increased together with the development of post-structuralism and postmodernism in social sciences and humanities, while hard and technical sciences found quite different ways to modify their research traditions.*

Technical sciences cooperate closely with strict and natural ("hard") sciences, but *these two cultural spheres differ essentially in their episteme: hard sciences are more paradigmatic,* see, e.g., (Kuhn 1962), *while technical sciences are more pragmatic than paradigmatic,* see (Laudan 1984). Some social philosophers of technology, such as (Latour 1987), speak about *technoscience,* but this is a mistake: it is true that technical and hard sciences closely cooperate, but they differ essentially in their values and episteme. Such misunderstanding of the epistemic character of contemporary technical sciences is characteristic of a large part of philosophy of technology; it will be discussed in more detail in final chapters of this book. *Both hard and technical sciences have understood for a long time already* (say, from interpreting the results about the uncertainty of measurements by Werner Heisenberg, 1927, or the epistemological theses of Van Orman Quine, 1964) *that knowledge can be only approximate and is a "fabric constructed by people that touches the reality only along its edges"* (as formulated by Quine), *but interpret this fact differently between themselves, and even more differently from humanities and social sciences.*

Even if representatives of hard sciences know that all knowledge is constructed by people and there are no judgments objective and true in the absolute sense, they nevertheless believe that scientific theories are *laws of nature discovered by people, not only models of knowledge constructed by them.* Truth and objectivity are higher, ideal values for them; metaphorically, *a representative of hard sciences resembles a priest.*

A representative of technical sciences is much more relativistic and pragmatic in its episteme, agrees without resistance that *scientific theories are only models of knowledge,* but requires these theories to be *as objective as possible,* tested in practical applications, are *falsifiable* (in the sense proposed by Karl Popper 1934, even if he did not perceive technical practice as falsification). Metaphorically, *a technician resembles an artist* (see also Heidegger 1954; Wierzbicki 2005) and similarly as an artist gives much attention to tradition.

Since social sciences and humanities are most diversified in their episteme, I shall comment here on the cognitive perspective of postmodernism, typical of only a part of soft sciences.[7] A postmodernist representative of social sciences and

---

[7] From my personal contacts I know that not all representatives of humanities accept postmodernism, thus I speak here only about a currently dominating approach in soft sciences, particularly in sociology of science.

humanities believes that all knowledge is subjective, constructed by social discourse, negotiated, relativistic, local. Such an episteme contains internal traps and
contradictions, see, e.g., (Kozakiewicz 1992); however, this internal crisis must be
overcome by social sciences and humanities themselves. Metaphorically, *a representative of postmodern social sciences and humanities resembles a journalist:*
all is admissible if it is interesting. A postmodernist also believes that if all (local)
worlds are chaotic, changeable and virtual (created by our imagination of them),
then he can use diverse words arbitrarily, without even checking whether the terms
used by him have quite different meanings in other cultural spheres or not.[8]

These differences of episteme can be illustrated by diverse examples of controversies between those three cultural spheres, but in this book I shall give only a
few of them, starting with the phenomenon of *science wars,* while further chapters
discuss also the differences in understanding the concept of *feedback* and the
conflict between *soft and hard systems analysis.*

Science wars occurred in the last decade of last century, when one of American
physicists, frustrated by the opinions of sociologists of science about the alleged
full subjectivity of science, wrote a pseudo-scientific sociological paper full of
complex terms. The paper was published; afterwards, the author confessed that he
wanted to prove full subjectivity, but of social, not hard science. This lead to huge
discussions and controversies. In the opinion of a postmodern philosopher of
technology, Val Dusek, about this phenomenon (Dusek 2006, p. 21): *"There are
scientists and technologists who believe that objectivity of their field is wrongly
denied by social, political and literary studies of science".* The postmodern attitude of the book of Dusek suggests that such opinions are represented by few
scientists and technologists, but a true humanist should know better. I believe, on
the other hand, that practically all representatives of hard and technical sciences
share such views, but not all are sufficiently frustrated to express them and take
part in *science wars.* The phenomenon of *science wars is a clear example of
controversies between different epistemai of these different cultural spheres.*

On this background, it is useful to stress the difference between *multidisciplinary* and an *interdisciplinary* approach, terms that are popular but often used
imprecisely. Multidisciplinary approach is one encompassing the knowledge of
several disciplines, interdisciplinary approach is a holistic approach which should,
nevertheless, take into account the differences in episteme of the three cultural
spheres discussed above, including results of social sciences and humanities, but
also strict and natural sciences as well as technical sciences, and the concept of
technology proper discussed in more detail in the next chapter.

---

[8] Particularly in fast and inadequate translation. See, e.g., (Agamben 2007), where the author
used actually the concept of *dispositif,* but postmodernist translators into Polish used the technical
word *urządzenie* (*assembly*). However, there are many more general examples, such as the
sociologist use of the concept of *network* (a set of nodes and connections between them for a
technician, a loosely defined set of people for a sociologist).

2.3 Main Elements of a New Conceptual Platform

Disintegration of the old episteme of the epoch of industrial civilization, called sometimes, not quite accurately, *modernism, scientism*, etc., was motivated by the emergence of a new *conceptual platform*, a set of new concepts inconsistent with the old episteme, contributed by the development of science or technology. We shall briefly list here commonly known concepts resulting from the development of science in twentieth century that contributed to the new conceptual platform:

- *Relativity theory and relativism (Einstein 1905),*
- *Logical pluralism (multivalued logics, their diversity, Łukasiewicz 1911),*
- *Indeterminism of measurements (Heisenberg 1927),*
- *Dependence of truth from meta-assumptions (Gödel 1931; Tarski 1933).*

Some of them were widely discussed and commented, some are known only to specialists. The relativity theory by Albert Einstein had a tremendous impact. Actually, it started from an attempt to modify classical theories assuming that the velocities of movements shall be simply summed up, in the face of the empirically established fact of a finite speed of light, independent from a coordinate system. However, it has eventually shown the relativity of perception of time and speed (lower than the speed of light) and further consequences, as the equivalence of mass and energy. It has had diverse interpretations, both in physics, especially quantum physics that until now cannot reconcile relativity theory with Heisenberg uncertainty principle, and in philosophy, or in common media interpretations that often went too far. However, there is no doubt that the relativity theory of Albert Einstein became a foundation of the critique and disintegration of the old episteme (modernism or scientism).

Logical pluralism started (and continued) with Polish achievements. The theory of multi-valued logics by Jan Łukasiewicz (1911) anteceded its epoch even more that the relativity theory, but was noticed at most by a few specialists in abstract logics. Its importance can be understood only today, when many variants of multi-valued logics were developed and many technical applications thereof occurred, together with the negation of the principle of excluded middle, with the broad application of the feedback principle where the effect reflexively influences the cause (while temporal logics necessary for describing the feedback are still not fully developed), etc. Lofti Zadeh (1965) invented anew multi-valued logics and their applications (see also Kacprzyk 2001), calling them *fuzzy set theory;* Zdzisław Pawlak (1991) has shown how a three-valued logic arises necessarily from the nature of big data sets, calling this *rough set theory.* Today, technicians know well that the classical binary logic is a great simplification in describing the world and that it is not adequate even for language analysis; thus, *it is necessary to select logic that is adequate for a given application.* Incomprehension of this fact even in philosophy has led to great mistakes that I discuss later while criticizing the logical errors of philosophical scepticism and analyzing the concept of feedback.

The indeterminism of measurements was discovered by Werner Heisenberg and concerns quantum level, thus its importance was noticed, beside physicists, more by technicians (particularly in electronic engineering), less by philosophers of science. However if, on the quantum level, the very fact of measurement distorts the measured variable, then this phenomenon influences and changes the concept of measurement: there are no absolutely accurate measurements, every measurement is vitiated by an error. Naturally, we can construct diverse models, statistical and others, of measurement errors, but in this way the technical science anticipated later opinions of philosophy and sociology of science that a measurement depends on a theory used to prepare it. It is obvious for a technician that theories substantiating measurements can lead to diverse errors, including gross (systematic) errors, if selected in an inadequate way. This does not mean, however, at least from a cognitive perspective of technicians (and also that of experimental hard science), that they should resign from striving for measurements as accurate as it is possible or needed for a given application. We should stress once more that until today the construction of a theory combining the indeterminism of Heisenberg and the relativity of Einstein is an open question, by some physicist considered to be the most important problem of contemporary science.

The importance of the observation that truth depends on meta-assumptions, the results of Kurt Gödel. and Alfred Tarski, was noted by philosophy of science (and also by information techniques), and especially by philosophy of mathematics. However, it was interpreted in a specific way because it encountered the critique of the paradox of infinite regress and an incapacity to resolve this alleged paradox (because of using an inadequate logic), as it is discussed in Chap. 6. Only a more precise definition of the concept of hermeneutic horizon proposed by Zbigniew Król (2005) allows, in my opinion, a correct resolution of this paradox.

Beside hard sciences, technical sciences and technology, and especially information techniques had also a great impact on the emergence of new concepts. The new conceptual platform was influenced by the following developments in informational techniques (beside mechanization, electrification, assembly line production organization, and other aspects characteristic of industrial civilization):

- The beginnings of telecommunications, the concept of a *network*;
- The beginnings of radiocommunication, *transmitter* and *receiver*;
- The beginnings of television and resulting *spectacle society;*
- The beginnings of automatic control and the concept of *feedback*;
- First *analog computers*, their development and impact;
- The concept of a *flip-flop switch*, its applications;
- *Digital computers*, their beginnings and impact on episteme;
- Transistors and integrated circuits, law of Moore, *digitalization* of technology;
- Spontaneous emergence of *software* from a *hardware* approach to computers;
- Nonlinear dynamics, *pseudo-random number generator, deterministic chaos, order emerging out of chaos, emergence;*
- *Computational complexity* versus *systemic complexity* (*holism, synergy and emergence*);

- *Computer networks, hypertext*;
- *Personal computers*;
- *Mobile telephony*;
- *Robotics and automated factories*;
- *Internet and web services*;
- *Human-centred computing:* the role of *emotions* and *intuition*, of *tacit knowledge* as opposed to *artificial intelligence*.

All these developments had a ground-breaking impact on the set of concepts typical for contemporary world, on the perception of this world. As already mentioned, computer networks and personal computers date the caesura of the beginning of knowledge civilization epoch (even if this is only a beginning, such as the invention of Watt was only a beginning of the industrial civilization). However, in the conceptual sense to the most significant I include:

- *Feedback as a concept with a specific temporal logic resolving ostensible paradoxes (vicious circle, infinite regress etc.)*
- *Software versus hardware as an example of spontaneous emergence*;
- *Deterministic chaos and emergence of order out of chaos as a mathematical and technical justification of emergence principle*;
- *Computational complexity as a cognitive limitation;*
- *Technical justification of the power and fallibility of intuition as a basis of multimedia principle.*

For example, I believe that a good understanding of the logics and dynamics of feedback is equally important for the understanding of temporary world as apprehension of relativism or indeterminism. I mentioned above the emergence principle and multimedia principle that will be formulated and discussed in next chapters, similarly as all concepts mentioned above. Here it is necessary to stress that a chaotic and emergent understanding of the world seems to be the foundation of a new episteme of the knowledge civilization era:

> *The epoch of industrial civilization perceived the world as a great clock, turning with the regularity and inevitability of celestial spheres; today, we see the world as a plurality of chaotic systems, in which everything can happen, and new forms of order are likely to emerge.*

This is not a postmodern, but rather post-postmodern or *informed* view of the world. Even if postmodernism considers the concept of chaos as its own concept, co-defining the postmodern view of the world, yet I personally participated, as it is presented in more detail in further chapters, in the emergence and the rationalization of the contemporary concept of chaos, and it occurred earlier than the beginnings of postmodernism. In further chapters I shall repeat and enlarge the above observations, starting with general epistemological observations, continuing with a recall of fundamental occurrences in contemporary history of informational techniques, and ending with further and more detailed epistemological conclusions and some warnings concerning future developments.

References

Agamben, G.: Qu'est ce qu'un dispositif? Rivages Poche (Polish translation (2010): Czym jest urządzenie. In: Agamben, Przewodnik krytyki politycznej, (eds.) M. Ratajczak and K. Szadkowski, Wydawnictwo Krytyki Politycznej, Warsaw) (2007)

Bard, A., Söderqvist, J.: Netocracy—The New Power Elite and Life After Capitalism, Reuters/ Pearsall UK (Polish translation (2006) Netokracja. Nowa elita władzy i życie po kapitalizmie. Wydawnictwa Akademickie i Profesjonalne, Warsaw) (2002)

Bell, D.: Coming of Post-Industrial Society: A Venture in Social Forecasting. Basic Books, New York (1973)

Castells, M.: End of Millenium: The Information Age, vols. 1, 2, 3. Blackwell, Oxford (2000)

Dusek, V.: Philosophy of Technology—an Introduction. Blackwell, Oxford (2006)

Einstein, A.: On the Electrodynamics of Moving Bodies. Ann. Phys. **17**, 891–921 (1905)

Foucault, M.: The Order of Things: an Archeology of Human Sciences. Routledge, New York (1972)

Gödel, K.: Über formal unentscheidbare Sätze der Principia Mathematica und verwandter Systeme. Monatshefte für Mathematik und Physik **38**, 173–198 (1931)

Heidegger, M.: Die Technik und die Kehre. In: Heidegger M. (ed.) Vorträge und Aufsätze, Günther Neske Verlag, Pfullingen (1954)

Heisenberg, W.: Über den anschaulichen Inhalt der quantentheoretischen Kinematik und Mechanik. Zeitschrift für Physik **43**, 172–198 (1927)

Kacprzyk, J.: Wieloetapowe sterowanie rozmyte (Multistage Fuzzy Control) WNT, Warsaw (2001)

Kameoka, A., Wierzbicki, A.P.: A Vision of New Era of Knowledge Civilization. In: Ith World Congress of IFSR, Kobe, Nov 2005

Kopczyński, M.: Ludzie i technika (People and Technology). Oficyna Wydawnicza "Mówią Wieki", Warsaw (2003)

Kozakiewicz, H.: Epistemologia tradycyjna a problemy współczesności. Punkt widzenia socjologa (Traditional Epistemology and Contemporary Problems. Perspective of a Sociologist). In: Niżnik J. (ed.) Pogranicza epistemologii (On Boundaries of Epistemology). Wydawnictwo IFiS PAN, Warsaw (1992)

Król, Z.: Platon I Podstawy Matematyki Współczesnej (Plato and Foundations of Contemporary Mathematics). Wydawnictwo Rolewski, Nowa Wieś (2005)

Kuhn, T.S.: The Structure of Scientific Revolutions. Chicago University Press, Chicago (1962) (2nd edn, 1970)

Latour, B.: Science in Action. Open University Press, Milton Keynes (1987)

Laudan, R. (ed.): The Nature of Technological Knowledge: Are Models of Scientific Change Relevant? Reidel, Dordrecht (1984)

Lubacz, J.: Czy potrzebna jest dalsza reforma szkolnictwa wyższego? (Do We Need a Further Reform of Tertiary Education?) Przegląd Telekomunikacyjny i Wiadomości Telekomunikacyjne, 2008-2 (2008)

Lucertini, M., Gasca, A.M., Nicolo, F.: Technological Concepts and Mathematical Models in the Evolution of Modern Engineering Systems. Birkhauser, Basel (2004)

Łukasiewicz, J.: O wartościach logicznych (On Logical Values). Ruch Filozoficzny I:50–59 (1911)

Mattelart, A.: Histoire de la société de l'information. Editions La Découverte, Paris (2001)

McLuhan, M.: Understanding Media. Ark Paperbacks, London (1964)

Naisbitt, J.: Megatrends: Ten New Directions Transforming Our Lives. Warner Books, New York (1982)

Norman, J.M.: In: From Gutenberg to the Internet: A Sourcebook on the History of Information Technology. historyofscience.com, Novato, CA (2005)

Okin, J.R.: The Information Revolution: the not-for-Dummies Guide to the History, Technology, and Use of the World Wide Web. Ironbound, Winter Harbor (2005)

Pawlak, Z.: Rough Sets—Theoretical Aspects of Reasoning About Data. Kluwer, Dordrecht (1991)

Popper, K.R.: Logik der Forschung. Julius Springer, Vienna (1934)

Quine, W.V.: (originally published in 1953) Two dogmas of empiricism. In: Benacerraf P., Putnam, H. (eds.) Philosophy of Mathematics, Prentice-Hall, Englewood Cliffs (1964)

Sadowski, Z.: Trapiące problemy przyszłego rozwoju świata (Grieving Problems of Future World Development). In: Kleer J., Mączyńska, E., Wierzbicki A.P. (eds.) Co ekonomiści myślą o przyszłości (What Economists Think About Future). Committee of Future Studies "Poland 2000 Plus" at the Presidium of P.Ac.Sc. and Polish Economic Society, Warsaw (2009)

Snow, C.P.: The Two Cultures. Cambridge University Press, Cambridge (1960)

Stehr, N.: The Texture of Knowledge Societies. In: Stehr, N. (ed.) Knowledge Societies, pp. 222–260. Sage, London (1994)

Tarski, A.: The Concept of Truth in Languages of Deductive Sciences. In: Tarski, A. (ed.) Logic, Semantics, Metamathematics. Hackett Publishing Company, Indianapolis, 1956 (1933)

Toffler, A., Toffler, H.: The Third Wave. William Morrow, New York (1980)

Wierzbicki, A.P.: Education for a new cultural era of informed reason. In: Richardson, J.G. (ed.) Windows of Creativity and Inventions. Lomond, Mt. Airy (1988)

Wierzbicki, A.P.: Finansowanie nauki w krajach rozwiniętych na progu gospodarki opartej na wiedzy a sytuacja nauki w Polsce (Financing of Science in Developed Countries on the Verge of Knowledge Based Economy and the Situation of Science in Poland). Przyszłość: Świat, Europa, Polska. 2: 105–119 (2008)

Wierzbicki, A.P.: Megatrends of Information Society and the Emergence of Knowledge Science. In: Proceedings of the International Conference on Virtual Environments for Advanced Modeling, JAIST, Tatsunokuchi (2000)

Wierzbicki, A.P.: Technology and change: the role of technology in knowledge civilization. In: Ith World Congress of IFSR, Kobe (2005)

Wierzbicki, A.P., Nakamori, Y.: Creative Space: Models of Creative Processes for the Knowledge Civilization Age. Springer, Berlin-Heidelberg (2006)

Zadeh, L.: Fuzzy Sets. Inf. Control **8**, 338–353 (1965)

Chapter 3
What is "Technology"?

3.1 Introduction: Basic Definitions

A technician, reading works on philosophy of technology, usually has the impression that the authors do not have any idea what technology is and write about an imaginary entity. There are many examples, starting with an evidently anti-technical book (Postman 1992) about *technopoly*, where the author does not explain in which sense he uses the word "technology" and actually writes about a socio-economic system of applications of contemporary technology, without clearly making this distinction. Another example might be a recent, excellent book of Darin Barney, *Network Society* (2002), correct in the conclusion that the thesis about a domination of network society might be premature. However, in these parts of *Network Society* where the author writes about his understanding of the essence of technology and the opinions of philosophy of technology about this essence, an absolute lack of understanding of the object of discussion is evident: whether he writes about technology as such, or about technological artefacts, or about a socio-economic system of production and utilization of technology products, or about fascination of people with the possibilities of technology. To express all these possible meanings, and there are more, a key-word "technology" is used.

Perhaps this is natural, or perhaps it is only an effect of the dissimilarity of hermeneutic horizons or perspectives of technology and that of social sciences and humanities, but the problem of providing a good definition of technology is fundamental. Clearly, it can be something else for a technician, something else for a sociologist, it can be different for ancient Greeks and in the time of industrial civilization, it can change with the turn of the informational revolution, but these distinctions should be made and clarified before anybody starts any farther-reaching discussion of the role of technology. An example might be a recent and excellent book of Arthur (2009) that takes up the question what

© Springer International Publishing Switzerland 2015
A.P. Wierzbicki, *Technen: Elements of Recent History of Information Technologies with Epistemological Conclusions*, Intelligent Systems Reference Library 71,
DOI 10.1007/978-3-319-09033-7_3

technology is,[1] but seeks an answer to it from the hermeneutic perspective of an economist. Also this chapter is devoted mainly to this question, but addresses it from an interdisciplinary perspective, perhaps with a domination of a technical perspective.

The word "technology" has many meanings. Some of them are as follows:

- For a philosopher of technology: *a socio-economic system of creation and utilization of products of technology* (often much more imprecise meanings);
- For a postmodern humanist or sociologist: *an autonomous force enslaving people, together with a "technocratic" way of perceiving the world;*
- For an economist: *a way of proceeding when producing artefacts, a techno-logical process;*
- In colloquial language (particularly in English): *a product of technology, an artefact;*
- For a representative of hard (strict or natural) sciences: *an application of sci-entific theories;*
- For a technician: *techne, meaning the art of creation of tools and artefacts, a fundamental faculty of people, motivated by the joy of creation:*

  - *Making people free from hard work;*
  - *Supporting brokers of technology* (capitalists, bankers, entrepreneurs) in making money, and if any effect of this *enslaves* people, then the socio-economic system or the brokers are responsible;
  - *Stimulating the development of science* by *inventions* and *instruments* that provide it with new ideas, concepts and theories together with new types of measurements.

If there are such diverse interpretations of this word and we do not agree on its usage, then the philosophy of technology can use it arbitrarily, choosing the implied meaning conveniently for the thesis currently being proved. It is signifi-cant, e.g., that Dusek (2006) in his book *Philosophy of Technology, an Intro-duction* admittedly discusses diverse meanings of the word "technology", but omits the meaning *techne,* the only one acceptable for a technician. Technicians or technologists believe, on the other hand, that they have fundamental rights to define the meaning of the word "technology".

I know that such a belief may cause objections; therefore, I propose to express the technical meaning of the word "technology" by *"technology proper"* or *"techne"*, which is consistent with the understanding of the word "techne" by ancient Greeks, and nearly consistent with the interpretation of Heidegger (1954). Nearly, because Heidegger, also without asking technicians for an opinion, maintained that technology had lost this meaning of *techne.* However, he sub-stantiated this opinion by the mass character of the socio-economic system in

---

[1] And gives an answer that is partly correct (and fully correct in Polish language, where the word "technology" should be clearly distinguished from the word "technique"): technology is a recipe for producing artefacts.

which contemporary technology is used, thus he adapted the meaning of the word *technology* to the thesis that he wanted to substantiate. This manoeuvre is repeated by many other philosophers of technology.

If asked, a technician would express an opinion that the art of creating tools certainly changes with a civilization epoch and tools typical for that epoch, but the mass character of usage of tools influences this art in a feedback process only to a small degree. We can speak, therefore, about $techne_1$ in some ancient epoch, $techne_2$ in the epoch of late middle ages, when gothic cathedrals were constructed, $techne_3$ in the epoch after Gutenberg, when we constructed clocks and telescopes, $techne_4$ in the epoch after Watt, with an enormous richness of constructed tools, $techne_5$, or rather $techne_n$, if we take into account the possible number of ancient civilization epochs using different types of tools, after informational revolution, when the most important tools are of software type. Clearly, it is difficult to define when a given epoch ends and a new one starts, because old types of tools are used along with the new types. Therefore, in the definition of caesurae we should rather follow historians; but the best historians assume that a new civilization epoch starts with a ground-breaking invention of a new type of tools, or rather with a beginning of socio-economic penetration of these tools. However, the essence of *techne* (with full understanding of the fact that an essence of a concept is never fully attainable cognitively, but we can empathize with it hermeneutically) does not change: it is a creative, intuitive art, naturally supported by science, but e.g. painting is also supported by science. Therefore, the hermeneutic perspective of a technician is similar to that of a *creator,* not an *observer.*[2]

For further discussions I recall here the definition of *technology proper* that was given in Wierzbicki (2005):

> Technology proper is a fundamental faculty of humans that makes it possible to create tools and artefacts. It serves human intervention in natural environment, but can be also used to limit such intervention to a necessary extent. In its essence, it is a creative activity revealing truth; in this sense it is similar to art. It consists, mostly, in solving practical problems. It relies on the results of hard and other sciences, if they are available; if not, technology proper finds new, own solutions, often creating new concepts and problems in that way, which are later assimilated by hard sciences or even social sciences and humanities.
>
> Technology proper is not an autonomous force, since it is a faculty of humans. It is, however, sovereign in the sense in which art is sovereign. Autonomous forces and processes emerge in the socio-economic system of technology usage.

The reasons which cause that the system of applications of technology might appear to be, or even really become, autonomous, uncontrollable, are at least two:

- Delays in technology development resulting from the fact that the system of applications of technology is complex, has many stages, and many years can pass from the moment of creation of a technical idea to its broad

---

[2] For this reason, technicians perceive the description "technocratic" as an offensive and imprecise epithet: in historical evidence, technocrats were brokers of technology, not proper technologists.

socio-economic penetration (which is discussed in detail in the next chapter and commented later with many historical examples). For an outside observer, such delays give raise to the conviction that the system is uncontrollable (in technical terms; philosophy of technology uses rather the term autonomous).

- Social fascination with certain technology products that can mask some deep dangers: disadvantageous side effects can appear for the first time when this fascination is very advanced, while technology brokers or political leaders can use it to pursue their excessive ambitions.

Technology proper, techne, or more generally, technology (together with its products, processes of producing and socially using them) we should also distinguish from *technical sciences.* They are similar to hard sciences, but differ in being less paradigmatic, more pragmatic; their episteme belongs to the cultural sphere of technology. They are distinct from technology proper, although they support it: *the art of creating tools cannot be taught by purely theoretical education; technicians learn techne in praxis.*

This chapter addresses also the question of the role of technology, and in particular of diverse informational techniques, in the civilization epoch after the informational revolution, the epoch that I call knowledge civilization, as well as the issue of dangers hidden both in technology proper and in the socio-economic system of technology usage. Such dangers are grave, but do not correspond to the dangers on which contemporary philosophy of technology concentrates. *This is the fundamental danger: incomprehension of what technology really is leads to false diagnoses and to focusing on apparent problems.*

3.2 Three Different Epistemai

Fifty years ago it was taken for granted (see e.g., Ritchie Calder 1962) that human civilizations developed due to the creation and utilization of tools, hence technology is an inherent faculty of humans, and many ancient civilizations collapsed because their leaders (pharaohs, kings, head priests) exploited technical abilities of their people for too ambitious purposes (transforming natural environment to a grand extent or building pyramids), so even if technology serves people to obtain control over nature, nature severely punishes civilizations that use their technical abilities without restraint.

However, during the last 50 years some representatives of social sciences and humanities (starting, e.g., with Marcuse 1964) accused technology (without differentiating its definition) of enslavement of people and devastation of natural environment. But at the same time, due to the approaching informational revolution and related *dematerialization of work* though automation, computerization and robotization, technology actually liberated people from the majority of hard occupations and works. This, on the one hand, enabled realization of the equity between genders, but on the other hand resulted in rising unemployment and in the emergence of a new class of *precariat,* discussed in detail in Chap. 14. Thus, technical development has led to a new civilization epoch, made it possible to solve many old problems and raised many hopes, but also created lots of new problems and dangers; unfortunately, these new dangers are not those predicted, e. g., by Marcuse (1964) or even by Scharff and Dusek (2003).

In such situation, a new reflection on the role of technology in the epoch after the informational revolution is necessary. Convinced about the fundamental character of contemporary civilization changes, I started to read books on *philosophy of technology* and the state of this field appeared to me to be deeply disturbing. Disturbing, because various authors of this field not only did not elaborate a common and convincing definition of technology, but often use definitions that are not acceptable for a representative of technical sciences. In particular, they do not discriminate between technology proper and its socioeconomic applications. Moreover, they do not seem to be even interested in what can be said on this topic by representatives of technical sciences. A more detailed critique of contemporary philosophy of technology will be presented in Chap. 13, here I stress only a fundamental difference in the ways of perceiving the world.

Therefore, it is worth first to consider why during the last 50 years, three evidently distinct epistemai emerged: different ways of perceiving the world by *social sciences and humanities,* by *strict and natural sciences,* and by *technology and technical sciences.* These three fields together with their distinct views of the world I called *cultural spheres.* As it will be shown in more detail in next chapters, these cultural spheres use different language and concepts (e.g., the concept of *linearity* is different for mathematics and technical sciences, different for sociology), often use different logics (e.g., when interpreting the impact of an effect on its cause as either a *vicious circle* or a natural example of *feedback*), different ways of constructing and justifying knowledge. How these spheres perceive each other, what are main differences in their epistemai? Only after discussing these fundamentals we can turn to the future role of technology after the informational revolution.

In the preceding chapter I already mentioned three recent civilization epochs that are characterized by gradually increasing globalization. These epochs are:

- *preindustrial civilization 1440–1760:* print, banks, geographical discoveries;
- *industrial civilization 1760–1980:* steam, electricity, mobile transportation;
- *knowledge and information civilization 1980–2100(?):* robotization, computer networks, mobile telecommunication, knowledge management and engineering.

Each of these epochs used a specific *conceptual platform,*[3] a set of new ideas and concepts that emerged earlier to be later transformed into a new episteme characteristic for a given epoch. Michel Foucault is correct in stressing that an

---

[3] I used earlier, in Wierzbicki (1988, 2000), the term *cultural platform* to describe a set of new concepts and ideas preceding a change of civilization epoch, but now I use a better fitting term, conceptual platform, instead.

episteme specific for a given epoch develops only after the beginning of this epoch. He dates the emergence of a preindustrial episteme at least a 100 years after Gutenberg and of the industrial episteme around 50 years after Watt, however, he does not analyze initial elements of episteme. Before Gutenberg and preindustrial episteme we had already the ideas of the Renaissance, before Watt and industrial episteme we had works of Isaac Newton and French encyclopaedists. The episteme of the epoch of knowledge civilization is not formed yet,[4] but the destruction of the episteme of industrial epoch and the construction of a new conceptual platform for the epoch of knowledge civilization started with *relativism* initiated by Albert Einstein, *indeterminism* initiated by Werner Heisenberg, with the development of the concept of *feedback* leading to the concepts of *deterministic chaos, order emerging out of chaos,* and finally to *the emergence principle.* I mentioned these concepts in the former chapter and will discuss them in detail together with the contribution of technology to their formation in next chapters.

The principle of emergence is not yet broadly understood. The episteme of industrial epoch was based on *the principle of reduction* of properties of a complex system to the behaviour of its component elements; this principle, however, is correct only if the level of complexity of the system is low. Today, for systems with high level of complexity, technical and information sciences have been forced, by increasing complexity of technical systems, to apply *the emergence principle, postulating emergence of new properties of a system resulting from increasing complexity, properties qualitatively different from the properties of component elements of this system, irreducible to these component properties.*

We should stress that the emergence principle expresses the essence of the concept of complexity and means much more than simple synergy principle,[5] which says that the whole is greater or more valuable than the sum of component parts, but does not imply irreducibility of the new properties of the system. The synergy principle, by the way, is obvious for technicians: because they are motivated by the joy of creation, they always consider a whole product as more valuable than the sum of its parts. Moreover, it were technical sciences that contributed to the development of the emergence principle, rationally and theoretically by their contributions to the theory of chaos and order emerging out of chaos, pragmatically by applying this principle as the only way of mastering the growing complexity of technical systems. The concept of emergence was,

---

[4] It could be said that people living in former epochs did not know their episteme consciously and Michel Foucault used this concept only in a historical sense, as a human mind construct describing past situations. On the other hand, since the publication of Foucault this type of interpretation of the term *episteme* has been a part of intellectual heritage of humanity, so we have right—or even duty—to use it also to forecast future developments.

[5] *Synergy* implies *complementarity:* two parts are complementary if they fit each other and create together more than a simple sum; in more general, descriptive terms, synergy is related to the concept of *holism*; but even this concept in systems theory did not imply emergence, although it should.

admittedly, older, used e.g. in philosophy, but until now it lacked theoretical or pragmatic justification.[6]

I discuss here the new understanding of emergence principle in order to show that all these new concepts and changes in the way the world is perceived or in creation of knowledge contribute to the *destruction of the episteme of the epoch of industrial civilization,* diversely and sometimes imprecisely called also *positivism, scientism, modernism.* There are also other processes that contribute to this destruction.

In the time of industrial civilization, technology, for the first time in human history, partly freed people form hard work, but also unveiled the possibilities of total destruction of life on Earth. Fast and inexpensive travel, mass media and mobile communication, robotization and automation, the beginnings of cosmic travel and landing on the Moon had a reverse side: the spectre of nuclear anni- hilation, catastrophes of nuclear power plants and other irreversible pollution of natural environment. And even worse, entire societies or social systems became fascinated and choked on their ostensibly unlimited power over the nature con- ferred by the technology of industrial civilization. This led to diverse threats to natural environment, especially in the communist system where official ideology boasted of transforming the nature, and praxis brought the biggest destruction of natural environment in the global scale, such as in the *black triangle* between Czech Republic, Poland and Germany. However, the same state of affairs con- tinues in the capitalist system, based on the trust in the functioning of markets together with the conclusion that the market should determine the applications of technology. After a deeper analysis such a conclusion is, nevertheless, extremely doubtful, particularly when it comes to issues related to the protection of natural environment, but also to the ethics of social technology applications in general (see Morawski 2011). This has led and continues to lead to many ideological and intellectual controversies and we should not wonder that *the ideological and intellectual crisis at the end of industrial civilization was especially deep.*

The basic conflict of the epoch of industrial civilization was about the own- ership of basic productive resources of that epoch, industrial resources. Together with the end of this epoch this conflict expires: the dematerialisation of work divests proletariat of importance, which in turn divests the communist system of legitimization. However, many intellectuals, even if disappointed towards com- munism, were deeply engaged in that conflict; the change of situation sharpened the intellectual crisis.

In epistemology, the beginning of the end of industrial epoch episteme was related to the end of neopositivism. Positivism was the peak effect of the episteme of industrial epoch and anti-positivism started as early as at the end of the

---

[6] The rational and theoretical as well as pragmatic justification of the emergence principle was presented for the first time in Wierzbicki and Nakamori (2006, 2007), although a parallel formulation of the emergence principle (without stressing the irreducibility of emerging properties and thus equating emergence with synergy) was given independently in Skyttner (2001), Cempel (2006).

Nineteenth century, but the episteme was still defended by neopositivism, or logical empiricism, as it is called in English tradition. Its internal critique was started by Kurt Gödel or Roman Ingarden; however, the final sign of its end was a paper by Quine (1964) proving that logical empiricism was internally logically incoherent, that *"All human knowledge ...is a man-made fabric that impinges on existence only along the edges"*. Despite this, Quine persuaded that an incompatibility with reality on the edges of knowledge should result in its modification if such a human-constructed knowledge should be useful in human evolution, and that implied that knowledge is objective in a limited sense. Almost at the same time, Snow (1960) formulated a thesis about *two cultures* of strict sciences and social sciences with humanities; he did not account for technology or technical sciences, and neither he could note that his observations concern actually not *cultures,* but distinct *epistemai* of these two cultural spheres. The concept of historically relative *episteme* was introduced later (Foucault 1972), although Foucault analyzed only historical changes of *episteme* and did not notice that his work contributes in fact to the destruction of the current *episteme,* its split into *epistemai* of three distinct cultural spheres.

Actually, a large part of social sciences and humanities went much further than Quine, maintaining that all human knowledge is only a result of *discourse,* is fully relativistic and local. This general belief has many variants. One of them is radical biological constructivism, represented e.g. by Maturana (1980), and earlier by von Foerster (1973): this variant postulates that the concept of truth should be replaced by evolutionary effectiveness. Another one is radical relativism in sociology, including the strong program of the Edinburgh school, e.g., (Barnes 1974; Bloor 1976), developed concurrently by postmodernism, e.g. (Foucault 1972; Derrida 1974; Lyotard 1984).

Part of humanities opposed to postmodernism stressed the social and humanistic importance of the concept of truth, e.g., in the writings of Gadamer (1960). However in this current already Marcuse (1964) originated the opinion (repeated later by many followers, until Postman 1992; Jackson 2000) that technology is an autonomous, dehumanizing force enslaving people and forming a *one-dimensional man.* Such opinion was very popular in social sciences and humanities in the end of the twentieth century, together with treating a technical (incorrectly called *technocratic*[7]) way of thinking as a *functionalist worldview,* see e.g., Habermas (1987, pp. 72–73) or Jackson (2000, pp. 107–210).

In all these discussions, however, *reductionist arguments were used:* the meaning of more complex concepts, such as truth and objectivity, was undermined

---

[7] As mentioned above, *technocratic* is an imprecise and abusive epithet, since the concept of technocracy originated from Henry Ford, a broker of technology, not a technologist proper, who tried to transfer a technological organization of production into broader society; technologists proper or creators of technology by no means believe in technocracy. Many social scientists and humanists use, however, the term *technocratic* as an epithet, ostensibly characterizing the behaviour of all technologists. Such usage of this word is an expression of cultural imperialism—a judgment of other culture without actually knowing it (see e.g. Levi-Strauss 1958).

by attempts to reduce them to simpler ones, such as discourse, negotiation, power and money. The result was clichés: that knowledge gives power and money is important for science, that absolute truth does not exist for many reasons. However, this is well known also for technologists: each measurement is distorted by the fact of measuring, words are only an approximate code trying to describe reality, etc. If we admit that truth and objectivity are concepts of a higher order of complexity, then the arguments of extreme relativism lose their validity. Truth and objectivity might be unattainable, but retain their importance as ideals, values.

Without trying to be objective, technology could not be effective e.g. in increasing the reliability of cars. Technology reduced the fallibility of cars during the last century to the extent that today few drivers would repair their cars themselves (admittedly, cars have become also too complex), while it was a normal phenomenon in the first half of the twentieth century. However, this increase in reliability resulted from multiple empirical tests, from seeking objective reasons for design improvements; we would not achieve such results if technology abandoned its striving for objectivity.

Thus we can reckon that these reductionist attempts to deconstruct fundamental concepts by postmodernism constituted in a sense only *a sign of the end of a civilization epoch, where a general uncertainty of values results in a general anarchy, a game in which anything passes. An inflated interpretation of chaos (without understanding the conditions of chaos emerging out of order and order emerging out of chaos), the fear of the speed of change, an inflated interpretation of virtuality etc. make a postmodernist to deny any objectivity, to reduce truth to money what actually serves the interests of large corporations trying to privatize the intellectual heritage of humanity with the use of slogans of "intellectual property".*

The reader will be right to judge that this is only an opinion of a technologist, but even this would signify a deep precipice that appeared between technology and social sciences at the end of industrial civilization epoch. I am not alone in seeing this precipice: e.g. Kozakiewicz (1992) considered the same problem from the perspective of sociology and pointed out its internal crisis, coming to a paradoxical conclusion that if science is (according to postmodernism) a social discourse, then it would appear that *sociology is a social discourse about itself.*

I have underlined already in the previous chapter that the cultural sphere of social sciences and humanities differs from the cultural sphere of technology, because they use different concepts and language, have different values, different episteme. The same, as observed already by C.P. Snow, concerns social sciences and humanities when compared to strict sciences. It is less evident that the same distinction applies to strict sciences and technology. Some sociologists of science, e.g. Latour (1987), write even about *technoscience*, which is, however, a great mistake, a sign of a deep misunderstanding of technology: strict sciences and technology have many common features, but differ essentially in their values and episteme. These differences will be discussed in detail later, but their essence should be indicated here: *the ideals of strict sciences are true theories, the ideals of technology are the art of constructing tools and solving practical problems even in cases when corresponding theories do not yet exist.*

Summing up the above considerations, I came to the conclusion that *at the end of the epoch of industrial civilization the old episteme disintegrated and a new epoch began; this disintegration consisted in divergent development of epistemai of three different cultural spheres, often without reciprocal understanding.* Cultural anthropology of the twentieth century developed an important principle of intercultural contacts: *you should not judge another culture without knowing it deeply.* Therefore, how should we assess the position of postmodern sociology of science, expressed, e.g., by Latour (1987)? By telling strict sciences that they do not seek truth, only power and money, a postmodern sociologist of science is similar to a party activist who tells a priest that he does not seek God, only power and money. By telling technologists that their products serve enslavement of people, a postmodern sociologist of science is similar to an activist who tells painters that their pictures serve enslavement of people. According to the principle of cultural anthropology mentioned above, the episteme of strict sciences should be discussed, subject to an internal critique and further developed by strict sciences; the same is applicable to technology. The same is also applicable to social sciences and humanities, but only when postmodern sociology overcomes its internal crisis, it can expect that its opinions about strict sciences and technology will be treated seriously.

A similar opinion was expressed by the representatives of strict sciences who spoofed the position of postmodernism in ostensibly scientific papers (send to and accepted by the editors of social science journals) in the so-called *science wars* in the end of the twentieth century. As a reminder: the philosopher of science Val Dusek tried to downplay this phenomenon, writing that "there are scientists and technologists who believe the objectivity of their field is being wrongly called into question by social, political and literary studies of science" (Dusek 2006, p. 21), while I believe that virtually all representatives of strict sciences and technology share such opinion.

On the other hand, if we treat the principle of cultural anthropology mentioned above too literally, these three cultural spheres become fully separated, which is neither desirable nor actually possible. It is necessary to find common language, and to do so, it is necessary to discuss fundamental assumptions and epistemai of individual spheres. We should start with existing differences.

The paradigmatic cognitive perspective of strict sciences was described in detail by philosophy of science in the second half of the twentieth century, e.g. in the works of Kuhn (1962) and Laudan (1977). A representative of strict sciences treats truth and objectivity as ideals, paramount values; old theories are equally important, because they were accepted as true before and can be abandoned only in a process of a scientific revolution when new theories, closer to the ideals of truth and objectivity, would emerge. Admittedly, science needs money to be developed, thus in some cases there might be certain compromises in the development of science, mostly in the form of inflated promises of socio-economic gains from the development of some scientific ideas; however, these compromises do not extend to such dependence of science on money as suggested by postmodern sociology of science.

A technician or rather a representative of technical sciences is much more ready to accept relativism, easily accepts that *scientific theories are only models of knowledge* and does not expect only one scientific theory to be true: if there are two competitive scientific theories, a technician can check the consequences of both (generally, a technician prefers to have two or more tools than only one). A technician uses such theories to solve practical problems, so it might be more useful in practice to test several ones. However, a technician is more demanding in one respect: the tools developed with the help of these models of knowledge, these theories, should be tested and prove correct in practice in possibly broadest conditions; these tools must be subject to *falsification*.[8] If there are no appropriate scientific theories, a technician does not wait until such theories have been developed, but tries to solve the practical problem creatively. (S)he is motivated by the joy of creation and appreciates tradition similarly as an artist: and old automobile might be a work of art. Again, technology needs money to be developed, hence certain compromises might occur (particularly for technicians employed at big corporations that try to appropriate every idea of an employee), but these compromises are similar to those a painter that makes portraits for money has to face.

A postmodern representative of social sciences and humanities believes that all knowledge is a subject of social agreement, results from social discourse, is constructed, negotiated, local, relative. There are traps in such an episteme, it will not stand up to an internal critique, e.g., such as suggested by Kozakiewicz (1992), but this crisis must be overcome by sociology itself. Not all fields of social sciences and humanities surrender to the fashion of postmodernism, and what is common for them is rather concentration on humanity and people—which is very noble but often leads to an excessive anthropocentrism, an issue that I discuss in Chap. 6.

Admittedly, the above characteristics might be treated as an overdrawn caricature, made from technological perspective. However, they show how far the disintegrating development of epistemai of these cultural spheres went and how difficult it will be to form a new, uniform episteme for the new civilization era.[9] Nevertheless, I will try to outline such a combined new episteme in one of further chapters.

---

[8] Technical sciences adhere to the postulate of falsification of Karl Popper (see e.g., Popper 1962, 1972) most closely. This fact is somewhat paradoxical, because Karl Popper, who did not know much about technology, maintained that technology does not use falsification (see Popper 1956), because allegedly does not reject unsuccessful constructions and does not use critical experiments (sic!). Possibly Karl Popper did not even hear about specific methods of developing critical destructive experiments in technology.

[9] The necessity of (only a relative) integration of the epistemai of the coming epoch appears obvious to me, because the existing divergence of the epistemai of three cultural spheres often leads to great interdisciplinary misunderstandings.

3.3 Selected Views of Philosophy of Technology

Since I do not want to strongly criticize here the philosophy of technology (I devoted a separate chapter to that), in this section I present only opinions favourable to technology, even if they are not frequent, often ignored or misinterpreted by other philosophers, but nevertheless important.

We should start with a classic and fundamental analysis by Martin Heidegger contained in *Die Technik und die Kehre* (1954), repeated in the anthology (Scharff i Dusek 2003) under a somewhat unfortunately translated title *The Question Concerning Technology*. A precise translation of the title of the paper by Heidegger is *Technology and Change*, since Heidegger, even if he used the term *die Kehre* in a specific sense (a turn or a change), recognised, however, that it is technology that enables civilization changes. Many philosophers of technology reject such interpretation, calling it "technocratic determinism[10]", criticize Heidegger for being too friendly to technology and prefer an imprecise translation of the title of his paper which makes it easier to infer that Heidegger did saw an autonomous and ominous force in technology. This diversity of interpretations is possible because in his heart Heidegger was a poet and played with words in search of good metaphors and *the essence of truth* opposite to *correct understanding*. Perhaps because of that he could empathically grasp *the essence and artistic nature of technology*, highlighting its three aspects:

Firstly, Heidegger stressed the human and artistic nature of technology:

- *Technology is a human activity aimed at transformation of nature;*
- *Technology is an art of solving practical problems, not an application of abstract theories;*
- *In its essence, technical act of creation reveals truth among many possibilities offered by nature.*

Some elements of the analysis of technology by Heidegger were already used in the definition of *technology proper* offered earlier in this chapter. We can interpret the conclusions of Heidegger as an opinion that *people cannot escape creating technology*, in the same way as a child cannot escape playing with blocks. Technical capabilities belongs to the fundamental characteristics of human kind, technology is an inherent capability of people.

> Regardless of how we define humanity, we would cease to be humans if we abandoned technical creation.

However, secondly, Heidegger maintained that technology of his contemporary times (in industrial civilization) lost its classical nature of *techne*. This opinion is both incisive and imprecise. The mass use of technology was clearly the

---

[10] This is because most philosophers are addicted to classic logics (either *a* causes *b* or *b* causes *a*) and the relation between technology and society is actually a feedback relation, see Fig. 3.1; see also Chap. 6 on the lack of understanding of feedback by philosophers.

foundation of industrial civilization, but this concerns *the socio-economic system of technology usage,* not *technology proper* that remained often a product of individual creation. *Techne, technology proper* or the art of constructing tools obviously changes with a civilization era because it is about other tools; but the mass use of these tools is a quite different socio-economic problem, clearly bringing both advantages and dangers, even if the latter relate more to the socio-economic system of the usage of tools than the art of constructing them.

Thirdly and finally, Heidegger recognised a fundamental fact: in human nature lies also the danger of fascination with technology, of trying to use it excessively. The formulation of this danger by Martin Heidegger remains a classical quotation:

> Meanwhile ... [as a result of fascination with possibilities of technology[11]] man exalts himself and postures as lord of the earth.

Heidegger noticed that the technology of industrial civilization changed qual-itatively when compared to earlier times and, at the same time, offered to man almost complete ascendancy over nature. *Such ascendancy, however, if exploited without consideration and self-restraint, endangers the very existence of humanity.* This warning was true, as we learned later at much cost; e.g., the catastrophe of the power plant in Chernobyl resulted not from a breakdown of technology, but from an irresponsible experimentation, a play of people in this power plant.

However, Heidegger did not blame technology as an autonomous force alienating and enslaving people; this denunciation emerged later, starting with the *One-dimensional Man* by Marcuse (1964). The warning of Heidegger concerned not technology proper, but the use of technology by people, in fact, the socio-economic system of technology usage, even if Heidegger does not stress this distinction.

Another philosopher of technology that shows an understanding of the *essence of technology* is Mesthene (1967), who writes about social impact of technological change. One of his fundamental questions is:

> At its best, then, technology is nothing if not liberating. Yet many fear it increasingly as enslaving, degrading, and destructive of man's most cherished values. It is important to note that it is so and try to understand why it is so.

Unfortunately, further analysis performed by Mesthene does not provide a clear answer to this question, because he did not differentiate to a sufficiently precise degree between technology proper and the socio-economic system of its use. I would answer this question as follows: many people fear that in the existing socio-economic system, other people, motivated by features of this system, e.g., competition or political ambition, would use technology to enslave or degrade them, that the system does not have enough checks and balances to restrain diverse possible uses of new technology.

---

[11] My own explanation in the straight parentheses, resulting from the context of this statement by Heidegger.

Distinctly neutral with respect to technology, and probably for this reason, ignored by most of other philosophers of technology,[12] is the monograph *The nature of technological knowledge. Are models of scientific change relevant?* by Laudan (1984). This monograph is devoted to the question whether the concept of scientific revolution by Thomas Kuhn is applicable also to technology; we shall analyze it in more detail in Chap. 13. However, the general conclusion of Laudan is that we should not use the concept of scientific revolution to technology literally, because technology has different mechanisms of qualitative change. It conforms with my personal conviction that technology is much less paradigmatic than strict science. Clearly, we can use the term *paradigm* also in technology, but in its popular meaning, not in the strict Kuhnian meaning as an exemplary theory reinforcing scientific beliefs. The monograph of Laudan presents, however, diverse definitions of technology, without asking the question which of them is acceptable for technicians. The definition accepted by Laudan is *technology as a faculty of solving practical problems* and it indicates a correct understanding of technology even if not fully representing its essence.

A deeper reflection is needed here. If technology consists of solving practical problems and, at the same time, is not a simple application of results of strict sciences (which is also stressed by Laudan and will be discussed in more detail below), then how it actually solves these practical problems? In my opinion, a correct answer to this question is: *technology is an intuitive creative activity, an art of creating tools and artefacts,* as perceived already by Heidegger and forgotten by almost all philosophy of technology. Technology must rely on rationality when checking and testing new tools, but in its essence it is an intuitive creative activity.

There are many similar misunderstandings in other texts on philosophy of technology, and here I will only comment the question of autonomy of technology. If technology is an art, then it is in a sense autonomous, similarly as artistic painting is. Moreover, since new technical solutions precede their mass socio-economic applications often by 50 years (which will be discussed in detail in the next chapter), then the autonomy of technology is far-reaching: technicians often develop new tools without a direct feedback from their users. However, technicians are people, live in a society and their actions are motivated by their perception of social needs, a perception that is justified to a greater or lesser extent. Therefore, a gross accusation that technology is unethical (see e.g., Schrader Frechette 1992) is an evident nonsense; a technician develops tools with the hope that they will serve society in a positive ethical sense (even in the case of defence technology); see a recent book on the ethics of technology (Morawski 2011). It is clear that there are no fully neutral tools; they stimulate a type of behaviour, a type of social fascination, but a final responsibility for their use rests with the users. A hammer can be used to drive nails or kill people, but

---

[12] For example, it is symptomatic that the anthology (Scharff i Dusek 2003) ignores the works of Rachel Laudan.

its hammer-likeness (a property that according to Heidegger decides of it being a hammer) does not decide its unethical use, even if it might fascinate people (and historically, has fascinated them).

3.4 The Feedback Relation of Strict Sciences and Technology

It is stressed here, and documented in more detail in historical chapters, that technology is not a simple application of theories of strict sciences, that technical solutions often precede the development of science, although technology, being pragmatic, uses scientific theories if they are available and helpful.

The first, rather obvious example is the invention of a wheel. Similarly as most inventions, it was at first used for war technology (chariots were meant originally as a dangerous weapon). However, a wheelwright created a wheel, before the invention of a lathe, as a polygon structure, cutting off more and more vertices until reaching an approximate circle and wheel. Therefore, the mathematical concepts of a circle and of actual infinity originated from technology, from the work of wheelwrights. Some historians and philosophers of mathematics (see e.g., Bronkhorst 2001; Król 2005) show that the majority of ancient mathematics before Greek mathematicians was oriented technically and, not knowing the concept of a mathematical proof, used only a pragmatic demonstration.

Another example, well known in the history and philosophy of science, is the impact of technical development of telescope on the discoveries of Galileo and astronomy. In the sixteenth century in the Netherlands, due to the improvements in cutting and polishing lenses, a telescope was invented, originally for naval applications. Galileo used the telescope to discover the moons of Jupiter which has supported the hypothesis of Copernicus that the Earth can be also a planet despite having a moon. The science of optics was developed later.

In the historical chapters I provide multiple, more contemporary examples. The improvement of a steam engine by Watt included a mechanical system of automatic control of the rotational speed of the engine, essential for its safety. This invention not only started the epoch of industrial civilisation, but also originated diverse scientific research. In general and mathematical terms, the invention of Watt concerned the stability of dynamical systems and resulted in research in this field by such great thinkers as William Kelvin-Thompson and James C. Maxwell, see (Thomson and Tait 1867; Maxwell 1868). Later, this led to a more thorough analysis of the dynamics of nonlinear systems and to the deterministic chaos theory. Other research concerned a very important phenomenon and concept of *feedback*, practically used by Watt. A complex history of this phenomenon and concept is discussed in Chap. 8, and the consequences of this phenomenon are not yet fully utilized by contemporary philosophy, as it is shown in Chap. 6.

A less known example of the feedback relation between strict sciences and technology is the *generator of pseudo-random numbers* in digital computers discussed in Chaps. 9 and 10, a practical precursor of the deterministic chaos theory, preceding this theory by over a dozen years according to the evidence of von Neumann (1951). There are many other such examples in the recent history of informational technology and sciences. The theory of relational data bases is an important and fundamental part of computer science, but with the beginning of the 1990s, the praxis of computational technology brought a new challenge, related to the storage of historical records of very large data bases (e.g., records of telephone connections) and a fast access to such records with the use of the so-called multi-dimensional addressing, as well as ensuring historical coherence of the data (non-removable dated data). In the face of these new needs, computational technology developed new types of data bases called *data warehouses,* independently from the existing theory, which in a sense surprised the theorists of data bases and delineated new directions of development in this field.

We see that technology often intuitively comes up with new practical solutions if it cannot find support in solving a practical problem in the theory of strict sciences or even technical sciences; and resulting concepts and tools stimulate the development of strict and technical sciences. This does not mean that the theories of strict sciences are not useful for the purposes of technology which clearly uses existing theories if they support the construction of new tools and artefacts. What it means is that there is *a positive feedback loop:* technology drives strict sciences, strict sciences drive technology.

3.5 Two Loops of Positive Feedback

Such feedback is not necessarily direct, often occurs via the intellectual heritage of humanity (e.g., Maxwell did not know Watt personally, but he knew the description of Watt inventions). This heritage can be treated as almost equivalent to the *world 3* of Karl Popper.[13] What is essential here, however, is that even if this reciprocal influence is not direct and often requires a long time (fortunately; otherwise the speed of change would be too fast), it has a character of a *positive feedback loop,* see Fig. 3.1: technical development stimulates the development of strict sciences, and new scientific theories are applied in technology.

We should stress here that the concept of feedback loop describes reciprocal influence of two time streams of effects and causes; it cannot be treated in terms of static cause-effect relationship, but dynamically or dialectically. Positive feedback is a feedback in which the stream of effects supports dynamically the stream of

---

[13] Almost, because Karl Popper limited the concept of *world 3* to rationally justified, explicit knowledge, while equally or even more important in the development of knowledge are its tacit parts, intuitive part and emotional pert of the intellectual heritage of humanity, which is discussed in more detail in Chap. 12.

Fig. 3.1 Two positive
feedback loops

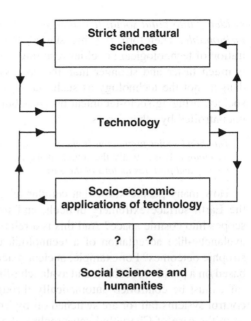

causes, which, if the process does not encounter constraints, results in an avalanche-like development.

Therefore it is good that the positive feedback loop between technology and
strict sciences functions slowly. The concepts and problems contributed by technology are analyzed by science with much delay, and technology does not respond
to new scientific theories instantly. Fortunately, because otherwise the resulting
avalanche of new theories and technologies could escape social control.

The second positive feedback loop concerns technology and its socio-economic
applications. The distinction between technology proper and its socio-economic
applications is not sufficiently stressed in social sciences, especially in postmodern
philosophy of technology, even if it should be clear at least from two reasons.
Firstly, technicians often work quite long on the solutions of technical problems
until these solutions find broader socio-economic applications. The second reason
is simple: *technicians do not make much money on technology development, as
profits from that go to technology brokers.* By *technology brokers* I understand
here entrepreneurs, managers, bankers and other organizers of processes of socio-
economic implementation of technology. If a technological product or service,
such as e.g. cellular telephony, brings much income, then these organizers find
more money to develop it further. This leads to truly avalanche-like processes of
socio-economic penetration of technological novelties.

However, these processes have peculiar dynamic properties. The socio-
economic acceptation of technological novelties is initially slow, sometimes we
observe substantial delays in the penetration of products and services as compared
to purely technical possibilities. Such delays have many reasons that shall be
discussed in the next chapter. One of the reasons is *social mistrust changing*

suddenly into blind social fascination at the moment a technical novelty becomes fashionable. Therefore, if it has started, the avalanche of socio-economic penetration of technological novelties resulting from the second positive feedback loop is much faster and stronger than the processes in the first loop. This is the real danger: not the technology as such, but its positive feedback with economy and society, bringing socio-technical progress but leading to a blind social fascination uncontrolled by other forces.

> *This blind social fascination is the actual autonomous force, erroneously attributed to technology.* It is actually the source of the danger noted by Heidegger that *man exalts himself and postures as lord of the earth.*

How many people know that cellular telephony has made radio astronomy on the Earth surface extremely difficult, and forced researchers to shift radio telescopes into cosmic space? And this is a relatively mild negative effect; what if an avalanche-like acceptation of a technological novelty will result in really catastrophic outcomes? For example, nuclear reactions in nuclear power plants are also based on a positive feedback and avalanche-like processes, but these, in order to be safe, must be controlled automatically through a negative feedback, and if such control systems fail (or are switched off by irresponsible people for entertainment, as in the case of Chernobyl catastrophe), the catastrophe might have no limits.

Such dangers intensify after the informational revolution, when knowledge, especially technological knowledge, becomes a fundamental productive resource, while postmodernism tries to reduce it to money and power, and neoliberalism recommends it's unlimited privatization (see e.g., Cellary 2011); these dangers will be discussed in more detail at the end of this chapter and this book.

> A response to the question of Mesthene: *why many people perceive technology as an alienating force, enslaving, degrading and destructive of man's most cherished values,* might be following: *the actual reason is an intuitive perception of dangers resulting from social blind fascination with technology that leads to avalanche-like processes of socio-economic applications of technological novelties and might result in a catastrophe.*

3.6 The Problem of Responsibility for Socio-economic Applications of Technology

As the social perception of dangers mentioned above is intuitive, it might not be rational and the related diagnosis might be erroneous.[14] Therefore, we should submit this perception to a critical diagnosis. Two questions arise:

- What mechanisms limit and stabilize avalanche-like processes of socio-economic applications of technological novelties?
- Who is responsible for overseeing the functioning of these mechanisms?

---

[14] More about a rational theory of intuition that is powerful but does not result in a priori truth and might be erroneous see in next chapters.

A mechanism that should at least counteract economic excesses is market economy; in the communist system people tried to replace market but without much success. However, market is only a robust mechanism that by no means solves all problems and often creates new ones. For example, since knowledge based economy strongly decreases marginal production costs (repeating knowledge does not cost much), hence today prices on high technology markets do not have any relation to marginal costs, and often are hundreds of times higher. Therefore a free, ideal market simply does not work for knowledge based economy and what is typical is an oligopolistic or monopolistic behaviour; see e.g. the opinion of Arthur (1994). But in that case, who might oversee such oligopoly on a globalized market?

The responsibility obviously rests with *technology brokers*. However, in order to be effective on the market they must be motivated by profit (whatever brings money is good). We can only hope that this motivation is somewhat mitigated by ethical concerns (see e.g., Morawski 2011); unlimited motivation by profit often leads to market degeneration, see e.g. (Salmon 2009). If ethics is a result of education, *then who educates* brokers of technology? Usually, they are not educated by technologists, their teachers are mostly representatives of social sciences and humanities who eventually also decide about the shape of education system. It seems that these representatives should not only include appropriate ethical aspects into education, but also help their students to understand socio-economic processes of demand for technology, of social fascination with technological novelties and related dangers.

Unfortunately, humanists, sociologists and economists often do not succeed in this task and shed the responsibility on technology, in addition technology not differentiated from the system of its socio-economic applications and censured as a technocratic tool of enslavement. It is expressed by question marks on Fig. 3.1: the role of strict and natural sciences as well as technology proper with respect to their socio-economic applications are well defined, but social sciences and humanities in this scheme seem not to fulfill their roles.

Thus, *the final responsibility for socio-economic applications of technology, for overseeing effective limitations of blind social fascination with technological novelties resides with technology brokers, but* social sciences and humanities are co-responsible.

This does not mean that technology proper is not co-responsible for dangers created together with technology; it should cooperate with social sciences to limit these dangers. However, a technologist is usually aware of the dangers that might emerge from what (s)he creates, analyzes possible results of application of (her) his tools and artefacts, and even if (s)he cannot be fully effective in limiting these dangers, (s)he knows that the blame for all misfortunes and misapplications of technology will be shifted to (her) him. At the same time we should not, for several reasons, expect that even the strongest feeling of responsibility on the part of technicians will result in avoiding all misapplications of technology. Firstly, in their attempts to collaborate with representatives of social sciences, technicians often face misunderstanding or even aversion. Secondly, the ingenuity of human stupidity is unbounded (*against stupidity, the gods themselves contend in vain*). The third reason is most essential: *the character of the epoch of knowledge civilization starting after the informational revolution will give people almost unlimited possibilities of selection in the scope of diverse technological options, including those that are suicidal.*

3.7 What will be the Technology of Knowledge Civilization Era?

We need today once again formulate the question of Heidegger concerning the character of *die Kehre,* the change in the nature of technology, in his text, the technology of industrial civilization era when compared to earlier times. The new question is thus as follows: *In which sense the technology of knowledge civilization era will essentially differ from the technology of industrial civilization era?* My own experience suggests the following answer, which at the same time is one of main theses of this chapter.

> The technology of knowledge civilization era will differ above all in terms of complexity. It will propose almost unbounded number of diverse technical possibilities, both concerning products and services, together with the creativity support service. However, only a small part of these possibilities will be actually introduced into socio-economic use.

In terms of complexity and diversity, this technology could be called postmodern, but the actual change will go deeper than the intellectual fashion of the end of industrial era; hence, it would be more correct to call it *post-postmodern technology.* We shall illustrate this change with several examples.

One of the most important possibilities created by informational technology is *the change of the art of recording intellectual heritage of humanity.* During last two civilization epochs, industrial and preindustrial, the dominant medium of that recording was print. Informational technology already partially enables, and soon will enable in full, a *multimedia record* of this heritage. In other words, instead of paper books (however much I personally regret this change) we shall have an electronic record including not only text, but also film, music, interactive exercises and virtual laboratories. Lectures of future Kant and Einstein will become available for future generations, but the change will go further. The multimedia form will support intergenerational transfer of intuitive knowledge and intuitive heritage of humanity,[15] and also make electronic and distance education more effective. However, the multimedia electronic transfer brings also diverse dangers, the increase of the danger of indoctrination, of plagiarism, of using the transfer for criminal purposes etc.

---

[15] See *Creative Space* (Wierzbicki and Nakamori 2006).

Another important possibility concerns the creation of *ambient intelligence,* called *AmI* earlier in Europe (today the name *Internet of things* is more popular), or *ubiquitous computing* or *wireless sensor network* in the United States; in Japan the same function is called *intelligent home* or *yaoyorozu* (eight million Shinto gods).[16] There is no doubt that the diversity of facilitations created by computational intelligence dispersed in sensors and processors hidden in our living space, in houses, offices, shops or vehicles, is tremendous, and people would buy such technology and services as soon they become truly accessible and inexpensive. However, the spread of *ambient intelligence* results also in an obvious danger, stemming not from technology itself, but from the way it is used by people. *Ambient intelligence* requires, for example, a recognition of a person coming into a room or a building. What will restrain too ambitious police from using this technology to develop a version of Big Brother? Full *ambient intelligence* is related to almost universal use of robotization. What will restrain creative criminals from using robots in mafia fights?

Finally, let us consider yet another possibility between the unlimited number of possibilities to be provided by future technology of the epoch of knowledge civilization. *Computer aided decision analysis and support,* developed in the end of industrial civilization, could further develop into *computer aided creativity support,* assisting the development of science and technology. To that end, we must understand creative processes better, especially the processes of knowledge creation, and not only in a historical macro-scale, as in the theory of scientific revolutions by Kuhn (1962), but especially in a micro-scale, with conclusions concerning the ways in which knowledge is created for the needs of today and tomorrow. Many such micro-theories of knowledge creation emerged already on the turn of the twentieth and twenty first century, in response to the demand of knowledge based economy, which is described in Chap. 12.

3.8 New Warnings: What Should We Pay Attention to?

The greatest danger results not from technology proper, but from us, people fascinated with technological possibilities; thus, we must repeat in new conditions and strengthen the warning of Heidegger. The seemingly unlimited technological possibilities can suggest to their users, in particular, to technology brokers, that human intellectual heritage is already rich enough to be used and privatized without limits, that market should decide what to select from these possibilities. This results, however, in the greatest threats. The unlimited use and privatization of natural resources in the epoch of industrial civilization had led to a pollution and endangerment of natural environment; similarly, unlimited privatization of human

---

[16] It is clearly a metaphor, much valued by the Japanese: eight million Shinto gods indicate their omnipresence.

intellectual heritage may lead to distortions of this heritage, which we already observe, e.g., on the pharmaceutical markets. Thus, the modified Heidegger warning is:

> In the epoch of industrial civilization people were blinded by seemingly unlimited dominion over nature, provided by technology, which has led to an excessive exploitation of natural resources combined with severe pollution of natural environment. In the epoch of knowledge civilization we must prevent being blinded by seemingly unlimited possibilities of products and services offered by technology, in particular, we should strive to preserve and develop our intellectual environment, the intellectual heritage of humanity.

The danger of leaving the selection of technical possibilities to market forces alone lies also in the positive feedback between technology proper and its socio-technical applications; the resulting avalanche-like development is not sustainable, it must lead to a catastrophe, and maybe even to self-extinction of human civilization, as suggested by the phenomenon of *eerie silence* (Davies 2010), discussed in more detail in Chap. 14.

3.9 Conclusions

There is no doubt that *technology contributed in a decisive way to the change of civilization epochs,* that the *informational revolution transforms industrial civilization into a new epoch which I call knowledge civilization.* This change concerns society, but results from the development of technology, from the emergence of computer networks and personal computers. Technology resulted also in *dematerialization of labour:* automation, computerization and robotization liberated people from most of hard labour, created conditions for full equality of genders, but at the same time increased unemployment.

These changes are positively evaluated by some sociologists, but, paradoxically, many of them, starting with Marcuse to Postman, censure technology as an autonomous force, alienating and de-humanizing, forming a *technopoly*, as a technocratic tool of enslavement, an expression of an instrumental vision of the world. The call of technology for objectivism is rejected as a remainder of positivism; such attitude is most frequent in postmodern sociology of science, but happens also in other humanistic disciplines. Such attitude is, however, dangerous, because it makes it difficult, if at all possible, to understand technology.

Technology is motivated by the joy of creation, it is an art of creating tools (the old Greek word *techne* meant both creative abilities and art). In order to be effective in creating tools, technology invokes *relative objectivism.* Technicians understand that there is no absolute knowledge and truth, such as there is no absolute measurement, but they must be as objective as possible. They cannot ignore uncomfortable or unpopular information because such omission can lead to a catastrophe.

The technical *relative objectivism* is not a positivistic belief in the existence of absolute truth based on empirical facts. Most technicians know that people construct their knowledge. Nevertheless, they assume that objectivism and empirical facts are useful, because until now they always helped in effective construction of tools and technical artefacts. *Postmodern sociology of science, and even a part of postmodern philosophy of technology, appear unable to understand this separate cultural sphere of technology.*

The definition of technology proper presented in this chapter stresses that technology is the most fundamental and specifically human ability, that it is an art of creating tools needed by people when dealing with nature. We cannot stop to create technology without stopping to be human. Technology enables human intervention in nature, but can also serve to limit negative effects of such intervention. As noted by Heidegger, *the essence of technology is revealing truth, thus technology is similar to art.* Technology is also an art of solving practical problems, even if recently, similarly to strict sciences, it also results in new perspectives.

The relation between technology and strict or natural sciences forms a positive feedback loop: technology discovers new problems for science and science creates new theories that might be used in technology. However, *technology is sovereign in this loop, solves problems and creates new concepts independently from the contribution of strict science.* In this sense we can say that *the technical discovery of a wheel motivated the development of mathematics;* reciprocally, the development of mathematics helps in technological development, although does not determine it.

Even more important is *the second positive feedback loop between technology and its socio-economic applications.* These applications are managed by *technology brokers*: entrepreneurs, managers, bankers etc.; our socio-economic system revolves around technology applications. This second feedback loop results in most socio-economic applications of technology but at the same time leads in dangers because of the avalanche-like character of resulting processes; if an additional stabilization of such processes is not effective, catastrophes may follow. *An intuitive fear of such dangers is the actual reason for many representatives of social science and humanities to condemn technology; but this intuitive fear is not founded on a correct diagnosis.*

A stabilization of avalanche-like processes in socio-economic applications of technology is provided by the market mechanism, which, however, ceases to work on high-tech markets, and allows monopolies or oligopolies to emerge. Moreover, the market mechanism by itself does not solve ethical problems related to socio-economic applications of technology; *thus, it is not technology but the market mechanism that results in the dangers of such applications.* Since brokers of technology are educated mostly by representatives of social sciences and humanities, *the ultimate responsibility for an effective limitation of a blind social fascination with technological novelties rests first of all with technology brokers, but social sciences and humanities are co-responsible.* In order to pursue their duties, the latter must not only stress ethics in the education of technology brokers, but also introduce technical themes, beside information technology, also robotics and biomedical engineering, into curricula of their studies.

The technology of upcoming epoch of knowledge civilization will differ from the technology of industrial civilization in terms of complexity, offering seemingly unbounded technological possibilities, not only in products, but also in services, together with a possible service of creativity support. I repeat and strengthen here the warning of Martin Heidegger concerning the fascination of people with technological possibilities: *in the epoch of knowledge civilization we must not be blinded by seemingly unlimited possibilities of products and services offered by technology,* and in particular, we should strive to preserve and develop our intellectual environment, the intellectual heritage of humanity.

References

Arthur, W.B.: Increasing Returns and Path Dependence in the Economy. Michigan University Press, Ann Arbor (1994)

Arthur, W.B.: The Nature of Technology: What It Is and How It Evolves. The Free Press, Simon and Schuster, New York (2009)

Barnes, B.: Scientific Knowledge and Sociological Theory. Routledge and Kegan, London (1974)

Barney, D.: The Network Society. Polity Press. (Polish translation (2008) Społeczeństwo sieciowe. Wydawnictwo Sic!, Warsaw) (2002)

Bloor, D.: Knowledge and Social Imagery. Routledge and Kegan, London (1976)

Bronkhorst, J.: Pāṇini and Euclid: reflections on Indian geometry. J. Indian Philos. **29**, 43–80 (2001)

Cellary, W.: Zasoby wiedzy dobrem ekonomicznym w społeczeństwie wiedzy (Knowledge resources as economic good in knowledge society). *Przyszłość: Świat, Europa, Polska, nr. 1/ 2011* (2011)

Cempel, C.: Teoria I Inżynieria Systemów (Theory and Engineering of Systems). Wydawnictwo ITE, Radom (2006)

Davies, P.: The Eerie Silence: Renewing Our Search for Alien Intelligence. Harcourt, Houghton Mifflin (2010)

Derrida, J.: Of Grammatology. John Hopkins University Press, Baltimore, MD (1974)

Dusek, V.: Philosophy of Technology—an Introduction. Blackwell Publishing, Oxford (2006)

Foerster, H.: On constructing a reality. In: Preiser, E. (ed.) Environmental Systems Research. Dowden, Hutchingson & Ross, Stroudberg (1973)

Foucault, M.: The Order of Things: An Archeology of Human Sciences. Routledge, New York (1972)

Gadamer, H.-G.: Warheit Und Methode. Grundzüge einer philosophishen Hermeneutik. J.B.C, Mohr (Siebeck), Tübingen (1960)

Habermas, J.: Lectures on the Philosophical Discourse of Modernity. MIT Press, Cambridge, MA (1987)

Heidegger, M.: Die Technik Und Die Kehre. In M. Heidegger, Vorträge und Aufsätze, Günther Neske Verlag, Pfullingen (1954)

Jackson, M.C.: Systems Approaches to Management. Kluwer Academic—Plenum Publishers, New York (2000)

Kozakiewicz, H.: Epistemologia tradycyjna a problemy współczesności. Punkt widzenia socjologa (Traditional Epistemology and Contemporary Problems. Perspective of a Sociologist). In: Niżnik, J. (ed.) Pogranicza epistemologii (On Boundaries of Epistemology). Wydawnictwo IFiS PAN, Warsaw (1992)

Król, Z.: Platon I Podstawy Matematyki Współczesnej (Plato and Foundations of Contemporary Mathematics). Wydawnictwo Rolewski, Nowa Wieś (2005)

Kuhn, T.S.: The Structure of Scientific Revolutions, 2nd edn., 1970. Chicago University Press, Chicago (1962)

Latour, B.: Science in Action. Open University Press, Milton Keynes (1987)

Laudan, L.: Progress and Its Problems. University of California Press, Berkeley (1977)

Laudan, R., ed.: The Nature of Technological Knowledge. Are Models of Scientific Change Relevant? Reidel, Dordrecht (1984)

Levi-Strauss, C.: Structural Anthropology. Allen Lane, The Penguin Press. Polish translation (1970) *Antropologia strukturalna*, Wydawnictwo KR, Warsaw (1958)

Lyotard, J.F.: The Postmodern Condition: A Report on Knowledge. Manchester University Press, Manchester (1984)

Marcuse, H.: One-Dimensional Man. Beacon Press, Boston (1964)

Maturana, H.: Biology of cognition. In: Maturana, H., Varela, F. (eds.) Autopoiesis and Cognition. Reidel, Dordrecht (1980)

Maxwell, J.C.: On Governors. Proc. Roy. Soc. 16(1868), 270–283 (1868)

Mesthene, E.G.: The Social Impact of Technological Change. In: Scharff, R.C. Dusek V.(ed.) op. cit (1967)

Morawski, R.Z.: Etyczne Aspekty Działalności Badawczej W Naukach Empirycznych (Ethical Aspects of Research Activities in Empirical Sciences). Wydawnictwa Uniwersytetu Warszawskiego, Warszawa (2011)

Popper, K.R.: Objective Knowledge. Oxford University Press, Oxford (1972)

Popper, K.R.: Three views concerning human knowledge. In: Lewis, W.H.D. (ed.) Contemporary British Philosophy, Third Series. Allen and Unwin, London (1956)

Postman, N.: Technopoly. The surrender of culture to technology, Knopf, New York (Polish translation (1995) Technopol. Triumf techniki nad kulturą. Państwowy Instytut Wydawniczy, Warsaw) (1992)

Quine, W.V.: Two dogmas of empiricism. In: Benacerraf, P., Putnam, H. (eds.) Philosophy of Mathematics, Prentice-Hall, Englewood Cliffs (1964)

Ritchie, C.P.: The Inheritors. The Story of Man and the World He Made, The Reprint Society, London (1962)

Salmon, F.: Recipe for Disaster: The Formula That Killed Wall Street. Wired Magazine 17.03.2009, Tech Biz: IT (2009)

Scharff, R.C., Dusek, V. (eds.): Philosophy of Technology: The Technological Condition. Blackwell Publishing, Oxford (2003)

Schrader-Frechette, K.: Technology and Ethics. In: Scharff, R.C. Dusek V.(ed.) op. cit (1992)

Skyttner, L.: General System Theory. Ideas & Applications, World Scientific, London (2001)

Snow, C.P.: The Two Cultures. Cambridge University Press, Cambridge (1960)

Thomson, W., Tait, P.G.: Treatise on Natural Philosophy, 2nd edn., 1883. Cambridge University Press, Oxford (1867). (reissued by Cambridge University Press, 2009)

von Neumann, J.: Various techniques used in connection with random digits. In: Householder. A.S., Forsythe, G.E., Germond, H.H. (eds.) Monte Carlo Method, National Bureau of Standards Applied Mathematics Series, 12, pp. 36–38. U.S. Government Printing Office, Washington, DC (1951)

Wierzbicki, A.P.: Education for a new cultural era of informed reason. In: Richardson, J.G. (ed.) Windows of Creativity and Inventions, Lomond, Mt. Airy, PA (1988)

Wierzbicki, A.P.: Technology and Change: The Role of Technology in Knowledge Civilization. Ith World Congress of IFSR, Kobe (2005)

Wierzbicki, A.P.: Megatrends of Information Society and the Emergence of Knowledge Science. In Proceedings of the International Conference on Virtual Environments for Advanced Modeling, JAIST, Tatsunokuchi (2000)

Wierzbicki, A.P., Nakamori, Y.: Creative Space: Models of Creative Processes for the Knowledge Civilization Age. Springer Verlag, Berlin-Heidelberg (2006)

Wierzbicki, A.P., Nakamori, Y. (eds.): Creative Environments: Issues of Creativity Support for the Knowledge Civilization Age. Springer Verlag, Berlin-Heidelberg (2007)

Chapter 4
Delays in Technology Development: Their Impact on the Issues of Determinism, Autonomy and Controllability of Technology

4.1 Introduction

This chapter presents a discussion of diverse delays in the processes of technology development. The concept of *delay* has a technical meaning here, related to the theory of automatic control. A delay in a dynamic system is the interval of time between the start of an action and the observation of its effects (first or advanced effects, this distinguishes pure delay from inertial delay). Delays in the processes of technology development explain, in my opinion, why these processes might appear autonomous, self-determining, uncontrollable, if they are regarded holistically, from outside. However, when regarded from inside, from a technical perspective, they are clearly controllable.[1] This apparent paradox might result from the fact that processes of technological development contain many delays, often summing up to above 50 years. Such a process might appear autonomous in the sense of philosophy of technology, or equivalently uncontrollable in the sense of technology, if it is seen from outside, without analyzing its initial stages known only to technicians. If we inquire into these initial stages, however, these processes can be controlled.

Therefore, the delusion of autonomy of technology, rather popular in the philosophy of technology, results from two reasons. Firstly, the delays in technology development are often unappreciated, often it is assumed that the interval "from an innovation to the market" is short and is further significantly shortening today; I will show that this is a wishful thinking. Secondly, the philosophy of technology assumes from Ellul (1964) on (erroneously, and this assumption might be the main

---

This chapter is based on the paper (Wierzbicki 2008), with modifications.

---

[1] The use of the concept of *controllability* suggests a technical perspective, since this concept was introduced over 50 years ago by automatic control theory and until now is seldom used by sociology or philosophy of technology.

© Springer International Publishing Switzerland 2015
A.P. Wierzbicki, *Techne$_n$: Elements of Recent History of Information Technologies with Epistemological Conclusions*, Intelligent Systems Reference Library 71,
DOI 10.1007/978-3-319-09033-7_4

reason of its inadequacy) that the development of technology might or even should be treated holistically from outside, from the socio-humanist perspective.[2] From such perspective one sees in technology only the socio-economic system of usage of technology products, and notices only its uncontrollability or autonomy, and focuses on *ex post* assessments of socio-economic use of technology products, without asking who is responsible for that use. This leads to accusations that technology is not ethical, while the responsibility, as discussed in the previous chapter, belongs to technology brokers and their teachers.

The socio-economic system of creation and use of products of technology is complex. A holistic approach to a complex system is often necessary, but without understanding the details, parts of the system and their relations, a holistic approach results in superficial opinions, e.g., a conclusion that the system is autonomous or in a superficial understanding of the issue of the so-called *technological determinism*. The last issue has two aspects. One of them is internal determinism: technology seems to be self-determining, which is equivalent to its autonomy. The second aspect concerns external determinism, the determination of socio-economic processes by technology: if technology is autonomous, and clearly influences socio-economic processes, then, if understood superficially, it determines them. Already Karl Marx in his historical materialism maintained that the development of productive forces determines socio-economic processes. However, such holistic approaches are too simplified for a technician who is a member of society and creates new tools in the hope that they will be socially useful; it is also clear for a technician that *techne,* including tools typical for a civilization epoch (and also tools used as productive forces), contributes to the socio-economic processes of that epoch but does not determine them fully, since people select tools and thus influence their further development in a feedback loop.

The complexity of a typical process of creation, socio-economic production and use of technology products can be illustrated with multiple stages of that process, which I discuss below. The construction of a new tool or other artefact requires time, from the idea through design, initial implementation, testing and evaluation, project of industrial production method, etc. The time needed, however, is small as compared to other stages, the development of new versions of the product that are effective but sufficiently inexpensive, the process of market penetration of new product, or stimulation of the demand and social acceptance of that product. At the same time, there are many reasons for a further increase of the delay time. New technical ideas often emerge in academic communities that create knowledge

---

[2] See Mitcham (1994), where Carl Mitcham clearly defends such paradigmatic attitude and maintains that an immersion into technical details hinders a humanist approach of the philosophy of technology. He writes (Mitcham 1994, p. 65) *"becoming mired in the specialized details of technology and its many processes tends to obscure relationships to nontechnological aspects of the human"*. Mitcham notes the understanding of technology as *techne,* but criticizes such understanding; his critique might be shortly paraphrased as follows: "If we, philosophers of technology, agreed that the word *technology* denoted both artefacts and their socio-economic use, we cannot agree that *technology* denotes creative activity of the human".

(including technology) in a different manner than industrial research organizations, see (Wierzbicki and Nakamori 2006, 2007), which results in additional delays in knowledge transfer. Applications of new ideas are often delayed by oligopolistic practices of large corporations that want to retain market positions of products already produced and often buy patents of new products in order to put them on the shelf; the reasons of such additional delays are multiple. Even if a product is ready for market penetration, its future users might be mistrustful; it takes time for the demand to develop and shape.

In historical chapters we shall see many examples of such delays, which I call civilization delays. Here I list them shortly: in the case of transistors it was about 26 years, but in the case of digital computers (from the patent of Konrad Zuse to Apple II) it was already 42 years, in the case of mobile cellular telephony it took slightly longer (it was invented in 1943), in the case of digital television, about 50 years, in the case of hypertext and WWW (if we count from 1945 paper of Vannevar Bush to Timothy Berners-Lee), 47 years. The applications of transistors were accelerated by their military use, for other important technical innovations a typical cumulative delay amounts to 40–50 years; even for less important innovations, such as home telefax, the cumulative delay amounted to 20–30 years (see Kameoka and Wierzbicki 2005).

The data illustrated in Fig. 4.1 show that the processes of social penetration of products and services of high technology have a specific dynamics that includes significant delays. The curves concerning the penetration of electric power and telephones in the USA are irregular, perturbed by the great crisis around 1930. Rather regular and relatively fast growing are the curves regarding colour television and video recorders (VCRs); the curves of penetration of cable television (Cable) and personal computers (PCs) are slower at the end of the twentieth century (in the twenty first century they accelerate) and a fast growth starts again with the penetration of the Internet. Such curves are usually approximated, see, e.g., Strużak (2006), Grzegorek and Wierzbicki (2009), with the help of a logistic curve, $y(t) = a/(1 + b \exp(-ct))$, or a Gompertz curve, $y(t) = a \exp(-b \exp(-ct))$, with appropriately selected parameters a, b, c, standing for the saturation level, shift in time and speed of growth, correspondingly. The shifts of practical beginnings of the curves when compared to the dates of invention of such products or services are even more significant than the speed of growth.

Let us consider the example of television. As described in more detail in Chap. 7, we can assume that it was invented either around the year 1880, or, when we consider a more practical version by Zworykin and Tihanyi, the years 1923–1928. The first public transmission of BBC took place in the year 1936. However, until 1960 the social penetration of colour television, in the USA, did not exceed single per-cents, see Fig. 4.1; the level 90 % was achieved around the year 1990. For a black-and-white television this process started about 12 years earlier, but it was replaced by one related to colour TV and not completed. Therefore, we can distinguish two types of delays in this process. The initial delay results from the works on TV receivers to make them more effective and less expensive, if we count from Zworykin this would amount to 37 years, including

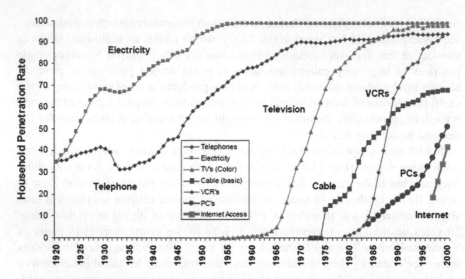

Fig. 4.1 Social penetration of selected services and high technology products in the USA in the years 1920, 2000. According to http://www.ntia.doc.gov/ntiahome/fttn00/chartscontents.html

the upgrade to colour television (if we count from 1880, this would amount even to 80 years). The second delay results from the dynamics of social penetration and exceeds 30 years.

Therefore, we must analyse delays from the perspective of a dynamic system. In dynamic systems, the effects of our actions accumulate and grow in delay to these actions, but the delays are of (at least) two types. One of them is called *inertial delay* and occurs commonly in all systems with accumulation; this concept is illustrated e.g. by the change of the level of tea when we pour water into a glass (it takes time to fill the glass, this is an inertial delay), another example is an electric iron that heats up after we switch it on. The other type of delay is called *pure delay* and also occurs commonly, e.g., in systems with transportation; it may be illustrated by the example of the delay of a letter from the moment of sending by post until its delivery, or the delay of our baggage at the airport (either total delay, when we wait for the transport of baggage from the aircraft, or final delay, when the baggage already appeared on the conveyor belt but did not yet arrive at the place where we are waiting). The delay mentioned earlier, corresponding to the time needed for the improvement and cheapening of invented products, is a pure delay; the time needed for socio-economic penetration of products is an inertial delay. In the second part, during the inertial delay, we can speak about a *co-evolution of technical solutions and socio-economic demand:* the constructors of new versions of the product react to the comments of users and improve subsequent versions. However, during the first part of pure delay we can hardly speak about any co-evolution: *the constructors of initial versions of new products are guided by their vision only;* this vision might depend on social opinions, but is usually deeply individual and much more difficult to control.

Dynamic systems with pure delay are generally much more difficult to control than systems with inertial delay only. This is related to the fact that in order to control the level of water in a glass it is enough to know only one quantity, the level of water in the glass; on the other hand, in order to control a system with baggage on a conveyor belt it is necessary to know the entire history of the baggage during delay time (whether the baggage did not fall off the belt or was not taken by mistake by another passenger). These facts are represented abstractly by the concept of *the state of a dynamic system*. The state of a glass can be represented by one initial state variable, the level of water (or by several variables, if we are interested e.g. also in the temperature of water in the glass, etc.). On the other hand, to describe the state of a system with baggage on a belt we must know entire trajectory of the baggage during the time of pure delay. And even worse, to optimally control a process with pure delay it is necessary to somehow compensate this delay through forecasting or anticipation[3]; this is explained in more detail several paragraphs below.

Here I would like to make a critical comment on the resulting philosophy of science. Karl Popper was only partly right in his critique of *historicism* (Popper 1962); he probably did not even know the concept of control of a time-delayed dynamical system and could not see essential gaps in his critique. It is clear for me that there are no absolute laws of historical and social development (such as there are no absolute laws of nature) and that great thinkers, enjoining people to act according to their interpretation of history, always committed errors. But historical and social (and also technical) processes include delays of diverse character. Hence the knowledge and interpretation of (especially recent) history is necessary, and we need a maximally objective version of that, without the contamination of "winners write history" slogan; otherwise phenomena that accumulate through a recent interval of pure delay will astonish us later by their inevitability.

That a time-delayed dynamic system (in particular a system with pure delays) seems to be autonomous or uncontrollable, if observed from outside, is well known to specialists in the theory of control of such systems, but might require an example and explanation for non-specialists. Let us consider the example of our suitcase on the conveyor belt at an airport. If we cannot come close enough to our suitcase on the belt, we lose control over it (it might fall off the belt or be taken away by another person); how can we regain at least partial control? There are several methods known to specialists, but all of them reduce to attempts to diminish the delay time, at least partly. In the case of our suitcase on the belt, we should select a place of observation enabling us to watch the suitcase from the moment of its appearance on the belt; we can then intervene if the suitcase is suddenly lost from sight.

How to use this analogy for regaining at least partial control over the processes of creation and socio-economic production of technology products? *If we would*

---

[3] Approximately 40 years ago I was intensively dealing with issues connected with the control of industrial time-delayed systems; I also published a *maximum principle*—a necessary condition of optimality of control—for such systems, see Wierzbicki (1970).

wait until a new technical solution found broad socio-economic demand, until its broad social penetration, and the philosophy of technology will then assess its social and economic significance, then such an assessment is obviously too late. Therefore, it is necessary to abandon the external, holistic approach: in order to influence the development of technology, we must concentrate on new technological ideas that did not yet find broad applications.

It follows from the foregoing that it is essential to abandon the old paradigmatic position accepted by almost all philosophy of technology after Ellul (1964) that the philosophy of technology should address only collective processes in (industrial) society, viewed holistically. This position is typical for the philosophy of technology, as evidenced by the opinion of Mitcham (1994) against "*becoming mired in the specialized details of technology*" quoted earlier. But one of main conclusions of this chapter is that the philosophy of technology has a difficult choice here: *either it will upkeep the tradition of not becoming mired in the specialized details of technology and it will continue to see in it only a dark, uncontrollable force, or it will decide to collaborate in technology control, but in such case it must penetrate into details of new technological ideas, not yet transferred to mass production.*

Let us illustrate this with another example, well known to specialists in software development for the purposes of new, multifunction mobile telephones. In Chap. 11, trends in the development of such telephones are discussed. One of such trends is an integration of the functions of a telephone, mobile television and personal computer, together with diverse additional functions, such as camera or a portable TV set. One of such functions is Global Positioning System (GPS) enabling a precise determination of the position of a phone with the use of satellites. Obviously, this has many advantages; the users are ready to pay for such service and the producers, to compete in providing improvements of such service. Together, it is an example of social fascination with the possibilities of technology against which Martin Heidegger warned. But if an ambitious minister of interior or justice appears (as it happened in Poland) who would like to use such a service for political purposes, we will become confronted with the Orwellian Big Brother.

Therefore, it is vital to penetrate technical details of a new technological products, perhaps not all of them, but at least the new functionalities. For this purpose, it is necessary to understand the stages of the processes in which technology is created.

4.2 The Stages of the Processes of Creation, Production and Use of Technology Products: Tools and Artefacts

We shall analyse possible stages of the development of a technology product in a rather detailed fashion. The list of such stages is given below. It should not suggest that these processes occur without recursion (in the language of technology; 'linearly' in the language of sociology); in reality, there are usually many recursions, returns to an earlier stage, in such a process. The list is used only to shorten the description, each stage might be divided into sub-stages, involve participation of various people in the process, require diverse choices and decisions, see e.g. Winner (1993), Arthur (2009). But even such a limited list is rather long:

(1) Motivation: a creative impulse or anticipation of social demand;
(2) *Techne:* actual construction or design of a prototype;
(3) Testing and evaluation;
(4) Transfer from a research environment to industry;
(5) Project of a mass production process;
(6) Initial marketing: promotion of demand;
(7) Mass production and marketing;
(8) Re-engineering of new versions based on opinions of users or customers;
(9) Designs of new versions based on technological progress and the nature of demand, including new functions.

These nine points are only an outline of main stages; some of them might be shortened or omitted, other expanded or repeated, with a possible recursion to earlier stages, some might be even carried out concurrently.

The stage (1) refers to an intuitive idea of a new tool, which may be called a creative impulse (actually, often of an artistic character), or, usually also an intuitive, sometimes ambiguous, inkling of a future social demand. This stage very seldom consists of simple instrumental application of results of strict and natural sciences, even if stimulated by such results. If a technician has a hunch of a new problem, (s)he usually does not wait until strict and natural sciences will produce premises of a solution and tries to find a solution (her-)himself. This stage, as well as other initial stages, *is motivated by the joy of creation,* and the lack of understanding of this fact, of a creative character of the initial stage, constitutes a fundamental error both of the philosophy of science and philosophy of technology (with rare exceptions, e.g. of Martin Heidegger). For example, in his otherwise excellent book *The Poverty of Historicism,* Karl Popper expresses in this context a conviction of a physicist (he was educated as a theoretical physicist and mathematician) that engineering is a simple instrumental application of scientific theories. He treats in that way the concept of "social engineering" (as an instrumental application of theory to social transformations) and writes (Popper 1999, p. 76)[4]: "As a technician or engineer, he will consider institutions from a functional or instrumental point of view". This allegation of functional or instrumental point of view, supposedly defining a technician or an engineer, is repeated many times by various authors, e.g. in Gadamer (1960), Habermas (1987), Jackson (2000). I do not suggest that such point of view is never used by technicians and engineers, but that if used, it is not what makes a technician; what makes a technician is rather creative solving of practical problems, expressed in stages (1) and (2).

---

[4] The quote here is after Polish edition; this book of Popper was published first as a series of papers in philosophical journals, as a book it was released for the first time in 1954, in an Italian translation.

However, this joy of creation may also be the reason of some dangers relating to technology. A creative engineer might underestimate possible threats resulting from the greed of technology brokers or ambitions of politicians and give them dangerous tools simply because it was a challenge to create such tools; from my own speciality I recall how the colleagues of mine approached the problem of constructing a pilotless aircraft as a challenge. This fact is an additional reason for the philosophers of technology to cooperate with technicians on the early, creative stages of technology development.

Stage (2) expresses the creative character of technology even stronger; as aptly noted by Martin Heidegger, *techne is a creative elicitation of truth out of many possibilities offered by nature.* An actual construction or design usually follows a specific methodology of a given discipline (differing, e.g., for architects and computer software developers), a codified experience of engineers gained from earlier examples of problems solved in that discipline. Nevertheless, *one cannot learn techne through theoretical studies, it can be learned only through personal practical experience.*[5] The best methodology and recipe for design do not suffice, personal creativity is necessary, as well as intuition based both on imagination and internalized experience. Such deeply intuitive character of technical creation is a main reason for the necessity of stage (3).

Stage (3), consisting of testing and evaluation, strongly depends on the character of the product tested and is diversified. If an inappropriate use of the product can endanger the safety or even life of people, then an essential problem is to design *critical tests,* such that will provide sufficient premises for the evaluation of product safety. This is not a simple problem, e.g., crash tests of cars have their specific methods, proven in many years of experience and supported by computer simulation, and even the partial problem of placement of sensors in the bodies of mannequins simulating the driver and passengers is not easy. Similarly, how to design critical tests for software that is intended to protect against cyber-assaults? Very often, critical tests lead to the destruction of exemplars of products; if these are costly, such as e.g. airplanes, initial tests are conducted in *virtual laboratories,* which means that real tests are preceded by tests with the use of complex computer simulations.

The difficult task of designing critical tests is not appreciated by many philosophers of science, starting with Karl. R. Popper who in his book (Popper 1956) formulated an opinion that technology does not use falsificationism since supposedly it does not use critical tests and (in Poppers opinion) does not abandon its products if the results of tests are negative. Personally, I believe that this opinion is biased by ignorance, since precisely stage (3) often results in abandonment or at least significant modification of such variants of products that do not pass the tests; and there are many variants of critical tests in technology. The opinion of Popper

---

[5] This concerns not only technology, but also many other fields or professions that require creativity. For example, we have many cookbooks, but it is not enough to study them in order to be a good cook.

can be acceptable only if we assume that he has thought about testing *scientific laws* that in his judgment could not be modified. However, soon afterwards such judgment was questioned by Kuhn (1962) who has shown that scientific laws can be modified or reinterpreted to defend a given paradigm of a field of science. Therefore, despite Popper's opinion, I think that it is technology that uses falsificationism, while according to Kuhn's opinion sciences, whether strict or social and humanistic, do not use it (at least in the short term; when we consider the long-term evolution of science we can treat scientific revolutions as an expression of falsificationism).

This perspective of technical falsificationism is misunderstood especially in postmodern sociology. I often encountered an allegation from sociologists: *why you, technologists, are so much positivists, the 19th century has passed, hasn't it?* I responded that we, technicians, since we are motivated by the joy of creation, must test the products of our intuitive imagination, even if we know well that from the time of Heisenberg (1927) we cannot expect absolutely exact measurements (which was known to engineers earlier, even if for different reasons) and thus an absolute truth.[6]

I requires time to test and modify technology products, often a longer one than in took to conceive the original idea, design and to construct them. Theoretically, one can imagine that a new product together with its test can be prepared within 1 year, but practice shows that it often lasts several, and in extreme cases over a dozen years.

Stage (4), the transfer from research environment to industry, exemplifies one of the biggest difficulties in technology development. Knowledge creation in research environments has quite different character than in business or industrial organizations, as it will be discussed in one of further chapters. This results in difficulties in transfer which are sometimes essential since both sides do not understand fully each other. The authors of new ideas try to patent them since they naively believe that, after buying rights to these patents, entrepreneurs will use them and pay the authors well. However, the reasons for buying rights to patents vary, which I know from personal experience (I patented ideas and sold them to industry): one of the main reasons is to protect oneself from the competition by buying a patent and storing it on a shelf to prevent others from using it.

The most telling example of this type of motivation is the history of radio frequency modulation (FM). The idea of frequency modulation was known already in the year 1922 (Carson 1922), but erroneously deemed inefficient. Edwin H. Armstrong patented a system of frequency modulation in 1934, and in 1935 he demonstrated publicly that frequency modulation results in much better quality of sound in radio transmission. However in that time, large radio corporations were interested in popularization of television and actually started to fight the frequency modulation (its adoption would result in large changes in the equipment and in the

---

[6] See next chapter concerning the absence of absolute truth even in the case of synthetic judgments a priori.

concept of development, including the development of television). After a dozen years of struggle some firms started to develop frequency modulation, but they did not respect the patents of Armstrong who sued them several times. In 1954, disappointed with the results, he committed suicide (see e.g. Lewis 1991); the widow of him Marion Armstrong finally won the legal battle, but it lasted until 1967 and required a decision of the Supreme Court of the United States. All this delayed (by 30 years, until about 1970) an universal use of frequency modulation which is fundamental in radio transmission of today; the total civilization delay in this case was more than 60 years.

Very long civilization delays of that type do occur, but are eventually overcome; nevertheless, a fast transfer of ideas from research environments to industry is also rare, and occurs in cases of rapport and trust on both sides. For this reason, large corporations prefer to use technical ideas of their own research organizations. Such industrial research organizations usually do not make breakthrough inventions (with some exceptions, as e.g. in the case of transistors described in Chap. 9), but they are efficient and quickly apply new results of strict and natural sciences. Academic research organizations are usually less efficient in this respect, but produce more breakthrough ideas. Because of such diversified competencies, it is difficult to agree with postmodern philosophy or sociology of science that postulates the concept of *technoscience* (see, e.g. Latour 1987; Ihde and Selinger 2003), in other words a synergy of science and technology motivated by money and power. Such synergy emerges perhaps in some industrial research organizations of large corporations, but it is not typical for academic research that contributes many breakthrough ideas.

Therefore, difficulties in cooperation between the sphere of research and development and market organizations can be observed in most countries. And they are especially strong in Poland, because of three reasons. The first one is constituted by the different manner of knowledge creation in academic environment and in industrial organizations, mentioned above and occurring everywhere, but especially evident in Poland. Secondly, there is a complicated question of conservatism of Polish universities, which manifests itself in curricula and fields of study. Finally, there is the domination of foreign corporations in Polish high technology industry.[7]

In the beginnings of the knowledge based economy these corporations were not interested (during the last 20 years it was so, now it is slowly changing) in cooperation with Polish science and technology. They preferred to treat Poland as a relatively large market and a source of relatively inexpensive and well-educated labour, implementing technical ideas developed in mother countries of these corporations; in return, they could avoid paying taxes in Poland by pushing the

---

[7] All high technology industry in Poland which before 1989 was not very advanced but in many cases very good, was sold to foreign corporations during the first phase of privatization without any conditions on the part of seller. In result, most of the takeovers were *hostile takeovers*—often, as in the case of ELWRO factory of computers, with the layoff of entire personnel and liquidation of equipment, and even of buildings.

profits towards zero with the use of arbitrary sums for technology transfer. For all these reasons it is a wishful thinking to believe that *the time from idea to industry is shortening*; in Poland, it is the opposite.

The delays related to stage (4) can, for all the reasons mentioned above, extend from 1, 2 to 30 years (as in the case of FM).

Stage (5) is the first of the stages implemented mostly by industry and market organizations. A design of industrial mass production, under conditions of its automation, robotization and computerization, would be a standard problem, hence a simple one and relatively fast to prepare, if the premises of the design were known a priori. However, it must be prepared before the mass production has been started, when the extent of demand is not yet known. For me, this example is a basic counter-example for fashionable neoliberal theories proclaiming that since it is impossible to forecast the future, one should not forecast (see e.g. Taleb 2007, where the ideas of Popper from the *Poverty of Historicism* are actually developed further). However, when we design a mass industrial production, we must forecast demand; even if we know that the forecast is uncertain, it is necessary. Therefore, a design of mass industrial production must be elastic, robust to changes in initial assumptions, easy to correct and modify; even a repetition of stage (5) should not introduce additional delays.

The same cannot be said about stage (6) devoted to initial marketing and demand promotion. If we see technology from the perspective of philosophy of technology, as a socio-economic system of applications of technology products, or even from the perspective of economics, as a process in which such products are manufactured, without any deeper insight into how they are created, then stage 6 appears as one of initial stages. But actually this stage is preceded by several stages described above and it may begin several, a dozen years, or even decades in delay to the initial idea of the product.

Let us analyze again, in more detail, the process of *market penetration* of a product signalized earlier, as a dynamic process. This problem was a subject of detailed research, particularly in Japan, where an attempt was made to accelerate such processes, using broad social processes of *technology foresight*[8] for this purpose. One of such processes concerned the production and sales of a small home tele-facsimile (telefax, see Fig. 4.2).

Using the pattern of the curves of market penetration of other products such as suggested, e.g., by colour television in Fig. 4.1, the vision proposed a similar

---

[8] *Technology foresight* is a social process whose aim is to predict technology development—or rather to construct a future of technology by testing and popularizing certain visions of technology development. Japanese experience—as exemplified by Professor Akio Kameoka who organized and participated in technology foresight for many years—indicates the necessity of an incisive elaboration of an initial vision (based on technical experience and intuitive illumination, called abduction by Akio Kameoka who followed American nomenclature in this respect) at the very beginning of the foresight process. Such vision is then tested, modified and popularized through the social character of foresight process that consists of several evaluations of subsequently corrected visions in the so-called *Delphi method*.

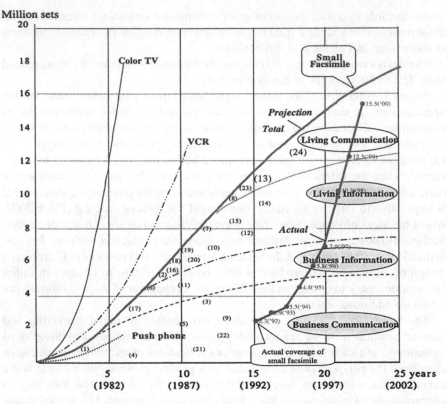

Fig. 4.2 The predicted (by a social foresight process) curve of market penetration of small facsimile as compared to that of colour television and of VCR; the actual curve of market penetration was additionally delayed, but faster. *Source* (Kameoka and Wierzbicki 2005)

though slightly slower curve of market penetration of the small home telefax. The participants of the foresight process were doubtful about the possibility of fast market penetration, hence the curve was slower and contained a delay, mostly an inertial one, as presented in Fig. 4.2.

As we can see in Fig. 4.2, the actual curve of market penetration of small facsimile has shown additional pure delay, initially about 10 years. Later the actual curve was much faster, almost as fast as for the colour television, but the total delay resulting from market penetration only was around 25 years. The initial stages (1),...(5) of the small facsimile demanded also about 15 years, thus the total civilization delay in this case was about 40 years, and this concerns a product of an important, but not a breakthrough character. In the case of digital computer or mobile telephony, the amount of civilization delay might be larger and was actually larger.[9]

---

[9] This was caused mostly by initial stages; stage (6), which is related to market penetration, takes typically around 30 years, for various examples, with maximal speed around 10 % per year.

Stages (7), mass production and marketing, (8), reengineering based on opinions of users and clients, (9), design of new versions based on technical progress and the type of demand, occur after or even in parallel to mass penetration of a new product. If we use only holistic evaluation of a new product, we must wait until these stages before we could influence e.g. reengineering.

Calculation and even short analysis of all these stages help to formulate additional questions: how to shorten this long way from an idea to industry? What to do to improve the cooperation between research and development field, and industry, e.g. in Poland? However, answers to such questions are not the objective of this book; an important conclusion is that some delays in technology development are inevitable and that the awareness of this fact results in a different view of typical assumptions of philosophy of technology.

Admittedly, the informational revolution changes the character of the tools typical for the epoch of knowledge civilization, which are now constituted by computer software. Typical processes of software development have also many stages, e.g. (Sacha 2010) lists six such stages in one type of software development processes, the so-called cascade process. These are: (1) Definition of requirements; (2) Analysis; (3) Design; (4) Implementation; (5) Integration and testing; (6) Application. Other types of these processes, e.g. the so-called iterative process, may have different stages, inclusive of iterative returns to earlier stages. However, the main objective of software engineering is acceleration of software development, e.g. through the introduction of new type of development processes called *nimble.* Therefore, new software systems might be developed in several years. This means that we can develop fast software for *everyday innovations,* say, adding new functions to a mobile phone in order to sell new versions of such phones every year. This does not mean, however, that the informational revolution annihilates the delays of more serious innovations based on *breakthrough inventions,* only that simpler *everyday innovations* are accelerated. This situation creates new dangers that are discussed in a further section.

4.3 Conclusions for Philosophy of Technology and Humanist Education

In which stages of technology development we should participate in order to influence technology development? No doubt, already the stage (3) of testing and evaluation should be influenced; this stage decides about the safety of technical products, and if the philosophy of technology believes that technology is not sufficiently ethical, philosophers should start with participation in that stage.

I can imagine the protests, e.g.: *but a philosopher of technology is not prepared professionally to understand the details of stage (3).* This corresponds to the question of Winner (1993): "where should a philosopher go to learn about technology?". My answer is: *philosophy studies will remain incomplete until they will encompass at least three technical courses: computer science, robotics and biomedical engineering* (perhaps also other ones; I give reasons for such a selection below). Without elementary knowledge of these subjects, a philosopher will not understand contemporary world. Moreover, we should expect a principle of reciprocity in this field. At least for 50 years, the education of engineers has included social and humanist courses (e.g., philosophy, economics, even pedagogy; so why we, engineers, do not encounter reciprocity in this respect?

Mitcham (1994) writes about this issue sufficiently clearly and objectively, but his opinion between philosophers is exceptional: "Engineering studies require engineering students to take humanities courses, but how many engineering courses are humanities required to take?"

Without at least elementary knowledge in computer science, a philosopher in the world of Internet will be crippled. Man will be soon accompanied by robots and using them might become one of fundamental ethical problems; moreover, without a deep, even intuitive understanding of a feedback system behaviour (fundamental for robotics), a philosopher is apt to fall into the trap of alleged paradoxes of vicious circle and infinite recourse, see Chap. 6; for a specialist in automatic control and robotics, these paradoxes are equally invalid as the ancient paradox that Achilles will never overtake a turtle. Biomedical engineering will soon accompany us almost everywhere, both as a cause as well as an effect of growing expected lifetime and ageing of societies. Without knowledge in these fields, philosophers, humanists, sociologists will not be able to understand the changes of new society.

It is necessary to account for new trends in the development of these disciplines, because a modification of inevitable consequences of delays in technology development is possible only if we sufficiently early recognize conflicts and threats that might result form an application of new technology; we see already today that such threats might be serious, see Chap. 14.

So how to control the system of socio-economic applications of technology, if it contains large, pure delays? In such a way as our luggage on the conveyor belt at an airport: *we should select such place of observation so we can watch the luggage from the moment of its appearance on the belt and intervene appropriately, if something wrong happens to the luggage; we must thus compensate the delay by pre-emptive observations and forecast.* In this respect I fully agree with Ihde (2002) who in last chapters of his book stresses the necessity of interdisciplinary involvement of philosophers of technology and of incisive consideration of diverse consequences, including environmental ones, at the early stages of design and development of new technical products. However, Ihde himself probably does not know what an "early stage of design" means in a situation where the delays might extend to 50 years. Therefore, the conclusions of Don Ihde should be modified as follows: *the philosophy of technology should learn in depth the details of recent history of technology together with the development of technical thought through the last 50 years, including not only the technology products that have already appeared on the market or generally in socio-economic application, but also ideas and stages of development of new products* in statu nascendi.

Here I imagine the response of philosophers: "But this is impossible, you say yourself that we represent a different episteme, how can we penetrate technology to such an extent?" I respond that, firstly, one cannot practice the philosophy of technology without understanding the episteme of technology. Secondly, philosophers of mathematics know the recent history of mathematics well. Thirdly, the approach in classical philosophy of technology boils down to an *ex post* evaluation of technology, not an evaluation *in statu nascendi,* so it might be too late to make corrections if something wrong happens. And such corrections are necessary in an obvious way; moreover, they might be crucial for the *survival of human race.* I mentioned earlier that one of possible interpretations of the *eerie silence* (Davies 2010), the lack of response from cosmos to radio signals indicating the presence of intelligence on Earth, which have been sent by us already for 50 years, is the possibility of self-annihilation of civilizations that develop in an avalanche-like manner as a result of the positive feedback loop between science and technology on the one side and their socio-economic applications, self-driven by market forces on the other. If such a threat exist, we cannot avoid responsibility: even if the visions of technologists stem from a fully positive motivation, they might have catastrophic consequences. For example, from its beginning robotics was motivated by the vision of liberation of people from heavy work; when this actually happened, we have problems with unemployment, but more dangerous might be drones, warlike robots; who will be responsible for controlling them?

4.4   Forecast and Foresight of Technology Development

The delays in technology development have also an advantage: if we know them well and know the dynamics of technology development, we can reasonably forecast the results of such processes. An analogy is constituted by a situation when we pour water in a glass; we can say in a chosen moment "in a second this glass will be full", or when you observe your suitcase on a conveyor belt: "in five second I will have the suitcase in hand". Of course, it is much more difficult to forecast technology development than in these examples. But if we limit our efforts to a specific type of technology products and if we know the dynamics of development of such products well, we can forecast the results of the process of development with large probability.

However, the above shall be considered against the background of general issues of *forecasting, predicting, constructing future.* By *forecasting* I understand any future-related statements of diverse character, including conditional ones, scenario-type ones, etc. By *predicting* I mean a specific forecast stating the most probable course of future events (for the author of the prediction, many of them are strongly subjective).[10] By *constructing future* I understand forming a vision of a

---

[10] The probability of a prediction depends on its specificity: a more general, less specific prediction has a larger probability of being correct. An absolutely specific prediction has zero probability of becoming true.

desired but possible future and then acting to realize this vision (e.g., we construct future when we build a house). Constructing future is actually a type of forecasting, only with our active participation.

Recently, it has become fashionable to question the possibility and point of all forecasting (see, e.g., Taleb 2007). We can suppose, however, that such position is politically motivated by a neoliberal attack on the role of the state in the economy (if it is not possible or reasonable to forecast, then any intervention of the state in the economy is irrational; compare such ideology with the actual state interventions after the great crisis of 2008). Another reason is postmodern belief in chaotic development; actually, it is true that after the informational revolution *everything flows* and *order emerges out of chaos,* but this mostly concerns such socio-economic events that are influenced by many causes and their dynamics is not well known. For processes of well known dynamics and specificity we can forecast or even predict with increasing probability (just realize how weather forecasts gradually improve). Some types of forecasts are much more precise than economic or general social ones. Such are e.g. demographic forecasts, where the character of fundamental dynamics is relatively simple and constant (in a year, people can be born, become a year older or die), only the impact of external and subjective factors (the average number of children born by a woman, the percentage of deaths etc.) is less precisely known.

Finally, *forecasting or predicting is a necessary condition of all rational human behaviour and the development of human civilization.* We would not organize airborne transportation if we could not predict the time of flight between airports. *Even if an absolutely precise forecast is obviously impossible, contemporary civilization cannot function without at least an approximate one.* This is the fundamental paradox of forecasting (see, e.g., Wierzbicki 2012) and, as most paradoxes, it can be resolved with the use of a triple-valued logic[11]: it is not important whether a forecast is absolutely right or wrong, but whether it is sufficiently precise and correct for a given application (and this is already a third logical value).

This concerns in particular the development of technology. Clearly, we cannot predict new, breakthrough inventions; but if we know well the state of development of a given field, we can forecast with the use of various methods (conditionally, by scenarios, etc.) the development of technology products relatively well and exactly. *The delays, both pure and inertial, if we know them well, make forecast easier* (when we notice our suitcase on the conveyor belt, we know what to expect). Thus, there is a specialized discipline called *technology forecasting* or *technology assessment,* together with its scientific journals etc. Nevertheless, it was noted that the danger of subjectivism for this discipline is significant, probably because of marketing attempts related to future technology products to dominate it. Due to the above, another method of forecasting, or, strictly speaking, constructing the future of technology, has been developed and called *technology foresight.*

---

[11] For more detailed comments see Chap. 6. Łukasiewicz (1911) justified his axioms of triple-valued logic by commenting that statements about future have a third logical value: "might be".

Technology foresight is a social process of constructing future, originally developed for new technological products,[12] started by RAND corporation more than 50 years ago in the USA but practically developed mainly in Japan. It begins with a creation of an initial vision (forecast or desired direction of technology development) and related scenarios, followed by a verification and modification of the vision and scenarios through a broad social consultation provided by many experts (including technologists, entrepreneurs, state and local government officials, etc.), while using special methods and techniques of information processing and groupware: panels, survey questionnaires, rankings, Delphi method etc. An important goal of such consultation is to build social consensus and support for the final version of the vision. Japanese experience in foresight was extensive and very positive; in Europe, first attempts to use foresight processes occurred over 30 years ago in Finland, and more recently this method is often used in the European Union.

Both *technology forecasting* and *technology foresight* are fields on which possible cooperation between philosophy of technology and technology proper can be stronger.

4.5 Breakthrough Inventions and Everyday Innovations

In one respect it is justified to question the possibility of more precise prediction: when it concerns *breakthrough inventions*.

During the informational revolution we observe a dissemination and acceleration of multimedia advertisements of new products, which results in an impression that the world suddenly accelerated. At the same time, the concept of *innovativeness* became a main slogan in the analysis and stimulation of economic competitiveness. Innovativeness is analysed in most new economic approaches (see, e.g., Arthur 1994), often in terms of *path dependence* (innovativeness is necessary for competitiveness, but does not secure optimal choice of new products, because the implementation of innovations depends on a cultural and socioeconomic path, oligopoly fights, selection of standards etc.). For a more detailed analysis of the conditions of innovativeness see e.g. (Wierzbicki 2010). We should stress here the fundamental difference between a *discovery, invention* and *innovation. A discovery* is often scientific, concerns a new scientific law or principle (but it can also have direct applications such as the discovery of vaccine against rabies by Pasteur). *An invention* is also a discovery, but aimed at broad applications. Nevertheless, it is usually technical, and later it stimulates both science and socio-economic applications (as the invention of a telescope stimulated both astronomy and optics as well as the captains of Dutch ships).

---

[12] The concept of *foresight* is now very popular in the European Union, and it has been used not only for technology development but also to deal with various socio-economic problems.

Finally, *innovation* commonly has a broader meaning that partly intersects with invention, but denotes also other improvements of products (or technological production processes) to ensure their better utility and larger market demand for them. Because of this ambiguity of the concept of innovation, it is necessary to distinguish its diverse types. There might be many of those, but here we shall distinguish two fundamental but extreme ones (there might be a gradient of intermediate types between them):

- *Breakthrough innovations,* which are usually *inventions* of large socio-economic importance; they are difficult if at all possible to predict, but have usually, as shown in this chapter, large delay times; their gradual dissemination or socio-economic penetration results in large economic, social, or even habitual or cultural changes; an example of such innovation is mobile cellular telephony.
- *Everyday innovations,* consisting of gradual improvements of products to increase their utility and competitiveness, hence the market demand; an example of such innovation is the service of positioning and navigation in a mobile telephone (equivalent to an integration of two earlier breakthrough innovations: cellular telephony and satellite positioning system, GPS).

It is understandable that market organizations concentrate on everyday innovations and introduce breakthrough innovations reluctantly (because the latter require fundamental changes in technological processes, new investments, new personnel etc.).

Sometimes we encounter a thesis that *in the past, breakthrough innovations occurred more often than today,* e.g. they were more frequent in the 19th century than in the 20th century. At the same time, we encounter a thesis that the distance from idea to industry has shortened. Both are incorrect and do not correspond to facts, but we can understand the reasons underlying their occurrence. Since market organizations exert pressure towards everyday innovations, there is not as much economic motivation and intellectual effort aimed at breakthrough innovations. On the other hand, everyday innovations require a fast path from idea to industry. If innovations had been determined by money, these theses would be correct. But as innovations are determined by money to a certain extent only, it is actually a wishful thinking.

As highlighted in this chapter, breakthrough innovations are improved and disseminated with large time delays; here is another example. A breakthrough innovation of *transitory computer memory* that retains its state long after switching off the power supply, based on NOR or NAND gateways with large capacities and entrance resistances, was discovered in 1984 by Fujio Matsuoka from Toshiba corporation, but the company did not use that innovation on the market. A decade later some young researchers from Singapore tried to market this invention, but it was difficult to start the market dissemination (the transitory memory was developed to replace magnetic floppy disks, then a dominating form of transferrable computer memory). Finally in the year 2000, a company from Singapore, TREK, introduced a transitory memory in the form of a *pendrive plug in USB standard* (originally called Thumb Drive by the company). This invoked interest of large corporations, hard competition and patent fights, etc. But today, pendrive memories have replaced floppy disks almost completely, even if it will take another 20 years (until around 2030, because inertial delays of high technology products do not shorten much) for a full market dissemination of pendrives as personalized computer memories. Thus, even if the pure delay of this very much needed innovation was very short, only 16 years, the total delay will anyway exceed 40 years.

We could indicate at least over a dozen of other examples of recent breakthrough innovations; some of them did not yet start a broad market penetration and remain at the stage of pure delay. Anyway, the thesis that breakthrough innovations are less intensive today is not confirmed by a more detailed scrutiny. Possibly this is because the main driver of innovations, in particular breakthrough innovations, is human curiosity[13]; profit expectation might be also important but it is secondary. Therefore, the focus of profit expectation on everyday innovations does not substantially influence the frequency or even the speed of penetration of breakthrough innovations.

Everyday innovations are necessary for knowledge based economy, but they rely on an only slightly delayed use of already existing knowledge (a year or two, otherwise the competition will outrun us). We should, however, deeply understand the determinants of everyday innovations resulting from the informational revolution. Since they are discussed in more detail in the closing chapters of this book, here I mention them shortly: *oligopoly on high technology markets, the diversity of software tools, conflict relating to the ownership of knowledge ("the intellectual property") versus the importance of the intellectual heritage of humanity, etc.*

In this section, I have stressed the necessity of a deeper understanding of the differences between breakthrough innovations (or inventions) and everyday innovations. The essential difference lies in the time delay related to them: usually very long one for breakthrough innovations, but rather short one for everyday innovations. Moreover, breakthrough innovations have quite different character, they result from human curiosity, not from market forces, and their socio-economic penetration is not only connected with large delays but it is also a slow process. Everyday innovations result from economic activity, they are motivated by the need to offer new versions of products to buyers again and again, to shorten the economic lifetime of products. *Thus, the impression that changes in the world accelerate results from the pressure of media advertising the everyday innovations, new versions of products, attempts to convince consumers that they should buy a new car every 3 years, a new computer every 2 years, and a new mobile phone every year.*

---

[13] For me, human curiosity is an attribute that co-defines the concept of humanity—together with tool making and speech.

References

Arthur, W.B.: Increasing Returns and Path Dependence in the Economy. Michigan University Press, Ann Arbor (1994)

Arthur, W.B.: The Nature of Technology: What It Is and How It Evolves. The Free Press, Simon and Schuster, New York (2009)

Carson, J.R.: Notes on the theory of modulation. Proc. IRE **10**(1), 57–64 (1922)

Davies, P.: The Eerie Silence: Renewing Our Search for Alien Intelligence. Harcourt, Houghton Mifflin (2010)

Ellul, J.: The Technological Society. Knopf, New York (1964)

Gadamer, H-G.: Warheit und Methode. Grundzüge einer philosophishen Hermeneutik. In: Mohr, J.B.C. (Siebeck) Tübingen (1960)

Grzegorek, J., Wierzbicki, A.P.: New statistical approaches in the systemic analysis of regional, intra-regional and cross-regional factors of information society and economic development. Mazowsze: Studia Regionalne **3**, 117–128 (2009)

Habermas, J.: Lectures on the Philosophical Discourse of Modernity. MIT Press, Cambridge MA (1987)

Heisenberg, W.: Über den anschaulichen Inhalt der quantentheoretischen Kinematik und Mechanik. Zeitschrift für Physik **43**, 172–198 (1927)

Ihde, D.: Bodies in Technology. University of Minnesota Press, Minneapolis—London (2002)

Ihde, D., Selinger, E.: Chasing Technoscience: Matrix for Materiality. Indiana University Press, Bloomington (2003)

Jackson, M.C.: Systems Approaches to Management. Kluwer Academic—Plenum Publishers, New York (2000)

Kameoka, A., Wierzbicki, A.P.: A Vision of New Era of Knowledge Civilization. Ith World Congress of IFSR, Kobe (2005)

Kuhn, T.S.: The Structure of Scientific Revolutions, 2nd edn., 1970. Chicago University Press, Chicago (1962)

Latour, B.: Science in Action. Open University Press, Milton Keynes (1987)

Lewis, T.: Empire of the Air: The Men Who Made Radio. E. Burlingame Books, New York (1991)

Łukasiewicz, J.: O wartościach logicznych (On Logical Values). Ruch Filozoficzny **I**, 50–59 (1911)

Mitcham, C.: Thinking Through Technology: the Path Between Engineering and Philosophy. The University of Chicago Press, Chicago-London (1994)

Popper, K.R.: Three views concerning human knowledge. In: Lewis, W.H.D. (ed.) Contemporary British Philosophy, Third Series. Allen and Unwin, London (1956)

Popper, K.R.: Open society and its enemies. Tłumaczenie polskie (1987) Społeczeństwo otwarte i jego wrogowie. Wyd. Krytyka, Warsaw (1962)

Sacha, K.: Inżynieria Oprogramowania (Software Engineering). PWN, Warsaw (2010)

Strużak, R.: Evolution of spectrum management concepts. 18th International Wroclaw Symposium and Exhibition on Electromagnetic Compatibility, EMC 2006, Wroclaw, pp. 368–373 (2006)

Taleb, N.N.: The Black Swan: The Impact of the Highly Improbable. Random House, New York (2007)

Wierzbicki, A.P.: Prinzip maksimuma dla processov s nietrivialnom zapazdywanijem upravlenija. Avtomatika i Telemekhanika **10**, 1970 (1970)

Wierzbicki, A.P.: Finansowanie nauki w krajach rozwiniętych na progu gospodarki opartej na wiedzy a sytuacja nauki w Polsce (Financing of Science in Developed Countries on the Verge of Knowledge Based Economy and the Situation of Science in Poland). Przyszłość: Świat, Europa, Polska. **2**, 105–119 (2008)

Wierzbicki, A.P.: On the Possibility of Prediction, pp. 61–78. Przyszłość: Świat, Europa, Polska 1/2012 (2012)

Wierzbicki, A.P.: Warunki kreatywności i innowacyjności dla Europy i Polski (Conditions of Creativity and Innovativeness for Europe and Poland). In: Kukliński, A., Pawłowski, K., Woźniak, J. (eds.) Kreatywna i innowacyjna Europa wobec wyzwań XXI wieku (Creative and Innovative Europe Facing the Challenges of XXI Century), pp. 63–72. Biblioteka Małopolskiego Obserwatorium Polityki Rozwoju, Cracow (2010)

Wierzbicki, A.P., Nakamori, Y.: Creative Space: Models of Creative Processes for the Knowledge Civilization Age. Springer Verlag, Berlin-Heidelberg (2006)

Wierzbicki, A.P., Nakamori, Y. (eds.): Creative Environments: Issues of Creativity Support for the Knowledge Civilization Age. Springer, Berlin, Heidelberg (2007)

Winner, L.: Social Constructivism: Opening the Black Box and Finding It Empty. Science as Culture 3(16), 427–452 (1993)

Chapter 5
Rational and Evolutionary Technical Theory of Intuition

5.1 Introduction

In this chapter, *rationality* of a theory is understood in a specific sense, not Platonian but close to the understanding by Popper (1934, 1972) and Quine (1964): a theory is rational if it can be deduced from abstract principles and reasonable assumptions, but is also empirically sound, according to Quine it *touches the reality at its edges* and according to Popper it can be *falsified experimentally* or, at least, allows to draw conclusions that can be checked in practice. The assumption of *evolutionary* character of this theory is more metaphysical, see the next chapter. Here, I assume that *all our knowledge of the history of the development of human civilization indicates that this development is evolutionary,*[1] and that intuitive abilities of humans may be explained in terms of this evolution. Finally, this theory is *technical* because I use elements of technical sciences, of telecommunications and computer science, to explain the power of intuition.[2] An additional, subsidiary conclusion *is an explanation of the reasons and an approximation of the extent of redundancy of human brain* together with its *inclination towards transcendental reasoning.*

In a common language of today, the adjective *intuitive* is used with a strong tint of *irrationality*, which has historical reasons. Intuition was a subject of philosophy from its beginnings, or at least from Plato (with the famous example of intuitive solution provided by an uneducated slave boy in answer to a difficult problem of

---

[1] This assumption does not imply belief in a primitive, neo-Darwinian (which usually means pseudo-Darwinian) individualistic evolutionism; see a discussion of diverse variants of evolutionary epistemology that usually follows the individualistic perspective of Western culture in Nowak (2007). Instead, I assume that the evolution of human civilization, similarly as for great apes, had a group character: in such development, innovations or new ideas were accepted if they had positive impact on the evolutionary survival of the group, an extended family or a tribe, not on the evolutionary survival of an individual; this concerns also metaphysical ideas.

[2] This chapter uses material from my several papers or book chapters concerning intuition, see Wierzbicki (1997, 2004, 2008), Wierzbicki and Nakamori (2006, 2007).

© Springer International Publishing Switzerland 2015   79
A.P. Wierzbicki, *Techne_n: Elements of Recent History of Information Technologies with Epistemological Conclusions*, Intelligent Systems Reference Library 71,
DOI 10.1007/978-3-319-09033-7_5

how to divide a square into two equal smaller ones[3]). From Plato to Kant, intuition was considered to be an internal source of sure and unfailing true knowledge, at least in mathematics. I recall here a quotation from John Locke (1690): "And this I think we may call intuitive knowledge. For in this the mind is at no pains of proving or examining, but perceives the truth as the eye doth light, only by being directed towards it. Thus the mind perceives that white is not black, that a circle is not a triangle, that three are more than two and equal to one and two. Such kinds of truths the mind perceives at the first sight of the ideas together, by bare intuition; without the intervention of any other idea; and this kind of knowledge is the clearest and *most certain*" (mine emphasis) "that human frailty is capable of".

However, it was mathematics that later discredited the infallibility of intuition, first through the development of non-Euclidean geometry, e.g. the geometry of figures on a sphere. Today we know that e.g. the verities quoted by Locke are not absolute: a circle is topologically equivalent to a triangle, and two plus one modulo two is not three but one, which is less than two. From the time of non-Euclidean geometry, philosophy contended with the concept of intuition, interpreting it diversely: differently in metaphysics (e.g. Bergson 1903); differently in mathematics (e.g. Poincaré 1913, Brouwer 1922); differently in phenomenology (Husserl 1973); differently as part of the concept of *tacit knowing* or tacit capabilities of humans (Polanyi 1966), differently by Lakatos (1976). An essential problem was the use of intuition in metaphysics and the issue of infallibility of intuition. Pierce et al. criticized the use of intuition in metaphysics (according to my opinion, unjustly, see Motycka 2010 and the next chapter) and the belief in infallibility of intuition (in my opinion, this criticism was correct, starting with the concept of fallibilism of C.S. Pierce with which I fully concur; see also the arguments of our *mesocosmic,* hence limited view of the world in Vollmer 1984). The most convincing is the approach of Król (2005, 2007) together with his use of the concept of *hermeneutic horizon.* However, here I present something different: an evolutionary, rational and technical explanation both of great power and fallibility of intuition, provided by me in Wierzbicki (1997, 2004) and derived from Japanese inspirations.

In contemporary mathematics, a strong impact on the understanding of intuition was exerted by Brouwer (1922) who reckoned that new mathematical theorems appear in the brain of a mathematician intuitively but they must be correctly proved, in a constructive way, *excluding the proof by contradiction,* because he believed that

---

[3] Here I shall answer as a technician: cut through a quadratic slab along its diagonals, from the resultant four triangular slabs compose (glue together) two quadratic slabs. A mathematician will not accept this as a solution of the mathematical problem, but this example illustrates the difference of a technical and mathematical episteme and indicates a loophole in the reasoning of Plato (or Socrates): it is technical experience and not an abstract reasoning which could suggest the solution to the slave boy (who might have been a carpenter). However, technical imagination bears some similarity to mathematical reasoning: sometimes they both have visual aspects, which is positively evaluated by technicians but often negatively by mathematicians (proofs should not be based on visual intuition).

the rule of excluded middle is faulty (in my opinion, later development of various types of logics has shown that Brouwer was fully right, see a more detailed discussion in the next chapter).

Despite or possibly because of all these discussions, the adjective *intuitive* is still commonly used today with a strong tint of *irrationality*. This might be the impact of Bergson (1903) who believed that intuition is a very important aspect of human behaviour but treated it as a mystical force, by definition not subject to any rational analysis. Even today, after over 100 years, I would decide better not to prove that intuition is fully rational; I only want to explain its functioning and power in rational terms, from the perspective of evolutionary naturalism and technology, while admitting it can be fallible. In this, I differ essentially from a philosophical tradition that treats intuition as a cognitive power which should "provide knowledge certain and necessary" (Motycka 2010) or interprets this cognitive power as a transcendental, extra-sensual ability (see also Jung 1971). There are many publications today (e.g. Gigerenzer 2007) that stress the importance of intuition, but I felt an absence of researches trying to explain and understand more deeply its functioning. For instance, this belief in the irrationality of intuition is so strong today that a very good book on human intelligence (Nęcka 2003) does not use the concept or even the word *intuition* at all. For me, however, intuition is an essential element of intelligence, hence it requires a scientific explanation. This need is the more urgent since contemporary approaches to the issue of knowledge creation, see Chap. 12, stress the interdependence of explicit, rational knowledge with *tacit knowing* (Polanyi 1966) or even *tacit knowledge* (Nonaka and Takeuchi 1995), while intuition is one of important elements of tacit, preverbal knowledge.

5.2 Technical and Evolutionary Explanation of Intuition

The technical explanation relies on a comparison of informational requirements for two types of human perception and reasoning: *verbal* and *preverbal*. The latter type is understood similarly to the immanent perception in phenomenology, using all senses but excluding verbal communication, with preverbal hearing, without words, included. Nevertheless, I am not only interested in perception, but also in the power of human reasoning. In order to compare the cognitive power of verbal and preverbal perception and reasoning, I use the results of technical sciences, in particular, telecommunications and computer science. An estimation, with some margin,[4] "at most", of the informational complexity of speech can be based on the frequency band needed to transmit sound, ca. 20 kHz. Another estimation, this

---

[4] This is because our hearing helps also in many kinds of preverbal perception: a sense of direction, recognition of a person speaking based on the timbre of their voice, recognition of a melody etc.

time incomplete, "at least", of the informational complexity of only one element of preverbal, immanent perception can be based on the frequency band needed to transmit an image, ca. 2 MHz. This band remains in almost no relation with the frequencies of light waves, but in a strong relation with the manner of the reception and interpretation of images by human eye.

Here we need a digression: there is an opinion that appeared in the philosophy of technology (Ihde 1976, 2002) that sound carries more information than vision, but it does not stand a technical critique. Don Ihde gives various justified arguments amounting to the fact that the sense of hearing relates to much more than understanding speech (the perception of direction, often better than visual, perception of music with all its complexity, recognition of the timbre of voice a person etc.). However, his main argument is faulty. He says: "The range of auditory perception with respect to the continuum of wave phenomena (for humans from 20 to 20,000 hertz) is vaster than for seeing within the optical range of light waves". This comparison contains puzzling or even absurd errors: the range of wave frequencies measured by the proportion of the upper frequency to the lower frequency is a physical, not an informational parameter and since we often use wave modulation for transmission, it has nothing to do with the informational content of a frequency band. It would be better first to ask information technology specialists about the wave bands they need, and only then try to present speech and music in comparison with vision, with the assumption of a similar quality of transmitted signals. According to the information theory (Shannon 1948) we transmit the same amount of information if the frequency band of sound is shifted by modulation to the frequency range 100–120 kHz (to long waves), or to the range 100–100.2 MHz (to ultra-short waves). Even if the ratios of the upper frequency to the lower frequency change drastically with modulation, the transmitted amount of information is the same and is characterized by the absolute breadth of the band, the *bandwidth* of 20 kHz (0.2 MHz) for sound.[5] The lower 20 Hz of frequency is meaningless, amounts to only 0.1 % in the informational sense (it is important only if one wants to secure a good quality of lower tone music in the loudspeakers), thus the argument about the ratio of upper frequency to lower frequency is absurd. However, 20 kHz for sound is an upper estimate of the complexity of verbal information, because (as rightly noted by Don Ihde) sound carries much more information than just verbs.

The situation is much more complex for vision, treated as a lower estimate for preverbal, immanent perception, but the range of light waves is of almost no importance: colouration is represented by three (or recently four) basic colours and signals about their intensity. Two issues are important: the number of light points (pixels) on the screen, and the manner in which changes in the informational description of a light point are coded. In the old black-and-white television, we used only ca. 20,000 pixels, each of several (say, four) degrees of greyness, and a frame rate of 25 times per second, which resulted in the need of a bandwidth of

---

[5] For speech transmitted by a telephone even a band of 3.4 kHz is sufficient.

$20,000 \times 4 \times 25 \approx 2$ MHz. Over the years of development of television this bandwidth was accepted as a basic standard, but the focus was on the development of image quality. Today, for the needs of colour television that requires encoding of three basic colours, we use the bandwidth of 6 MHz, although image quality requirements have much increased. We use much larger numbers of pixels, each basic colour has more degrees of intensity (more information for each pixel), image is repeated 100 or even 200 times per second (in order not to distress the eye of the viewer); how was it possible to increase the quality that much without changing the basic bandwidth? To achieve this, we use the possibility of encoding the changes in the informational description of a pixel in a similar manner as they are encoded in the human's eye retina and as human brain interprets signals it receives: we concentrate our attention not on a full blotch of colour, but only on the rand of the blotch and its changes over time. Therefore, instead of coding the information about each pixel independently, it is sufficient to code only how the information is changing over time and in comparison to adjacent pixels. Such coding is known as the compression of the image code; in contemporary television, it results in a much more efficient utilization of the standard bandwidth and in transmission of much better images, even if still of worse quality than the images perceived by a naked eye. The latter is evidenced by the development of digital photography industry, with its companies producing digital cameras competing in terms of image quality and achieving many millions of pixels, while colour television remains within a range of one million of pixels.

With all the complexity of this issue, we need here only a *lower estimate* of informational complexity of image transmission, once again treated as a lower estimate of the complexity of immanent, preverbal perception. For this lower estimate we can take the old standard bandwidth of 2 MHz, which enables taking into account the similarity of compression of image code to the functioning of human eye and brain. We should stress again that television signals are transmitted also with modulation, so only bandwidth is important, not the physical frequency band. To send an image we need *at least* 2 MHz of bandwidth, which is at least 100 times more information than when we transmit sound. *Therefore, vision is in informational terms at least 100 times stronger than sound,* and this is not any "traditional prejudice", as suggested by Don Ihde.

Therefore, if we take into account that immanent, preverbal perception uses all the senses (smell, touch, taste, preverbal aspects of hearing, and vision), then we again conclude that *the ratio of informational complexity of preverbal and verbal perception is at least 100:1.*

In order to further interpret this ratio we shall use achievements of another discipline, the computer science, and in particular the computational complexity theory. This theory is rather elaborate, but in great abbreviation it holds that computational complexity depends on the problem type and on the informational complexity of data, while the latter dependence is usually strongly nonlinear, often exponential or combinatorial, and polynomial dependence occurs only for really simple types of problems (linear dependence is exceptional). However, we need here only a lower estimate, of the type *at least,* hence we can assume a relatively simple nonlinear polynomial dependence, e.g. quadratic (since human brain or mind often solves rather complex information processing problems). It follows from that that the ratio of computational complexity of preverbal and verbal information processing in human mind[6] is at least 10000:1; *a picture is worth not a thousand, but at least ten thousand words.* A measure of this complexity can be e.g. the number of neurons needed for vision or speech processing. It is known that the number of neurons in human brain is about 10^{11}–10^{12}, but we do not know why we have that many neurons. Other apes have about the same number of neurons in their brain and developed diverse methods of communication, but not speech; perhaps they needed that many neurons because in their contacts with the environment they relied mostly on vision and on immanent perception?

We know today that our minds process signals in a parallel and distributed fashion, while the natural neural network is much more complex than artificial neural networks. For example, it uses rather multi-valued logics than classic binary one. Biological studies of a neuron suggest that a correct mathematical model of a neuron should be dynamic and nonlinear. For a precise modelling of a single neuron we would rather need a computational capability of contemporary personal computer, much more than a single binary flip-flop or a sigmoidal function (the latter is used in artificial neural networks to represent a single neuron).

What is the relation of the above to evolution and intuition? I shall explain it in an evolutionary thought experiment, which I later call shortly *dual thought experiment,* that concentrates on the question: *how people perceived reality and reasoned before the evolutionary invention[7] of speech?* In those times, people coped with their environment quite well, hence their preverbal perception and reasoning had to be strongly developed. Thus, *the invention of speech was a radical evolutionary shortcut,* it simplified the perception and reasoning at least 100, perhaps 10,000 times. It started (or rather co-defined, depending on what we count as the essence of humanity, whether only speech and communication, or also tool-making and curiosity) the evolution of civilization or rather of human communities. However, it also resulted in the concentration of rational reasoning within verbal communication, it suppressed to some extent the powerful possibilities of preverbal perception and reasoning, which, with no other words, we can call intuitive reasoning, even if a deeper analysis shows that it has three most important elements: *intuition, emotions* and *instincts;* the latter two can be treated together, because they are mostly inherited, while intuition is only partly inherited, mostly learned.

Since the evolutionary and rational, even if technical, theory of intuition will be further discussed and analyzed in the next sections, here I shall present some

---

[6] With all possible limitations concerning a comparison of human mind to a computer that should be taken into account, it should be noted that the computational complexity theory estimates complexity *in principio,* independently of a computer type. See further remarks concerning the concepts of brain and mind.

[7] The word *invention* is used here in a figurative sense; it was rather a revolution in the civilization development and, as any revolutionary stage in the punctuated evolution of civilization, the development of speech must have taken at least several generations.

comments on the above arguments. They explain rationally the power of intuitive reasoning, as well as the fact that historically, from Plato, through Descartes, Locke, Kant, many philosophers referred to *visual* aspects of intuitive perception of truth. On the other hand, none of above arguments suggests that intuition would lead to *infallible* truth; I believe, as a technician should, that even the best idea requires testing. Therefore, intuition is powerful, it is the source of our ideas, especially in technology that relies on intuitive creation of tools and artefacts, but it is fallible, new tools require testing or falsification.

The dual thought experiment suggests that *the invention of speech was an excellent shortcut that changed the nature of human evolution,* both in its individual and group aspects. It turned out that we could process signals from our environment at least 10^4 times faster than before that invention. This enabled the transfer of information and tradition between subsequent human generations, we started to build a *cultural and intellectual heritage of humanity;* the rational, verbal part of this heritage can be identified as *the world 3,* a concept introduced by Popper (1972). Biological evolution of people slowed down (some biologist even say that it stopped), we substituted this by an acceleration of cultural, intellectual and civilization evolution.

Many biologists consider *why* our biological evolution has stopped; this has led Alfred R. Wallace to formulate *Wallace paradox:* why primitive human tribes have practically the same brains as most advanced people in the world, if they cannot use those brains as intensively? It appears that the above discussions give a sufficient explanation, *why* it has happened: after the invention of speech it turned out that *we have an enormous excess of the capabilities of brain and mind.* This excess can be estimated as 10^4, ten thousand times, in other words, only 0.01 percent of our brain capacity would suffice for language processing and logical reasoning, hence it was not necessary for our brain to further develop biologically.[8] This tremendous surplus explains many problems, with some philosophical questions included, e.g. the question of an apparent *redundancy* of the concept of mind in relation to the concept of brain, see e.g. (Żegleń 2007). If we have such large excess of brain capacity over the rational, verbal needs, and we imagine the functioning of our brains in a simplified, cognitive verbal model, then we can under-estimate this excess and maintain that our minds are larger than brains. This concerns an under-appreciation of the actual capabilities of our minds, even if we treat them in a fully naturalistic sense as an ensemble of higher functions of our brains emerging from more elementary functions, according to the emergence principle, because of the high complexity of our brains.

---

[8] Some evolutionary epistemologists believe, see the discussion of this problem e.g. in Nowak (2007), that the invention of speech required the growth of our brains. However, such opinions do not take into account estimations of computational complexity. A modest increase of human brain did occurred, our brain is several times larger than in other apes, but this growth could be as well a result of using the excess (as compared to the needs of speech) capacity of our brains for the purposes of intuitive construction of more complex tools.

This tremendous excess explains also the functioning and the higher role of the *inward man* when compared to the *outward man* (see e.g. Jung 1971, Motycka 2010). It also provides a different understanding of the human predisposition to *transcendence,* understood here not as the existence of an absolute being above reality, but as our predisposition to abstract thinking, to transcend our *existential* experience. Evidently such abstract or even metaphysical reasoning was useful in the evolution of civilization, and it was possible because of the excess of our brains. It was well understood by Karl Jaspers who in his biography (Jaspers 1993, p. 131; quote after Polish translation) admits: "I write about an issue objectively, if I have to say something as a researcher; if I philosophize, I am guided by the sense of encompassing".

This tremendous excess of our brains and minds was utilized also pragmatically in the evolution of civilization, for the purposes of intuitive creation of tools and artefacts. There is no doubt that *the development of tools occurred concurrently with the development of language,* together with a self-supporting, positive feedback between both these processes. The development of language released this excess of brains and minds, enabling its use to create tools; new tools required in turn new words, at least in order to transfer, as part of the group evolution of civilization, the ability to construct and use the tools to next generations.

With all these successes and advantages of civilization development, the concentration on language had also certain disadvantages, as any simplification. When searching for better methods of convincing our interlocutors, we invented binary logic and the rule of the excluded middle (*"that must be true or untrue, there is no third way!"*). On one hand, binary logic brought in important civilization achievements, computers, telecommunication networks, then computer networks, the current informational revolution and the gradual advancement of the knowledge based economy and civilization. On the other hand, an excessive concentration on binary logic limits our understanding of the world, results in misunderstandings, also in ontology and epistemology (see the next chapter). The best example of such limitation is *cognitivism,* including the conviction that all our cognitive processes, together with perception, memory, learning, are based on a verbal medium, on a *language of thought,* see, e.g. (Fodor 1994); (Gardner 1985), and that all functioning of our brains (and perhaps of entire world?) can be modelled *computationally,* as the functioning of a giant computer. However, the dual thought experiment shows precisely where lies the error in the assumptions of cognitivism. *Cognitivism (in its computational version)[9] is a simplification of the same order as language is a simplification of functioning of the brain.*

---

[9] This does not concern all cognitivist research which includes very important experimental research of human cognitive processes and functioning of our brains. However, it does concern two paradigms present in cognitivism: *computational paradigm* that consists in belief that the functions of our brain and mind can be modelled as a giant digital computer, and the *neural network paradigm* that holds that artificial neural networks with their extremely simplified models of neurons can provide satisfactory similarity to natural neural networks in human brains.

If the language is only a code simplifying the processing of information about real world at least 10^4 times, then each word, necessarily, must have many meanings, and in order to describe our intentions more precisely we must invent new words. When we increase the number of words, we gradually describe the world more precisely, but even faster we discover new properties of the world, which is infinitely complex. If we must represent our knowledge in language, at least for the purposes of its interpersonal verification, and language is only an imperfect code, then absolutely precise, objective knowledge is not attainable. *This is not because of our limitations as cognitive subjects, but because we use limited tools for knowledge creation, starting with language.* The observation that language is only a very limited tool of description of reality was not treated seriously by all philosophy of the twentieth century, starting with logical empiricism. A large impact on such attitude had early works of Wittgenstein (1922) together with his famous (even if absolutely incorrect in the view of the above analysis) statements that "the limits of our language are the limits of our world" or that *"Wovon man nicht sprechen kann, darüber muss man schweigen"* (in a loose translation, if we cannot speak about something, we should be silent). This implied the rejection of all discussions about metaphysics and intuition. From this, a fully erroneous conviction arose that intelligence boils down to language, which drove many currents of thought in the twentieth century, starting with language structuralism and post-structuralism, and ending with cognitivism, constructivism, postmodernism.

5.3   Rational and Evolutionary Definition of Intuition

However, do we still possess our old capability of holistic processing of signals from our environment and memory? I call these capabilities *preverbal,* because we had them before the evolutionary invention of language and we still can observe them in animals accompanying us. The invention of language stopped the development of these abilities, pushed them off to subconscious or unconscious. Our conscious ego, at least its analytical and logical part, defined itself through language, verbal pronouncement. Because processing of words is at least 10^4 times simpler, out verbal, logical, analytic, conscious reasoning utilizes only a small part of tremendous capabilities of our mind developed before the evolutionary invention of speech. However, it seems unquestionable that *the abilities of preverbal, holistic processing remained with us and we call them commonly intuition,* even if we do not know how to rationally use them.

More precisely, I define *intuition as the ability of preverbal, holistic, unconscious (or subconscious, or quasi-conscious[10]) processing of signals from our*

---

[10] A quasi-conscious action can be defined as an action that we consciously start but do not concentrate our whole consciousness on it, such as walking, driving a car, etc. Abilities related to such actions were called *tacit knowing* by Polanyi (1966).

environment and memory, motivated by experience and imagination, an ability which is a historical reminder of the preverbal stage of human evolution. This definition I call the *rational evolutionary definition of intuition.*

The above definition puts weight on the processing of signals from our environment and memory, motivated by experience and imagination; in a sense, it separates diverse aspects of cognitive or decision-making processes that occur in our mind. Let us consider two following aspects: *intuition* and *instinct.* By *instinct* we understand something which is also a preverbal unconscious aspects, but motivated genetically; intuition is motivated rather by personal experience.[11] Therefore, a comment often heard during football matches, "the goalkeeper reacted instinctively" is incorrect, since it implies that the father of the goalkeeper was also a goalkeeper and the present goalkeeper inherited his goalkeeping instinct; actually, the goalkeeper uses his trained intuition.

Let us consider two other aspects: *imagination* and *emotions.* By *imagination* we understand the ability to create situations in our minds that actually did not occur, but could occur (therefore, dreams are elements of imagination); this is an effect of unconscious processing of signals from our memory, hence an important element of intuition. For young people, it can partly compensate their lack of experience in shaping intuition needed by them; hence it is very important to support the development of imagination in children, see e.g. (Piaget 1952). By *emotions* we understand some general states of our minds, such as love, anger, admiration, depression etc. These states are often unconscious or subconscious, they essentially influence the way signals are processed by our minds, hence also influence intuition. Moreover, contemporary neuro-physiological research shows that without emotions we could not make decisions. However, emotions constitute a different dimension than intuition, they are obviously genetically motivated and we shall treat them separately.

According to the rational evolutionary definition of intuition, *intuitive abilities should be related to a distinct part of brain.* This has been observed in the research on brain structure, in neuro-surgery and in research on hemispheric asymmetry of the brain, see e.g. (Springer and Deutsch 1981). These results suggest that a typical left hemisphere of brain (for right-handed persons; for left-handed we observe a reversal of the functions of brain hemispheres) is responsible for *verbal, sequential, analytic, logical, rational thinking,* while a typical right hemisphere is responsible for *non-verbal, figurative, spatial, parallel, analogue and intuitive* thinking. Note the contraposition of rational and intuitive thinking, typical as a result of the breakdown of the old philosophical belief in the infallibility of intuition. According to the research results, (Young 1983) defined intuition as the action of our right brain hemisphere. However, we cannot draw further conclusions from such a definition, e.g. how to stimulate and better utilize it.

---

[11] Even if intuitive abilities can be also inherited, but the use of them depends on personal imagination and experience.

Much more constructive is the rational evolutionary definition of intuition given above and we can derive such conclusions from it.

In order to illustrate these possible conclusions, consider the following finding: *memory related to intuitive thinking should have different properties than memory used in rational thinking.* Indeed, the foregoing is confirmed by the results of recent, independent research on memory functioning (see Walker et al. 2003): the phase of deep remembering occurs in sleep, when our consciousness is switched off.

Note further that every day each man makes many quasi-conscious, operational and repetitive decisions. It concerns well trained actions, e.g. during walking, we do not articulate (even mentally) the will to make next step. Such quasi-conscious *intuitive operative decisions* are common, obvious and simple, thus we do not attach any significance to them. However, we should investigate them in order to better understand intuition. Their quality depends on the level of experience. We trust in our operative intuition if we feel well trained in a given function. Dreyfus and Dreyfus (1986) had proved experimentally that the way of making decisions depends on the level of experience in a given field: it is analytic and rational for beginners and deliberative and intuitive for experts or masters.

We can also formulate an essential question: *does consciousness help or hinder the use of master abilities?* If intuition is an old way of information processing, suppressed by verbal consciousness, then *an expert using of master abilities should be easier after consciousness has been switched off.* This theoretical conclusion of the rational evolutionary theory of intuition is confirmed in practice. Every sportsman knows how important is to concentrate before a competition, and the best concentration is achieved through consciousness-switching off methods, e.g. Zen meditation that was practised by Korean archers before winning the Olympics in Seoul.

This theoretical conclusion can be also applied to creative decisions such as creation of scientific knowledge, formulation and proof of mathematical theorems, creation of new artistic ideas or new tools and technological artefacts. Creative decisions are similar to strategic political or economic decisions, they are usually non-repetitive, one-time decisions. They are reached deliberatively, based on reflection about the whole possessed knowledge and information. They are often accompanied by the *enlightenment effect* (which has many other names: *abduction, aha, illumination, brainstorm, eureka*), a sudden emergence of a new idea. This phenomenon is well known and it has been analysed many times in psychology, philosophy, and history of science, hence we will not analyse it here; it should be only reminded that it relies on unconscious or subconscious creation of a new idea.

5.4 A Model of a Creative Intuitive Decision Process

Before presenting such a model based on a rational theory of intuition, let us first recall the classical analytical model of such process. Simon (1957) defined essential phases of analytical decision process to be *intelligence, design and choice;* later, see e.g. (Lewandowski and Wierzbicki 1989; Wierzbicki et al. 2000) the fourth essential phase, that of *implementation*, was added. On the other hand, a different model of such phases was proposed in Wierzbicki (1997) for creative or strategic, intuitive decision processes:

(1) *Recognition*, which often starts with an intuitive feeling of uneasiness. This feeling is sometimes followed by a conscious identification of the type of the problem.

(2) *Deliberation or analysis*; for experts, a thorough deliberation suffices, as suggested by the Dreyfuses. Otherwise, any tools of analysis or an analytical decision process is useful, with intelligence and design but suspending the final elements of choice.

(3) *Gestation*; this is an extremely important phase, we must have time to forget the problem in order to let our subconscious work on it.

(4) *Enlightenment*; the expected eureka effect may come but not be consciously noticed; for example, after a night's sleep it is simply easier to generate new ideas (which is one of the reasons why group decision and brainstorming sessions are more effective if they last at least 2 days).

(5) *Rationalization*; in order to communicate our decision to others we must formulate our reasons verbally, logically, and rationally. This phase can be sometimes omitted if we implement the decision ourselves.[12]

(6) *Implementation*, which, after rationalization, might be conscious, or immediate and even subconscious.

Especially important are the phases of gestation and enlightenment. They rely on the enormous excess of the capabilities of our brain and mind on the level of preverbal processing discussed above. If conscious, rational thinking does not perturb this, our unconscious mind returns to the problem defined before as most important, not solved yet but forgotten by the conscious ego. There are many cultural institutions supporting gestation and enlightenment, particularly in the Far East. The advice to *empty your mind, concentrate on nothingness and beauty, forget the prejudices of the expert* originating from Zen meditation or Japanese tea ceremony can be precisely considered as useful methods of enabling the unconscious part of work of our mind.

Another important phase is *rationalization.* In classical metaphysics (see, e.g. Morawiec 2009; Piętka 2009) this phase corresponds to *metaphysical reduction*, intuitive substantiation of metaphysical conclusions, actually using deduction in a reverse direction than in logical deduction. Logically, this operation can of course lead to false conclusions, but this objection is countered by metaphysicians with the statement that in metaphysical reduction they seek a substantiation which can provide them with intuitive certainty. According to the evolutionary and rational

---

[12] The word *rationalization* is used here in a neutral sense, without necessarily implying self-justification or advertisement, though they are often actually included. Note the similarity of this phase to the classical phase of *choice.*

theory of intuition, this argument does have some value (intuition is ten thousand times stronger than logical deduction), but with the reservation concerning its absolute character: even intuition can be fallible.[13] The effect of enlightenment can bring a false idea and its rationalization might be erroneous. Logics can also lead to erroneous conclusions, if we use logics that are inadequate for a given problem, see the next chapter. Therefore, the arguments of metaphysicians about greater certainty of metaphysical reduction have some value, but conclusions obtained in that way are not absolutely true, but only well grounded intuitively.

5.5   Further Conclusions of the Rational Evolutionary Theory of Intuition

If we sum up that intuition is preverbal and based on imagination and experience, then it is clearly an important element of *tacit knowing* in the sense of Polanyi (1966). This concept, with wording changed, or rather meaning expanded, to *tacit knowledge* was used by Nonaka and Takeuchi (1995) to formulate a model of knowledge creation in organizations in the form of a *SECI spiral.* This spiral consists of subsequent transitions (conversions[14]) between group tacit knowledge, group explicit knowledge, individual explicit knowledge and individual tacit knowledge. The rational evolutionary theory of intuition might help to understand and analyse these processes, see Chap. 12.

In order to show the explanatory power of the rational evolutionary theory of intuition, consider first of all a simple conclusion concerning the phase of *rationalization* of a creative intuitive process: if we want to convey an idea to other people, we must rationalize it first, we must formulate it in words together with a justification. Therefore, in his famous discussion of seven possible meanings of the principle *nihil est sine ratione*, Heidegger (1957) omitted another, perhaps the most important meaning: *since an intuitive preverbal judgement must be rationalized when being formulated, it requires a rationale, e.g. by metaphysical*

---

[13] For example, the classical metaphysical theory of being, based on metaphysical reduction, did not note a fact that is fundamental for contemporary mathematics: the existence of a given mathematical entity, e.g. of a limit of infinite sequence in infinite-dimensional spaces, depends on the domain in which this entity was defined. Banach (e.g. 1932) noted that some infinite-dimensional spaces are *complete,* other *incomplete;* in the latter the limit of an infinite sequence of space elements might not belong to the space, thus might not exist in the domain. Therefore, before judging the properties of an entity, we must define the domain in which it will be considered; while this could not be noticed by classical metaphysics, by contemporary metaphysics, however, it should.

[14] The term "conversion" was used originally by Nonaka and Takeuchi, but it implies a consumption of the resource along with its transformation, a reduction of its amount; this leads to a misunderstanding since knowledge is not diminished when used, it can only increase in the subsequent transitions of the SECI spiral. This fact was noted by many authors, see e.g. (Jensen et al. 2003); hence I use here a more neutral term *transition*.

reduction. Equally far reaching conclusions can be drawn with respect to the old dispute of the schools of Popper and Kuhn, *paradigm* versus *falsification.* If people are social beings and language was a tool of civilization evolution of human tribes, individual thinkers had to present their theories to the group or tribe in appropriately selected words, even beautify and defend their theories, according to the concept of paradigm of Kuhn. However, in order to maximize the chances of survival of the group the evolutionary interests of the tribe required application of tested theories: too flowery ones might have been suspicious, and a falsification similar to Popper sense was needed. Therefore, both Popperian falsification and Kuhnian paradigms were necessary, in the civilization evolution they complemented each other.

If to the essential features of humanity we do not only include verbal communication, but also creation of tools and curiosity (the inherited tendency to explore), then the role of evolutionary interests of a group in civilization development of human societies has even greater importance. New tools not necessarily serve the immediate interests of the group or its leader, but they might serve the interests of their children. Research curiosity might be dangerous and not advantageous evolutionary for an individual, but of tremendous importance when it comes to the chances of survival of a group.

The rational, evolutionary and technical theory of intuition presented here in an outline allows many conclusions, in particular practical ones that can serve as a field of falsification. For example, in relation to individual intuition it suggests that the best ideas of intuitive decisions come after a long sleep, before we saturate our mind with the bustle and hustle of everyday life. This leads to a simple conclusion called *alarm clock method:* set the alarm clock for 10 min before the normal wake up time and just after waking up ask yourself *whether you found solution of the most pressing problem?* This simple experiment can serve as an empirical test, a method of falsification of the theory.

Another conclusion from the evolutionary rational theory of intuition concerns the distinction between one's own imagination and transmitted imagination (the latter transmitted, e.g. by a film in television); this distinction has something to do with creativity. Human mind has an undeniable, natural need to imagine. This need is satisfied by television, but if we saturate it, we might limit our own imagination. Therefore: *if you want to be creative, limit your contacts with television.*

Other applications of the evolutionary rational theory of intuition concern the theory and practice of negotiations. One of the conclusions is to use relaxation and sleep in order to let our subconscious work on previously defined but yet unsolved problems. A test of importance of such unconscious, intuitive problem-solving may be as follows: *organize a nontrivial, difficult and competitive test of simulated negotiations for students of the art and theory of negotiations,* but divide the students into two groups: *one group to solve the negotiation problem during one day, another group to solve it within the same time limit but in two sessions in two subsequent days, with a break for relaxation and sleep.* Then compare the results.

It is also worthwhile to mention the conclusions from the evolutionary rational theory of intuition in relation to the controversy between *soft and hard systems analysis* that will be discussed in more detail in Chap. 10. The theory of intuition stresses that people can have much better ideas than it stems from hard, mathematical model based systems analysis alone, thus soft systems analysis is also needed. However, the claims of soft systems analysis that hard mathematical modelling is unnecessary are erroneous, such modelling is necessary in order to rationalize and test our intuitive ideas that can be wrong.

Finally, it is worthwhile to summarize the conclusions from the evolutionary rational theory of intuition in the form of the following principle:

Multimedia principle: *words are only a simplifying code that serves to describe much more complex reality, the visual information and preverbal information in general is much more powerful,* it is connected with intuitive reasoning and knowledge; future record of the intellectual heritage of humanity will have a multimedia character that stimulates creativity.

This principle complements the emergence principle that is discussed in other chapters of this book. Both might appear to be common-sense, intuitive formulations; but it is important that even if both have a specific metaphysical character, they are rationally substantiated, relying on scientific facts. Moreover, both these principles exceed and in some sense correct the fashionable trends of post-structuralism and postmodern philosophy or sociology of science.

The multimedia principle is based on technical and informational knowledge: an image is worth at least ten thousand words. Post-structuralist philosophy stresses the role of *metaphors* and *icons*,[15] but reduces them usually to *signs*; the simplest argument against such reduction is presented in Fig. 5.1, where the Byodoin temple in Uji, Japan, is presented first as an icon (the 10 yen coin), later as a photograph of Byodoin.

Therefore, *the world is not constructed by us in a social discourse,* as post-structuralist and postmodern philosophy wants to convince us; *we observe the world with all our senses, including vision, using the immanent perception, then we try to describe the world with words, try to find words that are adequate to our preverbal observations in order to express them in language. Language is a shortcut in human civilization evolution, our original way of thinking was preverbal, often subconscious, evolutionary common with animals.*

This sheds a somehow different light either on the dualism of *transcendence and existence* discussed by Karl Jaspers and mentioned earlier, or on the dualism of *outward man and inward man* introduced by Jung (see e.g. 1971) and analysed deeply by Motycka (2010). The above conclusions, on the one hand, confirm the importance of the concept of *inward man,* but on the other hand they question the dichotomy of this distinction. Preverbal perception and cognition belong to a border set (with a nonempty interior[16]) between the outward man and inward man:

---

[15] Understood as simplified images.

[16] Similarly as in *rough sets* of Zdzisław Pawlak that can be depicted as a border of a set drawn on a piece of paper with a broad felt-tip pen. Therefore we have here not a dichotomy of *outward man and inward man,* but a trichotomy of *outward man* (with verbal perception and cognition), *preverbal perception and cognition* (in the border set), and *inward man.*

Fig. 5.1 An icon (*left*) and a photograph (*right*) of Byodoin, Japanese temple in Uji near Kyoto

even if this perception and cognition is an essential attribute of the inward man (without them it would be impossible to create archetypes, myths, metaphors), it also serves the outward man if he only can code the results of the preverbal cognition (enlightenment, illumination, abduction) verbally in order to convey them to the Other.

References

Banach, S.: Theorie Des Operationes Lineares. Chelsea Publishing Company, New York (1932)

Bergson, H.: Introduction to Metaphysics (English translation 1911). New York (1903)

Brouwer, L.E.J.: On the significance of the principle of excluded middle in mathematics, especially in function theory. With two addenda and corrigenda. In: van Heijenoort, J. (1967, eds.) A Source Book in Mathematical Logic, 1879–1931. Harvard University Press, Cambridge, pp 334–345 (1922)

Dreyfus, H., Dreyfus, S.: Mind Over Machine: The Role of Human Intuition and Expertise in the Era of Computers. Free Press (1986)

Fodor, J.A.: The Elm and the Expert: Mentalese and Its Semantics. MIT Press, Boston (1994)

Gardner, H.: The Minds New Science: A History of the Cognitive Revolution. Basic Books, New York (1985)

Gigerenzer, G.: Gut feelings: the intelligence of the unconscious. Viking. Polish translation (2009) Intuicja: inteligencja nieświadomości. Prószyński i S-ka, Warsaw (2007)

Heidegger, M.: Der Satz Vom Grund. Nachfolger, Stuttgart (1957)

Husserl, E.: Cartesianische Meditationen und Pariser Vorträge. [Cartesian meditations and the Paris lectures]. In: Strasser, S., Nijhoff, M. The Hague, Netherlands (1973)

Ihde, D.: Listening and Voice. Ohio University Press, Athens (1976)

Ihde, D.: Bodies in Technology. University of Minnesota Press, Minneapolis, London (2002)

Jaspers, K.: Philosophical autobiography. In: Schipp, P.A. (ed.) The Philosophy of Karl Jaspers. Tudor. Polish translation (1993) Autobiografia filozoficzna. Wydawnictwo COMER, Toruń (1957)

Jensen, H.S., Richter, L.M., Vendelø, M.T.: The Evolution of Scientific Knowledge. Edward Elgar, Cheltenham (2003)

Jung, C.G.: Psychological Types. Princenton Univeristy Press. Polish translation (2009) Typy psychologiczne. Wydawnictwo KR, Warsaw (1971)

Król, Z.: Platon I Podstawy Matematyki Współczesnej (Plato and Foundations of Contemporary Mathematics). Wydawnictwo Rolewski, Nowa Wieś (2005)

Król, Z.: The emergence of new concepts in science. In: Wierzbicki, A.P., Nakamori, Y. (eds.) Creative Environments, op.cit (2007)

Lakatos, I.: Proofs and Refutations. Cambridge University Press, Cambridge (1976)

Lewandowski, A., Wierzbicki, A.P. (eds.): Aspiration Based Decision Support Systems. Verlag, Berlin, Heidelberg (1989)

Locke, J.: An essay concerning human understanding. In: Stehr i, N., Grundmann, R. (2005, eds.) Knowledge: Critical Concepts. Routledge, Oxford (1690)

Morawiec, E.: Opis struktury bytu w metafizyce ogólnej (A description of the structure of being in general mataphysics). In: Motycka, A. (ed.) Nauka a Metafizyka (Science and Metaphysics). Wydawnictwo IFIS PAN, Warsaw (2009)

Motycka, A.: Metafizyka w oczach filozofa nauki (Metaphysics in the eyes of a philosopher of science). In: Motycka, A. (ed.) Nauka a Metafizyka (Science and Metaphysics). Wydawnictwo IFIS PAN, Warsaw (2009)

Motycka, A.: Człowiek Wewnętrzny a Epistéme (Inward Man and Episteme). Eneteia, Warsaw (2010)

Nęcka, E.: Inteligencja: Geneza—Struktura—Funkcje (Intelligence: Genesis—Structure—Functions). Gdańskie Wydawnictwo Psychologiczne, Gdańsk (2003)

Nonaka, I., Takeuchi, H.: The Knowledge-Creating Company. How Japanese Companies Create the Dynamics of Innovation. Oxford University Press, New York (Polish translation: Kreowanie wiedzy w organizacji, Poltext 2000) (1995)

Nowak, G.: (2007) Biologia i epistemologia (Biology and epistemology). In: Hetmański, M. (ed.) Epistemologia współczesna (Contemporary epistemology). Universitas, Cracow, pp 255–269

Piaget, J.: The Origins of Intelligence in Children. Norton & Co, New York. Polish translation (1966) Narodziny inteligencji dziecka. PWN, Warsaw (1952)

Piętka, D.: Status metodologiczny tez tomistycznej filozofii bytu (Methodological status of a thomistic theory of being). In: Motycka, A. (ed.) Nauka a Metafizyka (Science and Metaphysics). Wydawnictwo IFIS PAN, Warszawa (2009)

Poincaré, J.H.: The Foundations of Science (English translation, 1946). The Science Press, Lancaster (1913)

Polanyi, M.: The Tacit Dimension. Routledge and Kegan, London (1966)

Popper, K.R.: Logik der Forschung. Julius Springer Verlag, Vienna (1934)

Popper, K.R.: Objective Knowledge. Oxford University Press, Oxford (1972)

Quine, W.V.: Two dogmas of empiricism (original version in 1953). In: Benacerraf, P., Putnam, H. (eds.) Philosophy of Mathematics, Prentice-Hall, Englewood Cliffs (1964)

Shannon, C.: Mathematical theory of communication. Bell Syst. Tech. J. **27**, 376–405 (1948)

Simon, H.A.: Models of Man. Macmillan, New York (1957)

Springer, S., Deutsch, G.: Left Brain—Right Brain. Freeman, San Francisco (1981)

Vollmer, G.: Mesocosm and objective knowledge: on problems solved by evolutionary epistemology. In: Wuketits, F.M. (ed.) Concepts and Approaches in Evolutionary Epistemology. D. Reidel Publishing Co., Dordrecht (1984)

Walker, M.P., Brakefield, T., Morgan, A., Hobson, J., Stickgold, R.: Practise with sleep makes perfect: sleep dependent motor skill learning. Neuron **35**(1), 205–211 (2003)

Wierzbicki, A.P.: Modele i wrażliwość układów sterowania. WNT, Warsaw. English edition (1982) Models and Sensitivity of Control Systems, Elsevier-WNT (1977)

Wierzbicki, A.P.: On the role of intuition in decision making and some ways of multicriteria aid of intuition. Multiple Criteria Decision Making **6**, 65–78 (1997)

Wierzbicki, A.P.: Knowledge creation theories and rational theory of intuition. Int. J. Knowl. Syst. Sci. **1**, 17–25 (2004)

Wierzbicki, A.P.: Delays in technology development: their impact on the issues of determinism, autonomy and controllability of technology. J Telecommun Inf Technol **4**, 1–12 (2008)

Wierzbicki, A.P., Makowski, M., Wessels, J.: Model-Based Decision Support Methodology with
    Environmental Applications. Kluwer, Dordrecht (2000)
Wierzbicki, A.P., Nakamori, Y.: Creative Space: Models of Creative Processes for the
    Knowledge Civilization Age. Verlag, Berlin-Heidelberg (2006)
Wierzbicki, A.P., Nakamori, Y. (eds.): Creative Environments: Issues of Creativity Support for
    the Knowledge Civilization Age. Verlag, Berlin-Heidelberg (2007)
Wittgenstein, L.: Tractatus Logico-Philosophicus. Cambridge University Press, Cambridge
    (1922)
Young, L.F.: Computer support for creative decision-making: right-brained DSS. In: Sol, H.G.
    (ed.) Processes and Tools for Decision Support. North Holland, Amsterdam (1983)
Żegleń, U.M.: Kognitywistyka: czy nowa szata dla starych problemów epistemologicznych?
    (Cognitivistics: is it a new vestment for old epistemic problems?). In: Hetmański, M. (ed.)
    Epistemologia współczesna (Contemporary Epistemology), Universitas, Cracow, pp 271–303
    (2007)

Chapter 6
Problems of Metaphysics, Truth and Objectivity

Hermeneutic Horizon, Scepticism and Naturalism, Evolutionary Knowledge Creation

6.1 Metaphysics of Fundamental Naturalism

At the beginning I would like to explain my understanding of the word *metaphysics*. In the development of humanity, in all history of civilization, we observe both useful and inescapable tendency to *disciplinary specialization,* based on a closer contact with a selected fragment of reality, such as when carpenter specializes in wood processing and an astronomer in observations of stars and planets. This tendency, however, resulted in an opposite need for an *interdisciplinary reflection,* in at least two essential aspects. Firstly, there is a need to take into account or to complement knowledge belonging to many disciplines for the purposes of any broader practical application; this aspect is the subject of *engineering,* in particular, *systems engineering* that will be discussed in Chap. 10. Secondly, there is a more general need to rethink, to achieve a theoretical reflection on the general conclusions derived from the results of individual disciplines. Initially, in the time of Babylonian civilization for example, such role was attributed to *religion,* later to *mathematics,* but as both of them became specific disciplines, this role is fulfilled today by *philosophy in its interdisciplinary general sense,* and especially by *metaphysics.* Large parts of philosophy became also specific disciplinary fields, which resulted in the need of new approaches to interdisciplinary synthesis, such as *systems analysis.* But it turned out, as it will be commented in detail later, that an excessive specialization was dangerous for philosophy, so the latter detaches itself from the results of specific disciplines and needs a return to its general, metaphysical roots.

While consistently assuming here, as in the preceding chapter, an evolutionary position (which is a metaphysical assumption too, as it was observed e.g. by Popper, see 1972, see also Goldfinger-Kunicki 1993), I reject nevertheless all absolute interpretations, including absolute metaphysical verities. Next generations have the right to criticise, reject or refine our views, even those based on intuitive metaphysical reduction, "intuitive communing with transcendence,

© Springer International Publishing Switzerland 2015
A.P. Wierzbicki, *Techne_n: Elements of Recent History of Information Technologies with Epistemological Conclusions,* Intelligent Systems Reference Library 71,
DOI 10.1007/978-3-319-09033-7_6

transsubjective co-sensing of being" (Motycka 2009, p. 233). Gradually, evolutionarily we improve the intellectual heritage of humanity, with metaphysical verities included.

For these reasons, *I would like to understand metaphysics as a reflection on general, interdisciplinary questions, taking into account the results of specific disciplines but not reducible to them*, and not necessarily resulting from them.[1] For me, this also means that metaphysics *is naturally based on intuition*, that it is not only a logical reflection on the results of empirical studies. No wonder that logical empiricism rejected metaphysics, but it was its greatest mistake. I am aware that we can understand metaphysics in diverse ways: as reflection on being, time, on right and wrong, death and love, or in a Heideggerian way as a reflection on being, temporality and projective imagination, or as seeking for an ultimate base, or for an understanding who man is, see, e.g., Kołakowski (1988), Skarga (2007). Such a meaning, however, seems to me to be exceedingly anthropocentric, and this evaluation is related to the other concept, explained below.

The second fundamental concept that I want to explain here is *fundamental naturalism*. It can be treated as a variant of philosophical *realism* which is a conviction that we get to know reality in the first place and our own consciousness and cognitive processes only afterwards (see e.g. Krasnodębski 2010). But fundamental naturalism goes further, being based on two assumptions. The first one is the fundamental anti-solipsistic ontological assumption which is common with philosophical realism that *a man is not alone,* that there is also an external part of reality containing inanimate objects and animate subjects, microorganisms, plants, animals and people. This external part of reality can be called *nature,* although I am aware that strictly speaking the concept of *nature* can be limited to living subjects. The use of the term *subjects* might seem paradoxical, but I use it on purpose, as it results from the second assumption explained below. The first fundamental assumption means also that we humans improve our knowledge in relation with nature, even if this part of the assumption was many times contested by sceptical philosophy, which I shall critically discuss in a father part of this chapter.

The second basic assumption of fundamental naturalism is more subtle, even if also stronger and often unconsciously omitted or even consciously negated by Western philosophy: it is the assumption that *people are not lords of nature but only a part of it,* and other parts of nature can also be treated as cognitive subjects. I am aware that people have a dominating position on the evolutionary ladder, but this does not mean that we should treat other parts of nature as a field of unlimited exploitation. I am also aware that this assumption, after a deeper reflection, does not belong to the European tradition and might appear shocking. For example, Marcin Ryszkiewicz in his excellent book (2012) on evolutionism, subconsciously accepts this Western tradition and argues that man is lord of nature because he

---

[1] Such understanding of metaphysics does not directly correspond to its understanding by Aristotle—see e.g. (Blandzi 2009), but it does not contradict it either.

happened to become lord of nature, which is, for me, an example of how metaphysical reduction depends on pre-cognitive assumptions. However, after a longer stay in the sphere of Japanese culture, in which the said assumption of fundamental naturalism is fully natural and unquestioned, I would like to analyze some consequences of it.

After these initial remarks we should note that fundamental naturalism is not equivalent to materialism or atheism; e.g. Christian beliefs of St. Francis can be understood as an expression of fundamental naturalism, while Buddhist religion accepts fundamental naturalism as a basic part of its philosophy. Pantheism and some variants of agnosticism are close to fundamental naturalism. Fundamental naturalism has also an obvious ideological dimension, as it is close to the ideology of advocates of natural environment or to *green* parties. Fundamental naturalism can mean an acceptance of evolutionist assumptions (following rather the interpretation of Charles Darwin than of Alfred R. Wallace; the latter found in evolution a mystic message substantiating the dominant position of man) which, as it is known, are not contradictory to religion.[2] Fundamental naturalism is also not contradictory to technology, but suggests a moderation in its applications in order not to succumb to anthropocentric, technological pride (assuming that the destiny of humanity is to apply technology to subordinate nature). Fundamental naturalism is also not contradictory to humanism, but defies humanistic pride, *hubris,* either in the sense of theocentric humanism (assuming that God gave to people the dominion over nature and we have to use this dominion) or in the sense of anthropocentric humanism (assuming that only a representative of human sciences can understand the values of humanity). An example of the latter was also Marxist materialism that often openly proclaimed the domination of people over nature.

What is worse, such domination has been assumed by Western philosophy starting with Plato (see e.g. *Dialogs,* final parts of *Timaios*). Consider the example of Kant (1781, 1788): the assumptions of fundamental naturalism suggest that we should not ask *how human cognition is possible* or *how metaphysics is possible,* until we understand *how a tree, or a flower, or a cat learn,* or *what is the metaphysics of a dog?* In this light, even if it is obvious that "all forms of practice, tasks or activity are biased, ontically, formally and normatively", the conclusion that "stimulus, without interpretation, practice, without intentions, action, without design and ontical burden cannot exercise any epistemic functions" (Żurkowska 2008, p. 12, my translation) turns out to be doubtful; and how a flower opens its

---

[2] Even if we assume that God is an omnipotent creator (such assumptions are not represented by all religions, e.g., Buddhism does not use the concept of creator), then why should He bother with creating every microorganism, when He could use simple evolutionary mechanisms? On the other hand, the arguments of creationists against evolutionism arguing that an intelligent design was necessary to create an irreducible complexity are simply historically incorrect: the evolution of civilization during the last 50 years resulted in the emergence of irreducible complexity. Of such complexity is e.g. the concept and phenomenon of *software* that cannot work without *hardware* but is irreducible to hardware or technical equipment. See the discussion of the concept of irreducibility in Chap. 10.

petals? First of all we should ask what is common for a human being and for nature, and only later we can consider what distinguishes a human being from other natural subjects.

Yet a tree learns about its environment in order to survive and develop, it discriminates winter from summer (hence forms cognitive elements that could be recognized as conceptual categories), drought from rain. A flower discriminates a sunny day from a cloudy one and from night. A cat has obviously a much richer set of concepts and has an excellent spatial orientation, recognizes specific people. Clearly, in order to develop such an *epistemology of fundamental naturalism* a more detailed study is needed, taking into account botanical and zoological studies. However, we should note that all cognition of non-human natural subjects has a *preverbal* character, similar to human intuition. The rational but evolutionary theory of intuition described in the previous chapter is based on a similar assumption: it starts with the question of how much more important or stronger for us is preverbal processing of information (at least ten thousand times stronger), and proceeds to the question what has happened with preverbal capabilities of people after the evolutionary dissemination of speech. Therefore, intuition (and metaphysics) can be considered as common for people and other natural subjects.

Therefore, what is the metaphysics of dogs? In order to answer this question, we must first of all say what we mean by the concept of metaphysics when applied to dogs. There is no doubt that a dog has many practical abilities, such as digging in earth, retrieving, etc. But if a dog should be able of metaphysics, it would mean its ability of a broader reflection; is such an ability possible? After a long observation of dogs, I would answer positively, but most people would not. Why most people are convinced that a broader reflection is an ability limited to humans? Obviously, the reason is the usage of language to formulate the results of such reflection. However, if the broader reflection has an intuitive dimension, it contains preverbal elements. Therefore, we cannot rule out a general reflection in dogs; it might be only difficult to investigate the existence of such reflection. I have no doubt, however, that a talented zoologist may find methods of such investigation, hence the question of metaphysics of dogs might be cognitively stimulating.

Since metaphysics is based on intuition, then people use preverbal reflection to consider metaphysical questions. Thus it is not true that *the limits of my language mean the limits of my world,* or that *whereof one cannot speak, thereof one must be silent* (Wittgenstein 1922). Contrariwise, all focus of philosophy in the 20th century on language might result from an erroneous assumption that cognition is only verbal, continued in many works, see e.g. Foucault (1972) or Derrida (1974). However, if preverbal aspects of cognition are more powerful, then we must return to metaphysics, perhaps from a new perspective, and try to speak about that what cannot be said.

The paradox contained in the above sentence is obvious and equally obvious is the manner of overcoming it. If the intuitive and emotional aspects of human abilities have preverbal character, then we cannot speak about them; but if it is necessary to speak about them, then, as always in the development of language, we must find new concepts and verbs. The same is the case for metaphysical reflection: it creates new concepts and verbs, not necessarily empirically substantiated or resulting from linguistic tradition.

We can use an example here: the concept of *ontological pressure,* so correctly indicated as a metaphysical source of knowledge by Jan Srzednicki (see e.g. Żurkowska 2008), is a new concept, new verbal ensemble, although from the perspective of fundamental naturalism it might be interpreted as a pressure of preverbal information about beings and subjects surrounding us, leading to a preverbal creation of concepts. Another example is the struggle of Heidegger (1927) with words related to the concepts of *existence, being, becoming,* or Heidegger (1954) with giving a proper meaning to the concept of *technology.* We need to form new words or new meanings of words if we want to convey new content.

6.2 Truth Versus Hermeneutic Horizon and Perspective

I am aware that the question of *what truth is* belongs to the oldest problems of philosophy and human reflection upon nature, but I do not intend to present here a full history of this problem. I only stress one fact: today we understand that *there are no absolute verities,* because there are no absolutely precise measurements (Heisenberg 1927).

If there are no absolute verities, then how should we interpret the need of truth and human quest for truth? Let's first recall elements of the theory of truth in mathematics and formal languages. Gödel (1931) has shown that the question of truth cannot be decided inside a formal system; Tarski (1933) has shown how to use a metalanguage in order to rationally consider the question of truth in a given formal language. However, Król (2005, 2007), developed the ideas of Lakatos (1976) in order to stress the impossibility of creation and analysis of mathematics as a strictly formal game without meaning and interpretation; we can have no fully formalized mathematical theory, given as a formal system with a formal metalanguage. A strictly formal language requires formal metalanguage, the latter in turn requires a formal meta-metalanguage, etc., which results in an infinite regress. In further chapters I will show how we deal with an infinite regress in the praxis of robotics. Actually, Król suggests the same way: to stop the regress at a certain level and to accept some basic assumptions in an intuitive, abstract meta-environment. This idea stems from Plato and was also developed by Descartes and Kant, but unlike them, Król does not assume the infallibility of such basic assumptions, on the contrary, he shows their historical revisions.

An intuitive meta-environment of the assumptions about truth he calls *hermeneutic horizon* (also a concept with its own history[3]); but Zbigniew Król shows

---

[3] For example, Gadamer (1960) understood hermeneutic horizon as entire knowledge of a man, taken holistically. Zbigniew Król has given quite different and more precise meaning to this concept.

how such horizon was historically changing. For example, in the case of the so, called "Euclidean geometry", its understanding by ancient Greeks was quite different (taking into account the deepest interpretation of its axioms) than by Descartes, Newton and Kant, and today we understand these axioms in yet another way. I would like to make a remark here: if such a phenomenon occurs in mathematics, it can also occur, *mutatis mutandis,* in other scientific disciplines. Diverse paradigms are not only based on different, incommensurate (hence formally non-translatable) languages, but, more fundamentally, because languages are anyway only approximate codes, depend on different hermeneutic horizons, intuitive meta-environments of convictions about truth of fundamental assumptions, basic axioms that change in a long historical perspective. This phenomenon Król calls *horizontal change,* that not too frequently occurs in mathematics, but might be more frequent in other disciplines.

However, Król maintains that the perception of truth through a hermeneutic horizon is not subjective, even if it is historically changing; this perception is also not intersubjective: *we do not decide in a social discourse what constitutes a hermeneutic horizon.* Again, I would like to add my own comment here: in an obvious way the above concerns mathematics, technology and strict science. Nevertheless, the postmodern view of social science and humanities was different, but seems erroneous: even in social sciences the formation of an hermeneutic horizon is a process or *structure of long duration,* such as it was understood by Braudel (1979).[4] On the one hand, a hermeneutic horizon is intuitive, but on the other hand it is common, say, for all mathematicians or historians in a given era. The formation of a hermeneutic horizon might be subconscious, but it is connected to the canon of teaching in a given discipline, e.g., mathematics or history, in this era.

According to the above it follows, from the perspective of fundamental naturalism, that *people as social beings (not excluding similar phenomena e.g. between ants) need a relatively durable basis for the concept of truth and provide such a relatively durable basis through a punctuated evolution of hermeneutic horizons* or, speaking more broadly (because other disciplines than mathematics often do not specify their axioms, although they have certain cognitive beliefs), *hermeneutic perspectives* of individual disciplines. These horizons or perspectives are handed over from generation to generation through the teaching canon of a given discipline, they are structures of long duration, but they can change, revolutionary, similarly as the paradigms based on them, as a result of knowledge that amasses and modifies them.

I use here the slightly more general concept of hermeneutic perspective, because the disambiguation of the concept of hermeneutic horizon by Zbigniew Król defines it as a system of axioms intuitively accepted as true together with their intuitive interpretation, while many sciences other than mathematics, e.g. social, but also

---

[4] Since hermeneutic horizon is neither subjective nor intersubjective, it is an element of a long duration structure, then the postmodern concept of *truthness* (verity as a fully subjective feature resulting from an internal conviction) is rather poorly grounded.

technical ones, do not define their axioms precisely, but nevertheless accept them intuitively as obvious, true assumptions. Such precognitive assumptions depend on the discipline and might be quite different than in any other discipline, but often are not defined as axioms. For example, philosophy believes that its history has proven the inconsistency of any naturalism (I believe the opposite is true, that all such proofs are based on a basic logical error, see next point). It is important, however, that a reconstruction of an hermeneutic horizon is a difficult but solvable problem, as it was shown by Zbigniew Król who used historical sources to determine how e.g. ancient Greeks understood the axioms of Euclidean geometry.[5] I believe that with the use of similar methods we can try to reconstruct hermeneutic perspectives of various disciplines, e.g. asking what a given discipline believes to be fundamental verities for which there is no need of a proof.

The concepts of *hermeneutic horizon* or more broadly *hermeneutic perspective* can be tied to the concepts fundamental for the philosophy of science such as *paradigm* of Thomas Kuhn, *research programme* of Imre Lakatos, also *research tradition* of Larry Laudan, see an excellent discussion of these concepts in Michalska (2009). Even the most broadly understood concept of *research tradition,* including paradigms and research programmes, has, as it is stressed by Anna Michalska, a horizontal character, hence it might be based on such intuitive kernel as hermeneutic horizon or more broadly understood hermeneutic perspective.

My personal conviction mentioned earlier that after us will come new generations that will criticize, revise or improve our beliefs about the world, belongs to such hermeneutic perspective. This is, of course, a horizontal assumption, a firm belief which may turn out not to be true: we can perish in a cosmic catastrophe or self-annihilation of human civilization, or next generations might become so stupid that they will abandon human curiosity or *quest*, which is for me a feature co-defining humanity (see also Pasternak 2004); I have only hope and faith that we can avoid such catastrophes. I quote this personal belief in order to contradict those representatives of postmodern thinking who maintain that they do not believe in anything; does not it mean that they are too lazy in intellectual terms to check what they truly believe in?

6.3 Scepticism, Feedback and Naturalism

The tradition of scepticism and the critique of naturalism in epistemology is very strong. E.g. Kołakowski (1988, p. 13 of Polish edition) writes (my translation): "From the time of ancient sceptics it has been known that all epistemology, that is, any attempt to establish universal criteria of validity of knowledge, leads either to an infinite regress, or to a vicious circle, or to a not surmountable paradox of self-reference (not surmountable, of course, if not apparently resolved by

---

[5] They did not know the concept of infinity and for them, an elegant geometric proof was constituted by an analysis that did not use points outside the outline of the geometric object given originally.

reformulation to an infinite regress)." But for a technician, as I shall show in further parts of the book, a logically valid epistemology must rely on the concept of circular or spiral-like relations resulting from a positive feedback.

Recall that *feedback*, see further chapters on the history of this concept, means a dynamic circular influence exerted by a time stream of effects onto the time stream of causes, and it can be classified as *positive* (self-supporting) or *negative* (not self-contradictory in its essence, but regulating and stabilizing). A system with negative feedback used in each robot in order to stabilize its functioning needs a theoretically infinite time, an infinite recourse (regress in a positive direction of time, hence we can call it also progress) of the streams of effects and causes. In practice, we naturally assume that the system already achieved stabilization if we do not observe any changes for a sufficiently long time (it is like a limit of an infinite mathematical sentence: if the elements of this sequence are sufficiently close to the limit, we assume that it is practically achieved). Thus, a theoretically infinite time of action can be practically very short, if the stabilization is fast. Hence, according to the arguments of sceptics, each robot is a "hydra of infinite regress". Moreover, in each computer we have millions of elements with positive, self-supporting feedback and saturation (elements of electronic memory), millions of self-reference circles. However, if not stopped by saturation, positive feedback processes result in an avalanche-like development.

The main thesis of this chapter is drawing attention to the fact that a good understanding of behaviour of systems or artefacts with feedback explains rationally the apparent paradoxes of self-reference, of the hydra of infinite regress, of vicious circle, treating them as obvious features of the concept of feedback, broadly used in contemporary technology. Thus. these paradoxes are only apparent, similarly as it is clear today that Achilles will finally overtake a turtle, because an infinite sequence can have a finite limit (thus, it is not true that something is either finite or infinite, or either a cause or an effect, there is a third way, something can be both finite and infinite, or both cause and effect).

We must apply logics that is adequate for a specific field of applications. The relation between human knowledge and nature is obviously dynamic, not static. In a static approach, a vicious or self-supporting circle is a paradox: an effect cannot be at the same time a cause. In a dynamic approach, no paradox results: a time-stream of effects can be fed back as a time-stream of causes, naturally with some delay. Thus, the principle of excluded middle does not apply here: it is not true that something can be either an effect or a cause, there are many practical cases of feedback relation where something is both an effect and a cause. If it is a three-valued logics, such as the logics of rough sets by Zdzisław Pawlak, then all indirect proofs (by *reductio ad absurdum*) cease to be valid.[6] But all philosophical

---

[6] If there is a third way, then it can happen that something is at the same time infinite (say, in the sense of the number of elements) and finite (say, in the sense of a limit). This occurs in the paradox of Achilles and a turtle: the paradox suggests that Achilles will never overtake the turtle, because in the time when Achilles will come to the present position of the turtle, it succeeds to come away a small distance, etc. in an infinite regress. This was a paradox for ancient philosophy,

proofs of inconsistency of naturalism, which bases human knowledge on the observation of nature, consist exactly in *reductio ad absurdum,* in paradoxes sought and found in the relation between knowledge and nature, and in proclaiming inconsistency of this relation on that basis. Therefore, they are indirect proofs themselves.

Thus, *a good understanding of the concept and properties of feedback and the demand of using logics that is adequate to a specific field of application disprove the logical foundations of the entire tradition of philosophical considerations questioning the objectivity of cognition and criticizing naturalism from sceptical positions, while using paradoxical arguments related to vicious circle, infinite regress or self-reference.* I must admit that in static reasoning we should avoid e.g. circular definitions, but the development of human knowledge is a dynamic process.

The tradition of using the argument of vicious circle in order to question the objectivity of cognition is long, and should not be associated only with ancient sceptics. Edmund Husserl was afraid of drawing conclusions from Bergson theory of intuition, because, as testified by Ingarden (1974, p. 202 of Polish edition), he feared that "this threatens of a devilish vicious circle *(ein teuflischer Zirkel)*".

In his treatise *On Certainty,* Wittgenstein (1969) uses similar arguments in his justification of scepticism, when writing (thesis 130): "But isn't it experience that teaches us to judge like *this,* that is to say, that it is correct to judge like this? But how does experience *teach* us, then? *We* may derive it from experience, but experience does not direct us to derive anything from experience. If it is the *ground* for our judging like this, and not just a cause, still we do not have a ground for seeing this in turn as a ground".

Of course, the above remarks require a deeper analysis; such analysis was performed by e.g. Srzednicki (1995) and Żurkowska (2008). I quote after the latter (pp. 14, 15, my translation): "If in an explanation of the phenomenon of cognition what should be the subject of explanation becomes itself a tool of explanation (thus if the starting point is constituted by human praxis, language systems, scientific procedures, forms of cognition), we shall always be exposed to a *vicious circle.* That vicious circle in turn generates an *infinite regress* (in this theory, called a *hydra effect,* a hydra that regenerates in infinity "a hideous head of speculations"), since if we go in that direction, starting with the practical cognitive side,

---

(Footnote 6 continued)

but it ceased to be treated as a paradox when mathematicians proved that an infinite sequence can have a finite limit. The paradox of vicious or self-supporting circle is the next example of a phenomenon when a seeming paradox is explained away by the development of science and technology. Both these paradoxes are related to the exclusion of the middle. In the paradox of self-supporting circle it is assumed that something can be either an effect or a cause, *tertium non datur;* but we know today that effects can be fed back dynamically on causes. Some mathematicians and philosophers have long ago noticed the danger of using inadequate logics, even in treating the vicious circle as a paradox. This concerns especially the so-called *intuitionists* (e.g. Brouwer 1922), who questioned the principle of the excluded middle and indirect proofs, and maintained that only constructive direct proofs give a certainty of avoiding logical pitfalls.

we shall never identify such an irreducible ground to which no sceptics could knock. From obvious reasons Srzednicki does not want to go along that road questioned by Wittgestein, knowing well that those who head in that direction are thus exposed to scepticism..."

Even if I fully agree with the *solution* proposed by Jan Srzednicki, to base all cognition on the concept of *ontological pressure,* the thrust of beings surrounding us onto our consciousness, see next point, I nevertheless disagree with the explanation quoted above. My conclusions as a technician are clear: Wittgenstein had difficulty with understanding circular relations of the type of feedback which are fundamental for the processes of our learning, and has chosen to present his difficulties as a proof of general difficulty of the problem, a proof that scepticism is the only rational position, but this proof is based on the classical binary logics applied to a problem for which such logics is not adequate.

If Husserl could not know and Ingarden and Wittgenstein might not know that a contemporary computer includes millions of small devices, flip-flop switches, each using a positive feedback, self-supporting "vicious circle", then we cannot admit such excuse in the case of Bruno Latour who is a philosopher of technology and should know that fact. However, *the negation of objectivity by postmodern social sciences and humanities is based also on an apparent paradox, on a vicious circle found in the relation between nature and knowledge.* For example, (Latour 1987, p. 99) uses the following argument against objectivity: "since the settlement of a controversy is the cause of Nature's representation not the consequence, we can never use the outcome, Nature, to explain how and why a controversy has been settled".

The argument of Latour is perceived by a technician as a proof of ignorance as regards the logics of feedback, since for a technician the relation between nature and knowledge is not a vicious circle but an obvious example of a dynamic process with positive feedback. If the argument of Latour was logically correct then it would signify that computers memories (including millions of flip-flop devices using self-supporting positive feedback) and robots (using negative stabilizing feedback, hence infinite regress) cannot operate; but computers and robots do operate and contribute today to serious changes in technology and society. Arguments similar to those of Bruno Latour are used by Pobojowska (1996) in her criticism of naturalism.

Personally, I think that all such arguments indicate an inadequate logics, in this case a lack of understanding of the diachronic, dynamic character of the causal feedback loop. An argument based on the paradox of vicious circle indicates the use of logics that is not adequate for the analysed phenomena, and the recurrence of arguments based on inadequate logics indicates a disciplinary closure, self-limitation of philosophy. I fully agree with the remarks of Hetmański (2007) that philosophy should not isolate itself from other disciplines of science and knowledge; since (my translation) "The world has changed, and each theory explaining it should do so, also a philosophical one", then also "On epistemology, this imposes a duty... to go out of constructed fortresses, comprised of previous positions and concepts, and to participate in multi-disciplinary scientific research on cognitive phenomena" (Hetmański 2007, pp. 34, 67).

The concept of feedback changed essentially our understanding of cause-effect relations, explaining, among other things, the ostensible paradoxes of circular argumentation or vicious circle in logics, even if such paradoxes can obviously be explained only in a dynamic, not static approach to reasoning and models. *This fact was until now not fully understood by some philosophers who still have a tendency to use arguments based on paradoxes of vicious circle and an apparent impossibility of treating the same phenomenon as both the effect and the cause.*

Moreover, this error is often repeated by humanities and social sciences in their use of the concept of vicious circle.[7] There are too many such examples to quote them.

Unfortunately, philosophers, humanists and social scientists are not educated to understand the concept of feedback properly; that would require laboratory exercises or at least computer simulation of the behaviour of servomechanisms or robots. At most, they know the concept of feedback from popular works of Wiener (1948), popular, since the applications of feedback resulted first in the construction of analog computers (applied earlier than digital computers, by Vannevar Bush in 1931) and later, in the years 1930–1940, in an emergence of a distinct discipline of technical science, automatic control and automatics (much later supplemented by robotics); Norbert Wiener and his concept of *cybernetics* are secondary in relation to this development. This means, however, that the canon of humanist education is marked with essential gaps: it should teach both the concept of feedback together with computer simulations of the dynamics of feedback in servomechanisms, robots, or flip-flop devices, as well as the concept of *logical pluralism,* the multiplicity of diverse logics together with their manifold applications.

On the other hand, the search for a "irreducible ground to which no sceptics could knock" is doomed, because of the fallibility of our intuition. There are no absolute verities: for each example of a Kantian a priori synthetic judgment we can suggest conditions in which this judgement ceases to be true (an explanation of this seeming paradox is the fact that our intuition is based on our limited, mesocosmic experience). In metaphysics, we cannot search for the principles of existence of beings without defining first a domain to which these beings should belong. This was proven by Banach (1932). I expect the response of a metaphysician here: "But the classical metaphysics discussed real beings". True, but therefore it had trouble with the question *"Do ideas exist in reality?"*

Thus, we must first of all define the domains of beings: real and natural ones in nature, ideal in the intellectual heritage of humanity, including mythological beings (angels, devils, fauns, centaurs, Pegasus etc.) in mythology, belonging to the emotional part of this heritage; intuition of first ideas and synthetic a priori

---

[7] The misunderstanding of the concept of feedback between representatives of humanities and social sciences is so enormous that I encountered among them the use of the concept of *negative feedback* in the sense of a negative phenomenon—while negative feedback has usually positive effect of stabilization, such as the stabilization of the temperature of human body.

judgements in the intuitive part of this heritage; finally, rational models of laws of nature (called laws of nature by physics and chemistry) in its rational part, world 3 of Popper. Each of the domains can feature different rules of existence of beings, be complete or incomplete,[8] etc. Hence I would answer the question: *Yes, ideas do exist in reality, if we define their domains of existence well and in real terms* (e.g. as the pages of books explaining them or the universities where they are taught). Thus, it is difficult to speak about an *irreducible ground.*

Because of the principle of emergence discussed earlier, we should not concentrate on the search for an irreducible ground, we should rather ask for an ultimate goal and sense, as well as the dynamics of growing human knowledge.

6.4 Evolutionary Creation of Knowledge and a New Episteme

Because I regard the growth of human knowledge as a dynamic process, I have used the concept of spiral processes of knowledge creation (equivalent to processes with positive feedback) not only in *micro-theories* (or *micro-models*) of knowledge creation, which will be presented in later chapters, but also in a general explanation of evolutionary nature in which knowledge, the intellectual heritage of humanity, is gathered.

If we assume that we accept the existence of reality with its universal aspects[9] after all, we can describe the general relation between human knowledge and reality in the following way. Under *ontological pressure* (see above), motivated by curiosity, supported by intuition and emotions, with the use of the excess of our brains or minds, we, people, observe reality and construct hypotheses about the properties of nature, other people, human interrelations, and also construct tools that support us in contacts with nature or other people; together, we call all this knowledge.

People test and evaluate knowledge constructed by them through an application of this knowledge to reality: they conduct destructive, critical tests of new tools and artefacts, they invent critical empirical tests for theories describing nature, they try to apply and assess theories concerning social and economic relations, they perfect mathematics as an interdisciplinary (but intuitively grounded, according to the concept of hermeneutic horizon) language for models describing

---

[8] For example: is the limit of an infinite sequence of models of laws of nature an ultimate knowledge or the number 42? (we Poles think that this is a number, but slightly higher: 44).

[9] If we do not accept its existence, we can always falsify this belief through a *test of hard wall*, see Wierzbicki (2008): if we do not believe that reality with its universal properties exists, then we should place ourselves against a hard wall, close the eyes and try to convince ourselves that the wall does not exist or that it is not hard. If we cannot convince ourselves, then we accept the belief that reality exists and has some universal features. If we succeed in convincing ourselves, we can walk up with closed eyes....

Fig. 6.1 A general OEAM
spiral of evolutionary
knowledge creation (see e.g.
Wierzbicki 2008)

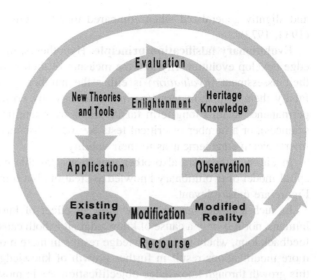

our knowledge. All this is possible because we have and utilize a tremendous excess and redundancy of our brains and minds, released by the development of speech.

Such a process can be described with the help of a general spiral of evolutionary knowledge creation presented in Fig. 6.1. The outer spiral in this Figure indicates an avalanche-like growth of knowledge resulting from a positive feedback: older effects (knowledge already gathered) become new causes (foundations of newly created knowledge).

We observe reality (in nature or in society) and its changes, we compare our observations with our intellectual heritage (world 3 of Karl Popper or also with intuitive and emotional elements of this heritage); this is indicated by the transition of *Observation.* Our intuitive and emotional knowledge helps us in generating new hypotheses (transition of *Enlightenment,* called also abduction, illumination, brainstorm, eureka) or in constructing new tools; we then apply them to reality (*Application*), usually in order to achieve a desired change (*Modification*); finally we return to observation.

The spiral is complemented by other transitions. The first one results from a natural evolution in time: a modified, new reality becomes actual reality (*Recourse*). The second one is what makes this spiral evolutionary it is the transition of evolutionary adaptation, of selection of tested knowledge (*Evaluation*): a majority of new knowledge can be recorded somewhere, but only its positively tested portion, resistant to falsification attempts, remains an essential part of the intellectual heritage of humanity. This can be also interpreted as an objectifying, stabilizing feedback.

This *objectification of knowledge* is especially important. It can be expressed in the following form of a *falsification principle,* which is in a sense metaphysical and slightly generalized when compared to the original formulations of Popper (1934, 1972):

Evolutionary falsification principle. Hypotheses, theories, models of knowledge develop evolutionarily and the measure of their evolutionary fitness used in their assessment (*Evaluation*) is either the number of unsuccessful attempts to falsify them in a short term (as in technology) by critical experiments, or their permanence against long term falsifications by scientific revolutions (as in strict sciences), or a number of critical tests by social discourse aimed at achieving an intersubjective agreement as to their validity.

In Fig. 6.1, there are also other transitions possible between the nodes of this spiral model of evolutionary knowledge creation, but transitions indicated on that Figure are most relevant.

In such a model *nature* is not only an effect of knowledge construction by humans, nor it is only a cause of knowledge, it is both cause and effect in a positive feedback loop, where more knowledge results in more modifications of nature and more modifications result in further growth of knowledge, but we try to control this growth through knowledge objectification. As in most positive feedbacks, the general result is an avalanche-like growth, and such a growth, if not stabilized by additional stabilizing feedbacks, brings beside great opportunities also various threats, often not anticipated but lurking in the future. Thus, the importance of selection of as objective knowledge as possible (I stress that there are no absolute verities, but we should strive for objectivity as an ideal) is related to the fact that avalanche-like growth results in diverse threats: *we must leave to our children the best knowledge possible in order to prepare them for threats of the future unknown to us.*

In this, the concept of *objectivity* is similar to the concept of *justice* as understood by Rawls (1971): because of the uncertainty concerning the future of our children, those social solutions are rational which are the most just ones, thus justice is interpreted as a concern about the welfare of children in an uncertain situation. Similarly, it is rational to strive for maximal objectivity of knowledge conveyed to our children. Thus, *an absolute justice just as an absolute objectivity is of course an unattainable idea, but according to the emergence principle they both arise as paramount ideas, irreducible to secondary ones, such as power, money or market.* Therefore, I do not believe that knowledge and intellectual heritage of humanity are a *reflection of nature* (see also Rorty 1980), but that they are an *image of nature* created by people and useful in civilization evolution.

At the same time I hope that a similar understanding of the growth and modifications of the intellectual heritage of humanity will become a foundation of a *new episteme of the epoch of knowledge civilization.* If after the debates of the Enlightenment a new episteme of the epoch of industrial civilization was formed (*modernism*), then after the discourses of the era of postmodernism a new episteme of the epoch of knowledge civilization, *post-postmodernism* or a broadly understood *informism* or rather *informed evolutionary objectivism,* or however we might call it, is likely to emerge. There is no doubt that the formation a new episteme is necessary: there are many negative features of the current separation of the epistemai of the three diverse cultural spheres. I stress this separation as an actually observed result of the break-up of modern episteme, but I do not consider this separation as a desired state.

What will be the new episteme then? There are at least two approaches to answering this question. The first is to deny any forecast: we shall live and see. Such intuitive approach can be diversely rationalized, e.g. by arguing that Michel Foucault treated his understanding of the concept *episteme* only ex post, historically, speaking only about past phenomena.

The second approach is ex ante, not precisely forecasting, but constructivist: we can try to construct one of possible futures, to help in the rise of a new episteme by speculating about it. I shall try to outline shortly a possible way of construction of such episteme. Between general principles, let us assume the evolutionary falsification principle described above is true, together with two general principles described earlier that I recall below:

Emergence principle. *Together with a growing complexity of a system, new properties of this system emerge which are qualitatively different from the properties of the elements of the system and irreducible to them.*

Multimedia principle. *Words are only a simplified code used to describe a reality which is much more complex; visual and preverbal information in general is much more powerful, and linked to intuitive reasoning and knowledge.*

We can also assume ten following specific principles to be true:

(1) *People are not alone in the world,* there is also a part of reality called *nature,* including (other) people, even if much of nature has been transformed by people for their convenience with the use of technology. Some parts of reality have local and diversified character, other parts can have more universal properties.

(2) People developed both *languages* for the purposes of communication with other people as well as *tools* in order to transform reality according to their needs. In both these endeavours, people were supported by their *curiosity* (an inherited aptitude for *quest*) that is positive in evolutionary terms of group development of civilization and has led to the growth of knowledge and science. Humanity can be defined only when taking into account these three basic features: *language, tools, curiosity.*[10]

---

[10] The contemporary evolutionary anthropology stresses the importance of a feature distinguishing us, people—even if encountered also in other examples of group evolution, e.g. in ants, from other species: in general, this feature is *curiosity,* even if it might have other names, such as *quest* (see Pasternak 2004), or *search.* For individuals, this feature is not positive in evolutionary terms (quest or excessive curiosity are dangerous), but it is strongly positive for groups; the foregoing applies especially to *cognitive curiosity,* essential for scientific development. Another feature, *creation of tools,* we share also with animals, e.g. great apes, but the development of

(3) According to *multimedia principle*, language is only a simplified code describing reality, while human senses (beginning with sight) perceive much more complex aspects of it. This more complete, immanent perception of reality is the foundation of human intuition; for example, tools have always been created on the basis of intuition, a more complete perception of reality than just verbal description. On the other hand, language is obviously necessary to rationalize preverbal observations and communicate them to other people.

(4) *The innate curiosity of people* (search, quest), aimed at other people or nature, together with the excess, redundancy of our brains and minds results in construction of hypotheses about reality, creates a structure and other models of the world. Until now, all such hypotheses, including the so, called laws of physics, at best turned out to be approximations; but we learn to improve them evolutionarily, using diverse tests, including the *evolutionary falsification principle*. Since we perceive reality as an increasingly complex structure, we construct new concepts on higher and higher levels of complexity in line with the *emergence principle*, and probably we shall always use only approximate hypotheses.

(5) *The sources of culture are both verbal*, such as stories, myths, symbols, *and technical*, as tools and artefacts used to improve human life. Both these aspects have rational features, which we describe with words, and also features that are intuitive, preverbal.[11] Both of them, together with their preverbal features, helped in the evolutionary development of *knowledge and science*, based on an accumulation of human experiences together with abstracting them and testing their credibility, a gradual development of appropriate models and theories. This is an evolutionary process, but *punctuated*, it includes not only gradual development, but also revolutionary periods.

(6) Accumulation of human experiences and culture occurs in the *intellectual heritage of humanity* together with its *emotional, intuitive and rational parts* (the latter resembles world 3 of Karl Popper). This heritage exists not only in human minds and in some sense independently of them, e.g., in libraries, museums, on the Internet and in other repositories of knowledge.

(7) Human thinking is based on imagination; it includes emotional, intuitive and rational aspects, stems from perception, sensual experiences, and also from interaction with human intellectual heritage together with its hermeneutic interpretation. Finally, it stems from human intuition and the excess or redundancy of our brains or minds. However, all these factors can be fallible.

(Footnote 10 continued)

languages and intergenerational transfer of knowledge helped people to develop tool creation ability much further.

[11] It should be stressed that we often use language intuitively, relying on a specific linguistic intuition—therefore languages, and obviously also myths and symbols, have also preverbal features. The creation of tools has, however, even more preverbal features.

(8) *Objectivism is a paramount value,* irreducible to money or power (how
     much money the life of our children is worth?). Objectivism helps in
     interpreting the intellectual heritage of humanity, in selecting its elements
     that better correspond to reality or are more suitable for the purposes of
     construction of tools, or describe social or economic behaviour in a better
     way.
(9) The recommendation of objectification is expressed by the *evolutionary
     falsification principle;* in the face of increasing complexity of the world we
     use the *emergence principle.* The sources of our cognitive power are related
     to the *multimedia principle.*
(10) The above principles are meant to be common for strict and natural sci-
     ences, for social sciences and humanities, and also for technical sciences,
     but they can be variously interpreted in these different cultural spheres.
     Strict and natural sciences can search for more universal theories, calling
     them *laws of nature* and using the falsification principle only within a long-
     term, historical perspective, social sciences and humanities might search for
     more local and diversified theories, and apply the falsification through a
     social discourse to them, while technology, motivated by the joy of creation
     of new tools, must use more short, term, experimental falsification.

I am aware that the current differences in the respective epistemai of these three
cultural spheres are profound and a common acceptance of the principles discussed
here will not be easy, will require time and discourse. But I formulate these
principles precisely as a ground for future discourse, since I strongly believe that
there is a need of future synthesis of a new episteme.

As an example of such need, it is sufficient to quote a recent opinion of a major
economic journal, *The Economist,* issue October 19, 25th 2013, on the theme *How
Science Goes Wrong,* more precisely, on the observed tendency to publish badly
documented scientific papers, precluding the possibility of verification by other
researchers. Under the influence of its neoliberal horizontal beliefs, *The Economist*
does not (or does not want to) note a fundamental reason of such a tendency: the
privatization of knowledge related to the strengthening of the positive feedback
between the results of science and technology and the market system utilizing
these results. If a scientist is rewarded accordingly to the amount of money her/his
achievements bring in result of market applications, she/he will not publish her/his
results in a way that allows for their repetition. Thus, there is a conflict between the
functioning of markets in knowledge based economy and the principles of
objectivity, and a resolution of this conflict requires a renewed synthesis of scientific episteme.

References

Banach, S.: Theorie Des Operationes Lineares. Chelsea Publishing Company, New York (1932)
Blandzi, S.: Arystotelesowska filozofia pierwsza a metafizyka (Aristotelian primal philosophy and metaphysics). In: Motycka, A. (ed.) Nauka a metafizyka (Science and Metaphysics). Wydawnictwo IFIS PAN, Warsaw (2009)
Brouwer, L.E.J.: On the significance of the principle of excluded middle in mathematics, especially in function theory. With two addenda and corrigenda. In: van Heijenoort, J. (ed.) (1967) A Source Book in Mathematical Logic, 1879–1931, pp. 334–345. Harvard University Press, Cambridge (1922)
Derrida, J.: Of Grammatology. John Hopkins University Press, Baltimore, MD (1974)
Foucault, M.: The Order of Things: An Archeology of Human Sciences. Routledge, New York (1972)
Gadamer, H-G.: Warheit und Methode. Grundzüge einer philosophishen Hermeneutik. Mohr J.B.C. (Siebeck), Tübingen (1960)
Gödel, K.: Über formal unentscheidbare Sätze der Principia Mathematica und verwandter Systeme. Monatshefte für Mathematik und Physik **38**, 173–198 (1931)
Goldfinger-Kunicki, W.J.H.: Znikąd Donikąd (from Nowhere to Nowhere). Państwowy Instytut Wydawniczy, Warsaw (1993)
Heidegger, M.: Sein Und Zeit. Niemayer, Halle (1927)
Heidegger, M.: Die Technik und die Kehre. In: Heidegger, M. (ed.) Vorträge und Aufsätze. Günther Neske Verlag, Pfullingen (1954)
Heisenberg, W.: Über den anschaulichen Inhalt der quantentheoretischen Kinematik und Mechanik. Zeitschrift für Physik **43**, 172–198 (1927)
Hetmański, M. (ed.): Epistemologia Współcześnie (Contemporary Epistemology). Universitas, Kraków (2007)
Ingarden, R.: Wstęp Do Fenomenologii Husserla (Introduction to Husserl's Phenomenology). PWN, Warsaw (1974)
Kant, I.: Kritik der praktischen Vernunft. Polish translation (1911) Krytyka praktycznego rozumu. E.Wende & Co, Warsaw (1788)
Kant, I.: Kritik der reinen Vernunft. Polish translation (1957) Krytyka czystego rozumu. PWN, Warsaw (1781)
Kołakowski, L.: Metaphysical Horror. Blackwell, Oxford (1988)
Krasnodębski, M.: Co nurtuje uczonych według filozofa? (What Bothers Researchers According to a Philosopher?) In: Misiak, V. ed. Co nurtuje uczonych (What Bothers Researchers), Difin, Warsaw, pp. 239–243 (2010)
Król, Z.: Platon I Podstawy Matematyki Współczesnej (Plato and Foundations of Contemporary Mathematics). Wydawnictwo Rolewski, Nowa Wieś (2005)
Król, Z.: The emergence of new concepts in science. In: Wierzbicki, A.P., Nakamori, Y. (eds.) Creative Environments, op.cit (2007)
Lakatos, I.: Proofs and Refutations. Cambridge University Press, Cambridge (1976)
Latour, B.: Science in Action. Open University Press, Milton Keynes (1987)
Michalska, A.: Niezbywalność założeń metafizycznych w nauce (Inalienability of Metaphysical Assumptions in Science). In: Motycka, A. (ed.) Nauka a metafizyka (Science and Metaphysics), pp. 207–227. Wyd. IFIS PAN, Warsaw (2009)
Motycka, A.: Metafizyka w oczach filozofa nauki (Metaphysics in the Eyes of a Philosopher of Science). In: Motycka, A. (ed.) Nauka a Metafizyka (Science and Metaphysics). Wydawnictwo IFIS PAN, Warsaw (2009)
Pasternak, Ch.: Quest: The Essence of Humanity. Wiley, New York (2004)
Pobojewska, A.: Biologiczne 'a Priori' Człowieka a Realizm Teoriopoznawczy (Biological 'a Priori' of Humans Versus Epistemological Realism). Wydawnictwo Uniwersytetu Łódzkiego, Łódź (1996)
Popper, K.R.: Logik der Forschung. Julius Springer Verlag, Vienna (1934)

Popper, K.R.: Objective Knowledge. Oxford University Press, Oxford (1972)

Rawls, J.: A Theory of Justice. Belknap Press, Cambridge, Mass (1971)

Rorty, R.: Philosophy and the Mirror of Nature. Princeton University Press, Princeton (1980)

Ryszkiewicz, M.: Homo Sapiens. Meandry ewolucji (Homo Sapiens. Intricacies of Evolution). CIS, Warsaw (2012)

Skarga, B.: Czło to Nie Jest Piękne Zwierzę (a Human Is not a Beatiful Animal). Wydawnictwo Znak, Cracow (2007)

Srzednicki, J.: To Know or not to Know: Beyond Realism and Anti-Realism. Kluwer Academic, Dordrecht, Boston (1995)

Tarski, A.: The Concept of Truth in Languages of Deductive Sciences. In: Tarski, A. (ed.) Logic, Semantics, Metamathematics. Hackett Publishing Company, 1956, (1933)

Wiener, N.: Cybernetics or Control and Communication in the Animal and the Machine. MIT Press, Cambridge, Mass (1948)

Wierzbicki, A.P.: Finansowanie nauki w krajach rozwiniętych na progu gospodarki opartej na wiedzy a sytuacja nauki w Polsce (Financing of Science in Developed Countries on the Verge of Knowledge Based Economy and the Situation of Science in Poland). Przyszłość: Świat, Europa, Polska. 2, 105–119 (2008)

Wittgenstein, L.: Tractatus Logico-Philosophicus. Cambridge University Press, Cambridge (1922)

Wittgenstein, L.: On Certainty. In: Anscombe, G.E.M., von Wright, G.H. (eds.) On Certainty. Blackwell, Oxford (1969)

Żurkowska, G.: Wprowadzenie do dyskusji: dlaczego metafizyka poznania? (An Introduction to Discussion: Why Metaphysics of Cognition?). In: Motycka, A. (ed.) Jana Srzednickiego Sapientia Restituta (Sapientia Restituta of Jan Srzednicki). Uniwersytet Warszawski, Warsaw (2008)

Part II
Elements of Recent History
of Information Technologies

Chapter 7
Telecommunication, Radio Broadcasting, Television

7.1 Beginnings of Telecommunication, Radio Broadcasting, Television: Important Events

Beginnings of telecommunication[1] relate to coding and telegraphy; the latter is becoming obsolete today, replaced by email. Codes, on the other hand, retain their fundamental importance because of the security demands related to network communication (see e.g. Lubacz 2001). Codes were of course used even in ancient times, but as the first practical code of the epoch of industrial civilization we can count the code of Luis Braille replacing print for blind persons. This code is not digital, but analog-digital: with the sense of touch, a blind person perceives the relative positions (analog information) of embossed points (digital information) on a page. In this sense, this code is closer to actual functioning of human brains than the cognitivist models of brain as a giant digital computer.

The code of Samuel Morse (broadly known "dots and dashes") was decisive for the development of telegraphy. In this code, dashes can be interpreted as a signal of a different digital value: today we would say that a lack of signal corresponds to the value 0, a short signal (dot) to the value 1 and a long signal (double dot or dash) to the value 2. Semaphore telegraphy towers were known much earlier, and broadly used e.g. in France in the end of the eighteenth century. The idea of electromagnetic telegraphy is attributed to Francis Ronald (year 1816). Many researchers worked on this idea, with Gauss and Weber around the year 1833 among them. Britons maintain that the electromagnetic telegraph was developed independently and applied in railroad industry in the year 1839 by Charles Wheatstone and William Fothergill Cooke. However, Samuel Morse invented a different, more efficient version of a telegraph in the year 1937, even if the first telegraph

---

[1] In Poland, we also use the old name "łączność" that was used for the first time around 1918 in Polish army to describe military telecommunication. The National Institute of Telecommunication in Poland still uses its old name "Instytut Łączności".

© Springer International Publishing Switzerland 2015
A.P. Wierzbicki, *Techneₙ: Elements of Recent History of Information Technologies with Epistemological Conclusions*, Intelligent Systems Reference Library 71,
DOI 10.1007/978-3-319-09033-7_7

transmission using this invention took place later than in Great Britain (Washington–Baltimore 1844).

After the invention of Morse code, telegraphy spread over the world: the first undersea telegraphic connection (between France and England, and then between Paris and London) was established in 1850–1854, the first transatlantic connection was finally achieved (after many failed attempts) in 1856–1866. The first connection of a telegraph with a typewriter was invented by Edward Calahan in the year 1867; after this invention, the use of telegraphy became universal even if it was many times enhanced later, leading to the concept of a telex, then telefax, then a computer printer that combines earlier functions of a facsimile.

Therefore, the time of a pure delay between the original idea of electromagnetic telegraphy and the beginning of its broad social utilization, which I call further civilization delay, amounts to circa 50 years even for this evidently useful invention. However, its usefulness was not evident for all people: even some great industrialists maintained that they must use traditional letters with signatures, not fashionable telegrams (hence the need of authentication of electronic signature is rather old). However, telegraphy was a precursor of globalization of information, a virtual reduction of the distance between people in various places over the globe, and also of instantaneous transmission of information for newspapers that initiated the era of mass media.

Telephony also belongs to telecommunications but it is younger than telegraphy. A well working version of a telephone was invented by Alexander Graham Bell in the year 1876. Also in this case it was not a fully original idea: it had been discussed and developed earlier by Antonio Meucci since 1849, and other but similar ideas were used by Thomas A. Edison also in 1876 but prior to Bell, in the construction of a gramophone. Social acceptation of Bell's invention was relatively fast: the first telephone line between Boston and Sommerville (a small city near Boston) together with a *switchboard* (a cross-connection board originally handled by female operators) was launched in 1877, the first telephone exchange in New Haven, Connecticut in the year 1978. Britons maintain that they were equally fast: the first telephone exchange in London started its operations in 1879. Again, however, there were great industrialists who commented that they could not imagine using such toys (telephones) in business. Nevertheless, during 20 years until 1895, ca. 240 thousand telephones were installed in the United States. Therefore, the civilization delay in this case was smaller, counting from Meucci to the exchange in New Haven it was around 30 years, even if it required much longer time until telephony arrived in all countries and was spread over the entire globe.

Radio was used at first in wireless telegraphy, first experiments are counted since 1832 (James Lindsay). However, the first practical version of wireless telegraphy was invented by Guglielmo Marconi in 1895, while the first wireless transatlantic connection was established in 1901. Soon it was also used in wireless telephony (Reginald Fessenden, in the year 1900). However, radio media transmission required yet the inventions of electronic vacuum tubes: diode (James A. Fleming in the year 1905), triode (Lee de Forest in 1906) and their further modifications (pentode etc.) to amplify signals, telegraphic, telephonic on longer distances, as well as radio signals. First broadcasting stations were established in 1920 (Westinghouse in Pittsburgh aired the first radio transmission of the results of elections); full socio-economic impact of the new media was achieved with the advertisements introduced just before 1930 with a commercial success of radio after 1935. Therefore, the civilization delay amounts in this case to at least 40 years (and counting from Lindsay, to almost 100 years).

The ideas related to television have also a long history. In 1873, Joseph May and Willoughby Smith invented the photocell. Stimulated by this, a Pole Julian Ochorowicz published in 1878 the idea of a monochromatic television camera. Slightly later, for in 1880 in Boston, George R. Carey published similar concepts. In 1884, Paul Nipkow invented (or augmented; it is an idea with a long history too) a rotating disc with apertures for scanning an image, the so-called Nipkow disc. In 1896, Aleksandr S. Popov constructed a duplex receiving and broadcasting antenna, essential for television transmission. In 1897, Karl Braun constructed a prototype of the cathode ray tube that enabled the construction of a television receiver; this prototype needed many improvements until it became the foundation of the works of Vladimir Zworykin and Kalman Tihanyi in the years 1923–1928.

In the years 1918–1925, John Logie Braid used Nipkow disc to construct the first working television system, demonstrated by him in London in 1925; hence in British sources Braid is treated as the inventor of television. However, fully electronic television started from the invention of an "iconoscope", the first functioning electronic television camera (Vladimir Zworykin, camera in 1923, and a complete TV system with a television receiver based on the works of Kalman Tihanyi in 1928). Therefore, American sources treat Vladimir Zworykin as the inventor of television (he was born and educated in St. Petersburg in Russia, but emigrated and worked in the United States). First experimental television transmissions occurred relatively fast after this invention, in 1931, in the United States, Germany, and Soviet Union. The first media transmission took place in 1936, through BBC in England.

The first experimental television station in Poland was established in 1937, through a cooperation of Polish Radio and the National Institute of Telecommunications. World War II stopped these works, but they were resumed and the first media television transmission was sent in 1952 from Instytut Łączności in Warsaw, which was a continuation of the works of the National Institute of Telecommunications. However, television receivers were more expensive than small automobiles, and the prices did not go down until after 1960. For all these reasons it is difficult to determine the civilization delay from the first idea to the start of a broad social utilization in the case of television. In Poland, if we count it from the idea of Julian Ochorowicz to the first medial transmission, it amounts to 75 years; but in Poland it often happens that we have groundbreaking ideas on the world scale, but their utilization is excessively delayed. Digital television was similarly delayed, the idea was discussed between specialists in Poland already before 1960, a practical transformation of television signals into digital form occurred around 1985, but a broad penetration of broadcasting digital television (not counting cable and satellite versions) was not yet finished in Poland in the year 2010.

If we count according to American data, either from George Carey or from Vladimir Zworykin, up to the start of broad social penetration in the United States, then, as already indicated in Chap. 4, see Fig. 4.1 and related discussion, we obtain a civilization delay of 80 years or only ca. 35 years. Indeed, television became so important in social terms that the delays in its further development were rather short (even if by no means negligible); today we use quite different TV receivers, cathode ray tube is a museum piece. We use flat screens made of light emitting diodes, up to four basic colours, with a frame rate up to 200 per second in order not to tire our eyes, etc. But these are innovations that penetrated the market rather recently, even if the related ideas were known long ago.

The great telecommunication revolution occurred after 1980 together with the development of computer networks and the Internet, which is discussed in Chap. 11, and after 1990 together with socio-economic penetration of radio telephony (called also mobile or cellular telephony). However, the history of radio telephony is older. As already noted, the idea of wireless telephony dates from Reginald Fessenden in the year 1900. The original mobile telephones were simply duplex radio devices combining the functions of a receiver and a broadcasting unit; the selection of an interlocutor was achieved by tuning the frequency of transmission. In military applications, this resulted in the need of coding. However, also for military applications, in the year 1943, the division of a terrain into appropriate cells was invented; these cells were equipped with corresponding antennas.[2]

The techniques of mobile telephony were intensively developed after World War II. For example, in the year 1947, Douglas Ring, Rae Young and Phillip Porter proposed to use hexagonal cells of terrain with antennas located in each corner of the hexagon and serving three adjacent cells. However, mobile telephones were massive and heavy, their miniaturization required the development of transistors and integrated circuits, which are commented in one of subsequent chapters. Thus, in the year 1973, CEOs of two leading corporations, AT&T Bell Laboratories and Motorola, demonstrated (in a TV broadcast) mobile telephones held (with some difficulty) in their hands. Despite great hopes related to them, they did not achieve a market success, both due to their large weight (they were called popularly brick phones) and their excessive prices. Only after the year 1990, Finnish firms Radiolinja and Nokia, and just after that other corporations, introduced market versions of light and relatively inexpensive cellular telephones which started their fast social penetration. Counting either from 1900 or 1943 to the beginning of this socio-economic penetration, the civilization delay for mobile telephony amounts to 90 or 47 years. Even if we take the lower estimate, we must remember that the process of socio-economic penetration of mobile telephony is not finished, even if quite advanced today. Thus we must account for additional

---

[2] Therefore, the word "cell" used sometimes in popular language to describe a mobile telephone device meant (and also means today) something quite different in technical language: it is an area of terrain served by an antenna of mobile telephony.

30–40 years for the actual socio-economic penetration. In total, even if counting only from 1943, this amounts to ca. 70 years.

Further on we shall distinguish *proper telecommunications*, telegraphy and telephony, even if the Internet today is also a means of telecommunication, from *broadcast telecommunication*, thus radio and television used as media of information transmission, but also as media of advertisement and a tool shaping public opinion. This is historical division, loosing its importance today because of the great *megatrend of digital integration* in the development of information society. A renewed integration of these individual domains is taking place, similarly as a renewed integration of larger fields of information techniques, such as telecommunication, control engineering and robotics, computer engineering (informatics), electronics. I shall use, however, the name of these fields in order to organize the material of this book.

7.2 Social and Conceptual Importance of Telecommunications

Today, *the social, economic, civilization importance of telecommunication* is obvious: telecommunication virtually overcomes the physical distance between people, it is a necessary (even if not sufficient) condition of globalization, changes our way of perceiving the world in such a way that the globe seems small to us. Even if telecommunication is becoming dominated by computer networks and the Internet, the latter can be nevertheless treated as only a next generation of telecommunication technologies. Globalization needs also a decrease of distance measured by the time of transport of people and goods, so also a development of transportation networks (airborne or maritime), but also these would not function today without telecommunication. Fifty years ago, the present-day easiness of mobile (radio, cellular) telephony, when radio communication was used mostly for military use, was difficult to imagine; but the changes of social habits resulting from this easiness are enormous.

What is much less broadly known, and unappreciated by, particularly postmodern, sociology, is the *conceptual importance of telecommunications* (its importance as the source of new concepts). And it is also tremendous. Various telecommunication inventions required, in order to better understand and analyze or improve them, new concepts and theories to be created, along with mathematical modelling.[3] For example, the concept of *input–output relation* was introduced by John Renshaw Carson in the year 1926 in order to describe and

---

[3] I use the concept of *mathematical modelling* not in the narrow sense of abstract logics, but in the broader sense used in systems engineering: mathematical modelling is the creation and utilization of mathematical models of diverse phenomena, in particular technical phenomena, for their computerized analysis by simulation, optimization, etc., in general, for *virtual laboratories*.

analyze telecommunication connections, or any electrical circuit, with the help of an aggregated operator. Such operator is called *transmittance* (or transfer function) and is defined as the ratio of Fourier transforms of the output signal and the input signal. Later this concept was generalized to include Laplace transforms, and much later, in the case of time discretisation necessary for transmission of digital signals, to Laurent transforms. Transmittance describes the dynamics of the input-output relation (because dynamics, as well as delays and other dynamical distortions of signals, are important in telecommunication connections) under the assumption of linearity of the connection and disregarding initial conditions. Today, the concept of input-output relation is broadly used in diverse disciplines of science, starting with economy (Leontief 1965), including sociology and philosophy, but usually in a static, not dynamic sense, and without even noting the technical origin of this concept and disregarding the fact that it can be applied only when assuming linearity in the technical sense.

Let me make a comment here on the difference between the languages of technology and sociology. In technical and strict sciences, the concept of *linearity* means the applicability of the *superposition principle:* the effect of sending a sum of two input signals should be the same as the sum of effects of sending these signals separately. This is equivalent to the assumption of linearity of mathematical models of a connection. Clearly, this is a simplifying assumption, since each real connection or circuit is nonlinear, e.g., it is subject to saturation when the amplitude of signals is too large. However, in sociology the concept of a *linear process* means something quite different: it is a process without recourse, without turning back to earlier stages of the process. The difference between these concepts results in diverse misunderstandings, particularly when we use their negations. For a technician, *nonlinear process* is such that the superposition principle does not apply; for a sociologist, it is a process that includes recourse. For example, *feedback*, another fundamental concept introduced by telecommunication, which has, however, the most important role in control engineering and robotics (hence it is analyzed in detail in the next chapter), is in its essence an infinite recourse, therefore the concept of a *nonlinear feedback* is not understandable for a sociologist, as in his understanding of linearity each feedback must be nonlinear. On the other hand, for a technician, the theory of linear feedback systems is an elementary foundation of control engineering, but especially interesting effects can be given by nonlinear feedback systems only: aperiodic oscillations, sliding motion, chaotic behaviour, even emergence of order out of chaos, see further chapters.

Another example: *network theory* was developed in order to analyze networks of telecommunication links, with the use of diverse mathematical tools, such as graph theory, queue theory, stochastic process theory. The classical understanding of network theory starts with Carson (1926), even if the concept of *network analysis* was introduced by Hendrik W. Bode in the years 1938–1945 (see Bode 1945). Even until 1970, the concept of *network theory* was defined by such basic technical monographs as *Active Network Theory* (Haykin 1970), or *Network Theory* (Murdoch 1970), describing such theoretical approaches as graph theory and topological network analysis, and such problems as the issue of network stability. Moreover, the classical problems of network combinatorics, such as routing, finding the most appropriate (in a given sense: shortest or least costly, or most safe etc.) connection between two network nodes, derive from telecommunication, and were only later transferred to operations research and management science.

However, the technical roots of network theory are nowadays forgotten and it seems that this attractive slogan has been appropriated by sociology and applied mathematics. The latter maintains that the network theory is the combinatory part of operations research (including problems similar to routing). On the other hand, sociology, especially postmodern and Anglo-Saxon,[4] goes as far as to state, invoking the works of Emil Durkheim and Ferdinand Tönies, that the network theory was devised in order to analyze interpersonal relations and contacts in social groups.

The attractiveness of the metaphor of a *network of interpersonal relations* for sociologic research is understandable. Manuel Castells (2000) maintains that the information revolution leads to *networking* in social relations, to *network society*. However, it is not true that the concept of networks of social relations is old: sociology always considered interpersonal relations, but only recently in a network sense. Actually, until around 1988 sociology used the concept of *social structure* rather than *social network,* see e.g. (Wellman and Berkowitz 1988). Therefore, we encounter here an example of a frequent phenomenon, an attractive slogan originating from technology is taken away and used it in a quite different sense, without even mentioning its technical source.

Also the concept of *automation* in contemporary civilization was introduced by telecommunications. As already mentioned, first telephone exchanges were served by telephone operators (usually female) making appropriate connections on demand of a client. Automation of these connections was one of first successes of *telecommutation,* a technique enabling the construction of automatic telephone exchanges. Such exchanges were a prototype of digital computers,[5] they were huge systems of electromagnetic *relays* (electromechanical switches) and *selectors* (originally, these were electromagnetic stepping switches selecting a connection between many output lines, and later, more advanced *crossbars*).

This trend towards automation of telephone exchanges resulted in the formation of a new technical discipline, control engineering, the technique of construction of automatic devices for various applications, not only telephone exchanges, described in more detail in the next chapter. It is control engineering that is responsible today for dematerialization of work with all its advantages and faults, but the first impulse for this megatrend came from telecommunications.

It was also telecommunication that introduced into a broader, both social and economic practice the concepts of *stability of a system* and of *conditions of wave*

---

[4] If we look on the Internet for the keywords *"teoria sieci"* (network theory in Polish), we will get pages where it is used either in technical or mathematical sense, but the keywords *"network theory"* result in answers providing mostly the sociological meaning.

[5] The patent of Zuse from the year 1936 describing the first digital computer is based on elements of automatic telephone exchanges.

generation. The stability of telecommunication devices, together with *stability of instability,* that is, stability of oscillations generated according to the needs of telecommunication engineering, is a fundamental demand. The whole development of telephony and telecommunications in general was related to the analysis of conditions of wave generation, especially in nonlinear circuits, which finally led to *chaos theory* and is commented in more detail in the next chapters. The abovementioned concepts introduced by telecommunication, more precisely, by theoretical analysis and mathematical modelling of problems emerging in telecommunication, do not exhaust the whole conceptual contribution of telecommunication. For example, *information theory* (started by Claude Shannon, publication in 1948) was even called originally *mathematical theory of communication* and clearly originated in telecommunications, even if its importance extends over computer science and other disciplines.

7.3 The Social and Conceptual Importance of Broadcast Telecommunication, Radio and Television

The social and economic importance of radio and television is perhaps even larger than of the telecommunication proper, and moreover, it is fully appreciated by sociology and political science: radio and television are fundamental tools shaping public opinion and fundamental media of advertisement and propaganda. Therefore, already in the year 1964 Marshall McLuhan formulated his theory of *media society* together with such metaphors as *the medium is the message*[6] or *the global village.* It should be stressed that the analyses of McLuhan were both deep and critical. He noted the impact of typed text on cognition (with the metaphor *typographic man*), noted and sharply criticized the strong dependence of media on advertisement, even made a bold attempt to intellectually attack fashionable (in his time, but even today) ethical judgements of technology as a force enslaving people. The arguments that he used in the latter case are telling: historically, the end of the seventeenth century was a period of a critique and ethical denunciations of printing technology, printers and booksellers who distributed too many "dangerous" books; now, at the end of the twentieth century, we have a period of anxiety about "the end of printed word" (lasting until today). From the above it follows that holistic ethical evaluations of technology are relative and unreliable, and that we must be well aware of both chances as well as threats of specific instances of technology. An ethical judgment must be well based on detailed knowledge.

---

[6] More precisely, *the medium is the massage,* not *the medium is the message*; McLuhan used the metaphorical play of words in order to stress the direct influence, and not only verbal, but rather a preverbal content, of a medium on the recipient.

However, the importance of media changes and grows together with digitalisation of radio and television, dissemination of satellite and cable television, with growing access to radio and television by the Internet and mobile telephony. At present, the said importance, especially of diverse forms of television, is much more immense than in the time of McLuhan (we live in a *spectacle society,* see Debord 1977). We also better understand why it has increased. As discussed in Part I, multimedia perception is a foundation of intuition and human intelligence in general, and the impact of multimedia messages on the unconscious foundations of human behaviour is particularly strong and only partly controlled by the conscious ego. Such state of affairs is exploited by advertisements and political propaganda in a ruthless way. In the face of future digital integration of media, we already observe political and economic struggle for the control of integrated media. Today, after the great crisis of the years 2007–2009, we witness also the growth of social resistance against instrumental treatment of viewers by television, against advertisements as tools of misinformation, aimed at increasing the *asymmetry of market information,* which will be discussed in further chapters.

The *conceptual importance* of radio and television broadcasting is in some sense smaller than that of telecommunication proper, but also profound. The basic concepts of broadcasting are *transmitter* and *receiver, broadcaster* and *viewer* or *listener,* together with initially assumed and subconsciously accepted asymmetry of their roles (the assumption that the broadcaster must have more to say than the viewer or listener, since (s)he is a broadcaster; this asymmetry was challenged for the first time in the era of the Internet). In social sciences, these concepts are broadly accepted in a metaphorical sense.

I would like to stress here, however, the importance of certain concepts generally related to telecommunications, but specifically to radio and television broadcasting, and with the information theory: the concepts of *modulation, broadband* and *information measure.* Modulation is an overlap (multiplicative, hence nonlinear in the technical sense) of two wave-like signals: *carrier frequency* and *operating band.* Radio and television transmission on various frequencies is possible because we use modulation in the transmitter, we mix these two signals, and transmit a modulated signal of high frequency (actually, operating frequency shifted by the carrier frequency), and in the receiver we reconstruct the operating band using demodulation.

There are two basic types of modulation: amplitude modulation and frequency modulation (apart from many variants and specifics; theoretically, phase modulation is also possible). Frequency modulation has many advantages; it serves better when we use larger carrier frequencies. The character of carrier frequency changes along with the frequency modulation: it is the base frequency transmitted when the operating signal is lacking.[7] The fact, however, that we use carrier frequencies to transmit a signal, does not change the measure of information sent

---

[7] The main advantage of frequency modulation is its noise tolerance, resilience with respect to perturbations of the amplitude of the carrier frequency.

by it. According to information theory by Shannon such information is measured by the operating bandwidth (actually measured in a logarithmic scale). Therefore, *modulation does not change information measure:* we could transmit voice and music on carrier frequency of 100 MHz (megahertz) plus-minus 10 kHz (kilohertz), or on frequency of 1 MHz plus-minus 10 kHz (kilohertz). In the former case we would use frequency modulation, in the latter, amplitude modulation, but in both cases operating bandwidth is 20 kHz which is sufficient for a good transmission of voice and music.[8]

In this respect, some philosophers of science (e.g. Ihde 1976) who want to compare the role of voice and vision in human cognition, commit a fundamental error of physical interpretation, they base their judgment on physical, not informational measures. Ihde derived his judgments from the ratio of the upper and the lower limit of a frequency band. In the example of modulation quoted above such ratio amounts to 1.0002 in the former case and 1.02 in the latter case; but these numbers say nothing about the amount of information transmitted by operating band, with bandwidth of 20 kHz in both cases. This bandwidth corresponds to the properties of human ears, but only with respect to the upper tones usually perceived by people. The lower tones, ca. 20 Hz, perceived as a lower limit by human ear, have an impact on the quality of music we listen to (e.g. when listening to a bass singer), but they practically do not influence the informational content of a human voice. While lower tones do have some informational meaning, e.g. a good automotive technician intuitively analyzes the tone of an engine, 20 Hz is only 0.1 % of the operating bandwidth of voice and it is not important in the measurement of informational content of voice.

For the purposes of transmission of colour television the standard operating bandwidth is 6 MHz (2 MHz for each of three basic colours). Beside using three colours, this has almost no relation to the frequencies of light waves, hence a physical interpretation is a rough error here. The accepted standards are historical, they first resulted from a rather blurred transmission of black-and-white television that used the operating bandwidth of 2 MHz. Later on, diverse improvements occurred, based mainly on an imitation of the properties of image analysis in a human eye that does not concentrate on the details of large coloured blotches, but only on their contours and changes. By analogy, we can transmit the data which does not concern every coloured point, but only changes between them on the image and over time. This results in a substantive *compression* of encoded images, a decrease of information needed for transmission. In practice of television techniques, this compression of codes was used conversely, not to limit the operating bandwidth, but to send images of much better quality, with a better resolution (number of light points called pixels), with the same historically preserved standard bandwidth. Obviously, for television transmission we also use modulation, hence carrier frequencies are much larger than 2 or 6 MHz.

---

[8] For the transmission of voice in telephony even today we apply less stringent requirements, using operating bandwidth of only 7 kHz or even 3.4 kHz.

All this discussion I recapitulate (it was already presented in Chap. 5) in order to stress how excessive simplification or reduction of reasoning to purely physical parameters can be misleading, can even lead to gross errors if the evaluated problems are essentially informational. This is precisely a part of the informational revolution: the emergence of new, informational concepts, typical for information society or knowledge civilization, and often just unaccountable within the physical episteme of the epoch of industrial civilization.

7.4 Some Disciplines Related to Telecommunications: Radar Technologies, Biomedical Engineering, Laser and Optical Fibre Technologies

The development of telecommunications resulted in the rise of many derivative, important disciplines. Some of them, such as control engineering and robotics, informatics and teleinformatics, are presented in detail in the next chapters. There are, however, many other disciplines, such as mechatronics and fine mechanics, which I shall not discuss in this book. I will limit this short description to only three of them:

- *radar technologies,* initiated in telecommunications already before World War II, but developed mainly for military applications;
- in a sense at the other end of the spectrum, but in some aspects closely related, is *biomedical engineering* that accompanied telecommunications since the end of the nineteenth century, but is currently especially intensely developing, not only because of the informational revolution, but also in result of the general trend of ageing societies and the need to intensify some forms of health care;
- finally, *laser and optical fibre technologies* which are decisive for a broadband access to information of the present day.

7.4.1 Radar Technology

The history of *radar technologies* can be dated back to the year 1904, Germany, when Christian Huelsmayer patented the first device detecting radio signals reflected from material objects, in particular metallic ones. The application that motivated Huelsmayer was the detection of ships in fog in order to increase the safety of navigation. Relatively fast, in the year 1922, it was noted that also wooden ships cause an interference of waves and can be similarly detected. In 1923 in England, Robert Watson-Watt applied a rotating directional antenna to detect the direction of approaching storm. The works on the development of such techniques continued (and were mostly strictly classified) and lasted over 30 years in many countries, Germany, Great Britain, the United States, Soviet Union, but also in Japan, France, Italy and many other countries. It is estimated that a conclusive development occurred in the years 1934–1939, when at least 8 countries independently developed operable radiolocation systems. The name RADAR was used for the first time in 1940 in the American navy; it was an abbreviation of Radio Detection And Ranging. *Ranging,* the determination of the distance from a detected object, was clearly a fundamental requirement for a naval ship aiming its guns.

Just after 1940, a *magnetron* was developed at the Birmingham University in England. A magnetron generated radio waves of high (for that time) frequency, 3–30 GHz (gigahertz), which resulted in much better parameters of allied radar systems and helped to give them advantage in World War II. The development of radar technologies in World War II was of enormous importance especially in aviation, where radar imaging helped not only in aiming, but also in landing at night or in fog (today it is universally used, but obviously not always unfailing). The civilization delay from the first idea to an universal (even if only military) use was in this case rather short, about 35 years.

A universal civil utilization of radar occurred for the first time after World War II, together with declassification of radar technology, hence another estimate of civilization delay in this case increases to ca. 45 years. The achievements of radar technology are commonly used nowadays in many disciplines, from radio astronomy and biomedical engineering (described below) to safety systems in airborne transportation (where, even if these systems are fallible, the achieved safety standards are much better than e.g. in road traffic).

Several problems were crucial for the development of radar technologies. The first of them was to increase the frequency of radio waves used (to improve the resolution, range and precision of a radar). Already in the years 1934–1939, the frequency was increased from ca. 30 MHz to ca. 200 MHz, but further frequency increase with the use of standard vacuum tube amplifiers was impossible, and it was only the invention of a magnetron that resulted in a substantial increase. Another problem was the application of pulse signals in order to achieve more precise ranging. Yet another one was the coordination of movements of the transmitting and receiving antennae; the duplex antenna, combining transmission and reception, was re-invented and improved several times; in radar technology it took place around 1935. All the time, an important problem was *imaging technology,* that is, presentation and interpretation of reflected signals received. In early solutions, imaging of reflected signals utilized cathode ray tubes, like in early television receivers, but if the signals were presented without additional processing, the obtained images were blurred.

The work aimed at the improvement of imaging technologies resulted in the idea of using computers, more precisely, sets of specialized image processors that processed in parallel parts of received images, in order to improve their quality and interpretation, which led to the emergence of *virtual reality*. Therefore originally, virtual reality was not artificial, but rather actual reality enhanced for better interpretation. These studies have a long history, they are dated back to a radar engineer Douglas Englebart[9] in the fifties of the twentieth century. They were accelerated by the development of *real time computing* for radar technology, parallel computer systems processing large amounts of information in time of reception of that information in a radar system, in the sixties of the twentieth century. At that time, it was the first real time computation, but on the other hand it started the development of many other applications of computer graphics.

For military applications, it was used in *flight simulators* that imitated real flying situations for the purposes of training of pilots. In the seventies of the twentieth century, similar techniques were used in the film industry to create virtual reality, starting with the *Star Wars* (1976) movie, and later broadly used in animated films. In this case it turned out how much more complex is image processing than language processing, how important is an interactive participation of the author of the movie in the preparation and correction of subsequent film frames of an animated picture. Therefore, first in the eighties of the twentieth century, the development of super-computers facilitated a substantial improvement of imaging techniques, used in film industry and in scientific research as well, and also in medical diagnostics.

7.4.2 Biomedical Engineering

We can say that *biomedical engineering* is as old as medicine, because it supplies working tools needed by a physician. In subsequent civilization epochs, such tools changed essentially and their further development is inevitable, at the very least because of the global ageing of societies, see e.g. (Tadeusiewicz 2008). At the same time, the best measure of socio-economic progress[10] is the expected life time of participants of a society. Therefore we observe here a positive feedback loop typical for all civilization development (see Chaps. 3, 6): the improvement of tools supporting the work of a physician prolongs human life, and a further prolongation of life requires further development of such tools.

The use of techniques similar to telecommunication and computer science in biomedical engineering can be dated in various ways: either from the research of Galvani (1786, see e.g. Kresse 1985) demonstrating the impact of electric current on living organisms, or from the discovery of Wilhelm Röntgen in the year 1895 of the so-called *x-rays* and their use for tissue imaging in medical diagnostics, see e.g. (Pałko 2009). Perhaps equally important was the discovery of electrocardiography by Willem Einthoven in the year 1903. Due to these discoveries (and neither of them

---

[9] Today, Douglas Englebart is more commonly known as the inventor of computer mouse; but actually he also started work on virtual reality.

[10] I am aware of the critique of the concept of *progress* in contemporary sociology and philosophy of technology, but I reject this critique outright, asking its proponents a simple question: If your critique is substantiated, then perhaps you would like to live, say, 200 years earlier, when the expected life time was two times shorter than today, because of, among other things, more primitive medical technology?

was a simple application of results of strict sciences, each was a discovery combined with an intuitive, to a certain degree engineering-related invention), all twentieth century was a time of fast development of biomedical engineering, stimulated by close cooperation with telecommunication and computer science.

Methods and appliances for tissue structure imaging, described as classic today, employ three main techniques: x-rays, isotopes and ultrasounds. The development of x-ray technology consisted in a reduction of the radiation dose and diversification of specialized appliances, such as for mammography, tomography, angiography, etc. All of these appliances used computerized imaging, electronic amplification of signals and computer aided visual tracks before displaying the results on a monitor. This aspect concerns all methods of tissue imaging, including echography (based on ultrasound penetration) and various variants of tomography (resulting in a layered image of tissue) using x-rays, single photon isotopes, positrons, magnetic resonance. Methods of computerized processing of images, developed in diverse fields, such as radar technology or virtual reality movies, found almost immediate applications in biomedical engineering. For example, contemporary biomedical imaging of the present day uses *multimodal imaging,* that is a combination of several methods of imaging provided by computer aided image processing in order to obtain a better medical diagnosis. Note that a physician, when using such tools, actually observes virtual reality; however, reality constructed in order to improve diagnosis about actual, biological reality. Therefore, sociologists who sometimes attach a great importance to the *artificiality of virtual reality* forget that the most important applications of virtual reality are real.

There are also efforts taken to use the methods of *artificial intelligence* in medical diagnostics, even if the classical artificial intelligence (see Chap. 11) concentrates too much on the automation of diagnosis and disregards the intuition of a physician, so the problem of interaction between a computer and its human user, discussed in other chapters, also arises here. Other efforts include mathematical modelling and computer simulation of the progression of various diseases, while taking into account the natural resistance of human body; the so-called *mathematical immunology.* All this indicates that diverse findings and methods of biomedical engineering, supported by other informational techniques, will be decisive in the coming decades for the improvement of health, especially of older people.

This is not yet an actual biotechnical revolution, which would mean radical changes in human evolution. Speculations about *radical evolution,* a vision of a cyber-man as an essentially new, to a large extent artificial product of biotechnical revolution, are quite frequent today, see e.g. (Garreau 2008); but they are far from being realistic and such evolution will not occur in the next forty, perhaps even 100 years. Already nowadays we observe a significant resistance to an excessive or hasty automation of human activities, kind of an appropriation of human subjectivity by computers and robots, or, in extreme cases, a dominance of computers over people. Will such experiences foster a radical biotechnical revolution that will include an implantation of microprocessors into human bodies? To the contrary: we can expect a significant psychological, social resistance that will significantly delay such a radical evolution.

The biotechnical revolution and radical human evolution will start at the point of a significant socio-economic demand: in biomedical engineering and in health care for older people. A microprocessor implant to stimulate heart action or an artificially developed bone tissue used in order to rejuvenate the bones of the elderly does not encounter substantial psychological resistance. Hence such techniques will meet a broad economic demand which will help in their further improvement, reduction of costs, and finally, making them broadly accessible. Together with techniques of *ambient intelligence* (called also *Internet of things*, discussed in Chap. 11), used for a non-intrusive monitoring of the state of health of elderly people, or even with mobile robots as companions of life for the elderly, the biotechnical revolution will become a natural enhancement and continuation of information revolution, but its broader social impact will be related with elderly people and will occur with significant delay. Nevertheless, even the socio-economic demand for the health care of elderly people might become, because of the growth of their share in society, a vehicle of economic development, especially in developed countries.

7.4.3 Laser and Optical Fibre Technologies

Today, *laser and optical fibre technologies* are decisive for a broadband access to information. It is true that new coding and processing techniques enable also a wireless broadband access and that most mobile telephony operators try to convince broad public that the access offered by new generations of cellular telephony will suffice. However, it is a typical example of the economic phenomenon of *asymmetry of information:* the seller propagates information that are advantageous to him, but not necessarily fully correct, and hopes for ignorance of the buyer. And only specialists in two specific disciplines of telecommunications, magnetic spectrum management and optical fibre technology, realize how much the information provided by mobile telephony operators serves their commercial interests. The spectrum designed for mobile telephony becomes quickly saturated (if all users of mobile telephony start to watch television on their tablets and to surf the web, the quality of wireless access will drop substantially); even today we observe various symptoms of an excessive use of radio spectrum. On the other hand, optical fibre telecommunication has practically unlimited band of information transmission, particularly if we assume a further development of optical amplification and signal processing.

The history of two fields, optical fibre technology and laser technology, is strongly interconnected, but the beginnings belong to optical fibre technology. It started with the discovery of physicists (Collodon and Babinet in the year 1840; John Tyndall in the year 1854), who have shown experimentally that light can travel along a curve (contrary to the theory that light travels at a straight line or is subject to an angular reflection or refraction). They observed light in streams of water in a fountain, or in a stream of water flowing out of a level pipe: light was bended together with water. This phenomenon can be easily rationalized by explaining that light is reflected from the boundaries of the stream and bends its course in that way.

With the use of a prototype form of optical fibre, bended glass rods, this phenomenon was utilised first in medicine and dentistry (Roth and Reiss in the year 1888, Smith in the year 1898). Around 1920 John Logie Braid, who worked on a television receiver, constructed and patented a matrix of glass fibres in order to show an image; this idea was further developed by many authors and in many directions.

However, optical fibre telecommunications developed in full only after the inventions of *maser* and *laser* in the United States. Maser (microwave amplification by stimulated emission of radiation) uses actually, similarly as laser (light amplification by stimulated emission of radiation), the phenomenon of a resonance of waves or photons. In the case of a laser, this concerns photons emitted by atoms during transitions to a lower energy state and the exciting the emission from other atoms by such photons; the atoms must be restored to the higher energy state with an external energy source, and photons must be repeatedly reflected in the same medium.

Maser was invented experimentally by Charles Townes in the year 1954. In 1958, Charles Townes and Arthur Schawlow have shown that masers can emit not only radio waves of very high frequency (as they did originally, if gas particles were excited by radio waves) but also light waves (if the excitation of atoms occurred in response to light repeatedly reflected in gas medium or in a specific crystal). The first practical lasers using gas medium (actually, neon or helium) and in a crystal (ruby) were constructed in 1960 by Theodor Maiman.

The combination of laser with glass optical fibre was achieved by Elias Snitzer in 1961. The core of the glass fibre must be sufficiently thin to conduct a specific monochromatic and single modal light wave in the best possible way. The results of Snitzer were sufficient for medical applications, but the attenuation of the light wave in the glass fibre used by him was too large for applications in telecommunication. Glass light fibres with attenuation less than 20 decibels per kilometre were developed for the first time in 1970. Efforts were also made aiming at the construction of semiconductor lasers integrated with a glass fibre; this was achieved in 1970, concurrently in the USA and in Soviet Union. In 1973, Bell Laboratories developed a technological process of drawing glass fibres with very low attenuation of light; in 1977, the first glass fibre telecommunication connection was established in Long Beach, California.

The first light amplifier (integrated with optical fibre and eliminating in this way opto-electronic transformations necessary for signal amplification in earlier versions of long light fibres) was invented in 1986 by David Payne and Emanuel Desurvire. Another 5 years were necessary in order to practically demonstrate, in 1991, the fully optical light fibre systems that originated a broader socio-economic application of light fibre technologies. Therefore, the civilization delay from the works of Townes in 1954 to fully optical telecommunication systems amounted to 37 years, which is rather a short time in comparison to other breakthrough inventions of information technologies.

Today, it is the fully optical fibre technology that makes the transmission of signals of high transfer rate possible, either in wide-area or in international networks, by telecommunication operators, cable television, Internet etc. As opposed to radio spectrum, there are no practical limits to the amount of information send by optical fibre; maximal transfer rates of a single fibre amount nowadays to terabits (10^{12} bits) per second. Thus, when we will saturate radio spectrum, which will happen soon because of the demand for mobile telecommunication and the commercialization of spectrum management, we will have to return to optical fibre networks with direct fibre connection to each house or apartment. Such access is not very costly, only the legal aspects of the ownership of such access must be regulated: it would be best if the ownership of such access was held by the owner of the premises (paying for it when investing in their development) who can later use this access for diverse services provided by competing service operators.

The laser and optical fibre technology has a much broader area of applications than just telecommunications. Lasers are used in medicine and biology (in bloodless chirurgical or ophthalmologist interventions, etc.), in broadly used digital disc-reading media players, in various mechanical and chemical technologies, in military technology, in precise measurements, in robotics, almost in every field of high technology. There are also high hopes of using lasers for a controlled thermonuclear synthesis, for 3D television and films or in other applications. Laser technology has also high conceptual importance, mostly for further development of hard science (it caused the discovery of diverse phenomena at the quantum level, such as optical bi-stability, generation of harmonic light waves in nonlinear optical media, etc.).

References

Bode, H.W.: Network Analysis and Feedback Amplifier Design. Van Nostrand, New York (1945)
Carson, J.R.: Electric Circuit Theory and the Operational Calculus. McGraw-Hill, New York (1926)
Castells, M.: End of Millenium: The Information Age, vols. 1, 2, 3. Blackwell, Oxford (2000)
Debord, G.: The Society of the Spectacle (English translation in 1970; rev. ed. 1977) Black & Red, London (1977)
Garreau, J.: Radical Evolution Random House. Polish translation (2008) Radykalna ewolucja, Pruszyński i Ska, Warszawa (2005)
Haykin, S.S.: Active Network Theory. Addison-Wesley, Reading (1970)
Ihde, D.: Listening and Voice. Ohio University Press, Athens (1976)
Kresse, H.: Handbook of Electromedicine. Wiley, Chichester (1985)
Leontief, V.: Input-Output Analysis. Sci. Am. **212**, 25–35 (1965)
Lubacz, J.: Telekomunikacja (Telecommunications). In: Morawski, R. (ed.) Wczoraj, dziś i jutro Wydziału Elektroniki i Technik Informacyjnych Politechniki Warszawskiej 1951-2001 (Yesterday, Today and Tomorrow of the Faculty of Electronics and Information Technologies of Warsaw University of Technology). Politechnika Warszawska, Warsaw (2001)
McLuhan, M.: Understanding Media. Ark Paperbacks, London (1964)
Murdoch, J.B.: Network Theory. McGraw-Hill, New York (1970)
Pałko, T.: Technika a postęp w medycynie (Technology and the Advancement of Medicine). Przyszłość: Świat Europa Polska **20**(2), 25–39 (2009)
Shannon, C.: Mathematical Theory of Communication. Bell Syst. Tech. J. **27**, 376–405 (1948)
Tadeusiewicz, R.: Inżynieria biomedyczna. Księga współczesnej wiedzy tajemnej w wersji przystępnej i przyjemnej (Biomedical Engineering. A Book of Contemporary Arcane Knowledge in a Nice and Accessible Version). Wyd. AGH, Cracow (2008)
Wellman, B., Berkowitz, R.D. (eds.): Social Structure: a Network Approach. Cambridge University Press, Cambridge (1988)

Chapter 8
Automatic Control, Analog Computers, Robotics: The Concept of Feedback

8.1 Introduction: Important Events

The possibility of automatic control was known and used even in ancient times, see e.g. (Bennet 1979). In the years 1769–1783, James Watt improved the already existing (but prone to exploding) steam engine, thus initiating its broad socio-economic utilization. A crucial part of this improvement was a rotating speed controller for such engine (presented schematically on Fig. 8.1a): two weights located on flexible arms of the rotating part of the engine would rise up in result of the centrifugal force, and this movement could be transferred in order to slightly close the valve supplying steam to the engine, preventing thereby the rotation speed from increasing.

This was an application of a phenomenon or principle which was much later called *feedback*: an increase of the rotating speed caused, in a feedback, a reduction of its cause, steam supply: *the effects reacted backwards on their causes.* Even today this principle is not sufficiently understood by many philosophers, which was discussed in Chap. 6. Nevertheless, this principle fascinated many researchers, especially physicists and mathematicians; e.g. James C. Maxwell was interested in automatic control and introduced the concept of a *governor,* which we today call *controller.* Maxwell addressed also the problem of *stability* of a control system, analyzed later by many mathematicians and discussed in further sections of this chapter.

The automatic control of diverse physical quantities, rotation speed of engines, temperature, the level of a fluid in a container, pressure, etc., led to many inventions, often based on the feedback principle. For example, the need to support human muscles when steering large ships resulted in the invention of a feedback controller of rudder position by J. Macfarlane Gray, patented in 1866, see e.g. (Bennet 1979). In the same time Jean Joseph Farcot in France introduced a similar solution and the name *servo-moteur* (today, after Házen (1934), we use the name *servomechanism*) for a system controlling the position of a large or heavy object

© Springer International Publishing Switzerland 2015
A.P. Wierzbicki, *Techne_n: Elements of Recent History of Information Technologies with Epistemological Conclusions*, Intelligent Systems Reference Library 71,
DOI 10.1007/978-3-319-09033-7_8

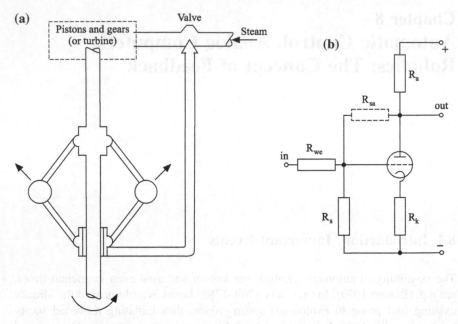

Fig. 8.1 A schematic illustration of the rotation speed controller by Watt (**a**) and the cathode based negative feedback by Black (complemented by an anode based negative feedback, **b**)

which actually uses the feedback principle; Farcot did not use this name although the principle was utilized by him, Házen knew this name already.

However, until 1930 the new discipline, *automatic control engineering,* dealing with the issues related to the control of diverse objects independently of their physical nature, did not emerge, because the habit of disciplinary specialization was too strong, even if appreciation of universality of automatic control principles had gradually increased.

In the year 1926, Harold Black invented, and first patented, then described using the concept of *feedback* and published (Black 1934), the negative cathode based feedback (addition of an additional resistor in the cathode circuit of a vacuum tube) in a vacuum tube amplifier for telecommunication applications. In Fig. 8.1b two feedbacks are presented, both *negative* (counteracting a strong increase of the output signal by feeding it back negatively on the input signal): cathode based feedback using the resistor R_k and anode based feedback using the resistor R_{sa}. The name *feedback* is slightly older, it was used to describe parasitic feedback (interconnection between circuits resulting from bad arrangement of cabling wires, or observed even today when microphones in a sound system are improperly situated). Such parasitic feedback might be called *positive,* since an increase of the output signal supports then a further increase of the input signal, although obviously *the positive feedback results usually in negative consequences of a sudden growth, and the negative feedback is used for positive purposes of*

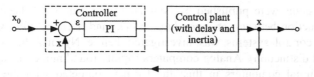

Fig. 8.2 A general scheme of a control system with the use of a *controller* (e.g. PI), controlling a *control plant*, with an indication of a *negative feedback*, the comparison of the *set value x_0* with the *controlled variable x*

stabilization. Harold Black used his negative feedback, among other things, in order to counteract parasitic feedbacks and generally to stabilize the properties of an amplifier.

In the year 1931, the Foxboro corporation introduced an automatic controller called Foxboro Stabilog, a prototype of a *PI-controller (proportional-integral controller)* that eliminated the disadvantages of a proportional controller of the type used by Watt (the disadvantages were known before, and their elimination was a problem addressed through the entire 19th century). If the controller acts only in proportion to the controlled variable, or actually to its deviation from a desired value (this deviation is called technically *control error,* denoted here by ε or $\varepsilon(t)$, and the desired value is called *set value,* denoted here by x_0 or $x_0(t)$), then in a *control system* consistent with the general scheme as in Fig. 8.2 we shall observe a *static error,* a discrepancy consisting in stabilizing between the controlled variable and its set value. In order to eliminate the static error, it is necessary to equip the controller also with an integral action, enabling it also to act proportionally to the control error integral. The introduction of the Foxboro Stabilog controller was a breakthrough event, however due to another reason: it was the first *universal controller,* suitable for the control of various physical quantities (under the assumption that it is possible to convert the result of their measurement to a common input signal for the controller). Therefore, from this event on we can count the history of contemporary automatic control engineering as a separate discipline.

On the other hand, the theory of *PID control (proportional-integral-differential)* was developed by Nicholas Minorsky even earlier, but still for a specific area of applications, steering of ships (Minorsky 1922). After these events, the theory of automatic control as a general discipline, not depending on a specific area of applications, developed fast and was mature until 1940, even if the basic monographs (Smith 1942; Hall 1943; Oldenbourg and Sartorius 1944) were published just after that date.

At the same time, in the year 1931, Vannevar Bush in the Massachusetts Institute of Technology applied the principle of negative feedback to construct an *analog computer,* originally called *differential analyzer,* since it served to provide a fast and relatively accurate solution of a system of differential equations. After these events, the theory of feedback systems developed quickly, stimulated by the results of Blake and Bush; e.g. the fundamental results concerning the stability of feedback systems were published by Nyquist in the year 1932: he introduced a crucial distinction between an *open-loop system* and a *closed-loop system*[1] (the stability of control systems was investigated before Nyquist, but without this fundamental distinction). Analog computers rapidly found numerous applications, preceding digital computers in this; the former competed with the latter until around 1970 (when software for digital computers emerged as a field fully independent from hardware, see next section, while programming of analog computers remained hardware-oriented and much more difficult). Today, solving differential equations on digital computers, even if more exact, is still tedious and time-consuming. Together with the invention of a *microprocessor* (in 1969, see next Chapter), also control systems switched to *digital control,* where the functions of a controller (either PID or its further digital modifications) were executed by a specialized digital computer working in the *real time,* according to the requirements of the dynamics of a control system, not a good will of a computer programmer or in a virtual time determined by the operating system of the computer.

In the year 1948, Wiener popularized a thesis (see Wiener 1948) that the principle of feedback is very general and applicable not only to diverse technical systems, but also biological systems (e.g. the stabilization of temperature in human body is based on a negative feedback) or even social systems (all processes of avalanche-like social changes are based on positive feedback). This thesis was not new, it was anticipated in earlier monographs on automatic control (Smith 1942; Hall 1943; Oldenbourg and Sartorius 1944; Oppelt 1947) which stressed the generality of the principles of feedback; besides, the emergence of automatic control engineering from telecommunications and other technical disciplines was based on the conviction about generality of its methods. But the contribution of Wiener was a broad popularization of this thesis, exceeding technology; he introduced a new name, *cybernetics,* to denote a new *science of control in technology, living organisms and social systems.* In result, some people started to call automatic control engineering *technical cybernetics,* but this is a misuse, since control engineering is older and it is the source discipline for cybernetics. While telecommunications and control engineering contributed to the development of *technical systems theory,* the reflections of Wiener initiated thinking on *general systems theory,* developed since (Bertallanfy 1956) and encompassing social and economic problems, which we shall discuss in further chapters. But again, the technical systems theory was an earlier discipline (see e.g. the title and publication dates of *Bell Systems Journal*) and the source of the general systems theory.

Today, the history of *robotics* is usually presented from the perspective of *artificial intelligence,* stressing the intelligence of technical artefacts. Such an idea is old, we could find its beginnings in ancient times and in historical thinking about computers. However, robotics actually developed as a part of automatic control

---

[1] Quite different than the similar names *open system* and *closed system* used later to describe different concepts in the general systems theory: in the technical meaning introduced originally by Nyquist, a *closed-loop system* is an entire system with negative feedback, an *open-loop system* is the system after disconnecting the feedback.

engineering, because to construct a robot we need to use servomechanisms to control the movements of its arms, manipulators and other moving parts. The very concept of a *robot* originated in literature, from a Czech theatre play *R.U.R., Rossum Universal Robots* written by Karl Čapek (first public premiere in 1921 in Prague). The possibility of robots fascinated also science fiction literature; already in 1950 Isaac Asimov published several volumes under a common title *I, Robot,* containing, among other things, *three laws of robotics,* expressing in general terms the idea that a robot should not harm people with its actions, or lack of action. However, the ideas of constructing a robot were actualized for the first time by the Servomechanism Laboratory of MIT just before 1956, when the first robot construction company, *Unimation* was established by former co-workers of the Laboratory, George Devol and Joseph Engelberger. A year later, the company constructed *Unimate,* a programmable universal machine to operate an assembly line; however, an actual application of *Unimate* on the assembly line of General Motors took place only in 1961. The first robots replaced people in heavy and dangerous work in industry.

On the other hand, the idea of artificial intelligence was indeed stimulated by the beginnings of robotics, but was started with the formation of the laboratory of artificial intelligence at MIT by John McCarthy and Marvin Minsky in the year 1959, two years after the first robot was constructed. After the construction of the first *mobile robot* (called S*hakey,* in 1970), applications of robotics have considerably widened, including areas of high risks for people, such as cosmic exploration (e.g. space probes sent to Mars, Venus, Mercury). The development of *humanoid robots* (under a telling name *Asimo*) was started by the Japanese Honda corporation in the year 1986. The development of robotics is very fast nowadays; however, after the year 2000 it has led to the development of *drone robots* for military purposes, the three laws of robotics by Asimov were disregarded in this case.

Robotics, together with automatic control engineering, provided the prerequisites for a broad *automation* of industrial production. The feasibility of the idea of CAD-CAM (*computer-aided design, computer aided manufacturing*), or a broadly taken automation of both design activities and industrial manufacturing, was demonstrated by the Servomechanisms Laboratory of MIT already in 1959.[2] CAD-CAM is based on a broad use of robots and other automatic control systems in the area of manufacturing and assembly. The aim of automation was always to liberate people from hard and dangerous work, but this noble goal, as it often happens, had also symptomatic, often disadvantageous side effects. An obvious one is the loss of

---

[2] Note a telling difference between the goals of the laboratory of artificial intelligence and the laboratory of servomechanisms at MIT: the former aims at constructing a fully intelligent machine while assuming that applications of such machine will be found in the future, the latter aims at the best utilization of the limited but already available intelligence of machines for contemporary automation of design and manufacturing. Both goals are noble, but both are related to serious dangers, more evident in the latter case, more hidden but perhaps bigger in the former case.

employment by a part of working force and resulting unemployment, which will be discussed in Chap. 14. Generally, we can say that we now know how to build depopulated factories with almost full automation, robotization and computerization of production, but the fundamental problem is what to do with the people employed until now in similar but less automated factories; the growth rate of automation is limited by socio-economic factors, not by technical concerns.

8.2 Analog Computers and Automatic Control

The invention of Vannevar Bush consisted in using the concept of feedback to equip an *operational unit* (later, an *operational amplifier*) with desired dynamic properties. These desired properties might include the properties of an *integrating element* (in which the output signal is an integral of the input signal), or a *proportional element* (in which the output signal is a strict proportional replication of the input signal), or an *inertial element* (in which the output signal is after some time like in a proportional element, but with a specific inertial delay). An operational element was based originally on a servomechanism, an electromechanical system amplifying the energy of movement with an electrical engine and a feedback to control this engine; in their first applications, such operational elements were used to compute trajectories of shells fired by naval guns. Relatively quickly, such servomechanisms were replaced by *electronic operational amplifiers,* direct current vacuum tube amplifiers *with phase inversion* (the output signal should change in the direction opposite to the input signal, which is, by the way, natural for most vacuum tube or also transistor amplifiers). See the illustration of an integrating element, proportional element and inertial element, constructed with the use of such amplifiers, in Fig. 8.3.

A system of appropriately connected operational amplifiers along with relevant feedbacks, see e.g. Fig. 8.4, was used to a fast and relatively accurate simulation of solutions of differential equations, hence the name *differential analyzer*, substituted later by *analog computer*. However, it is a fact that in their practical applications, analog computers preceded digital computers by over a decade, and there were many such applications, to start with solving differential equations describing the dynamics of a flight of an artillery shell in order to aim naval guns, or other differential equations describing the dynamics of a heavy object (say, transported by a crane), when designing an appropriate servomechanism. In fact, analog computers served for the analysis of dynamics of complex systems, including the dynamics of control systems, hence the concept of *systems dynamics*, which is ascribed by sociology to Forrester (1961), is actually 30 years older and was originated by Vannevar Bush.

With all their advantages, analog computers had an essential disadvantage that decided about their failure in the competition with digital computers: *they were programmed at the hardware level.* The programming process was based on block diagrams such as presented in Fig. 8.4, which however had to be translated into

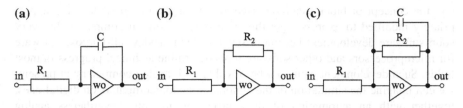

Fig. 8.3 Schemas of an integrating element (**a**), proportional element (**b**) and inertial element (**c**), simulated with the use of operational amplifiers with phase inversion (OA)

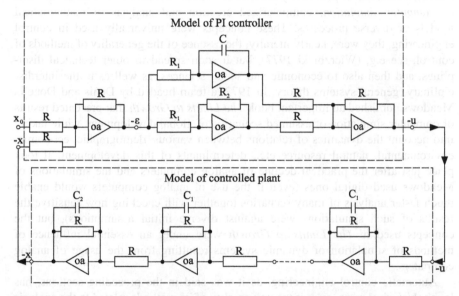

Fig. 8.4 An example of a block diagram of a system of operational amplifiers (simulating the dynamics of the control system from Fig. 8.2, with the control plant with an even number of operational amplifiers connected in a series and with a controller comprising an odd number of operational amplifiers connected in a series)

physical connections between subsequent operational amplifiers, executed with the help of special connection boards. For this purpose, interchangeable connection boards were used, but it did not change the fact that analog computer programming was strictly related to hardware.

On the other hand, higher level programming languages have been used in digital computers since very early, from the year 1952, when Grace Hopper developed the first compilation code (*compiler*, see the next chapter) and published the first textbook of computer programming. These high level programming languages enabled a gradual but spontaneous separation of programming (software) from hardware. Nevertheless, it was significant that the fundamental concepts of digital computer programming imitated originally analog computer programming, e.g. the concept of block diagrams used earlier in analog computers and appropriately modified to express algorithmic schemes in digital computers, but very soon software development became a separate field. Today, all-purpose software for microprocessors and other software tools determine technical progress in most fields. Such development could not be possible with analog computers; at present, we can imagine the simulation of the actions of analog computers on digital ones, together with an automation of their programming, but nevertheless analog computers already went down in history.

However, analog computers had a great conceptual importance. They originated the concepts of *process dynamics* and *systems dynamics,* as well as methods of *computer simulation,* computerized creation and analysis of mathematical models of diverse processes. These concepts were universally used in control engineering, they were, nearly literally, the essence of the generality of methods of control, see e.g. (Wierzbicki 1977), but it soon spread to other technical disciplines, and then also to economic and social sciences, as well as to the interdisciplinary, general systems theory. In 1972, a team headed by Denis and Donella Meadows published an important book *The Limits to Growth* that presented results of computer simulation of assumed scenarios of global development, while taking into account the dynamics of relations between various, demographic, economic, environmental, natural resource etc., determinants of this development. It took place just after the practical decline of analog computers and the simulations of Meadows used digital ones (even if the use of analog computers would enable much faster analysis of many scenarios together with checking how sensitive the results of such simulations were against diverse initial assumptions), but the concepts used in *The Limits to Growth* were under an essential influence of methods of simulation of dynamic systems resulting from the usage of analog computers.

Automatic control engineering, as it was said in the preceding sections, has longer historical roots, even if the universality of its methods related to the analysis of the dynamics of control of arbitrary physical quantities and arbitrary control plants started to be appreciated around 1930, concurrently and in a close relation to the construction of analog computers. In fact, the theoretical foundations of automatic control are much older, we can count the development of mathematical control theory from Airy (1840), a British astronomer who constructed a control system to automatically track the movements of stars on the sky and observed that, depending on the parameters he selected, the system might generate wild oscillations; hence, he tried to analyse this phenomenon mathematically. But Airy considered this to be a specific problem, did not suppose that this phenomenon of instability could arise in any system with feedback (since the concept of feedback, although practically used by Airy, was not yet defined as a general concept). Later works on the stability of systems described by linear differential equations included the studies of Maxwell (1868), Routh (1877), Vishnegradsky (1877), Hurwitz (1885), the works of Heaviside (1892) introducing operator calculus, and of Carson (1926) using this calculus, together with its relations to Fourier and Laplace transforms, in order to describe linear systems dynamics. However, the introduction of the concept of feedback systems resulted in the ground-breaking work of Nyquist (1932) describing the conditions of stability of a *closed-loop system* (with a functioning feedback loop) on the basis of the dynamic characteristics of the *open-loop system* (that is, the system after breaking the feedback loop) contained in the former one; these conditions will be described in one of the further sections of this chapter, where the importance of the concept of feedback is discussed. For a long time, this theory concentrated on the problem of stability, specifically of linear systems, even if stability of nonlinear systems was addressed relatively early, already in 1893, by Alexander Liapunov (and raised again by the automatic control theory only in the middle of the 20th century, after a renewed publication of Liapunov studies). I would like to stress the contributions of Polish telecommunications and electronics to the theory of stability of nonlinear systems, and especially the works of Groszkowski, e.g. (1947), and Kudrewicz (since 1964, see 1996).

The automatic control theory explored the most important problems related to the dynamics, optimization and intelligence of systems. The analysis of dynamics of nonlinear systems led to the emergence, as it will be described in further chapters, of the concept of *deterministic chaos*, perceived by automatic control specialists rather as a way of obtaining desired but unexpected properties of a dynamic system through the use of its strange behaviour resulting from the combination of nonlinearity with feedback, see e.g., (Wierzbicki 1963); (Lucertini et al. 2004). An example of such strange behaviour was also observed in the analysis of the dynamics of a *digital servomechanism,* that is, an engine moving a heavy object to a desired position, with a control of this engine executed by a microprocessor with a specified time cadence (fixing the value of measured signals during a time span needed for signal processing) and with a digital measurement that does not distinguish changes smaller than a given quantum of measuring. This example will be discussed in detail in Chap. 10, together with an analysis of the history and concepts of deterministic chaos.

The automatic control theory dealt with many other problems, starting with the optimization of control. Bellman (1957) introduced a fundamental concept of *dynamic programming,* remaining current also today, which, beside a complex mathematical theory, represented by the so-called Hamilton-Jacobi-Bellman equation, includes a fundamental rule of dealing with optimal planning or programming of the course of complex dynamic processes: *in order to have a good plan, one has to start with the end time of the process and its desired final state, then to step back stage-wise until the current time.*[3] Pontryagin et al. together with his co-workers (1962) formulated the so-called *maximum principle,* a fundamental condition of optimality of control; to prove this condition, it was necessary to enhance the classical calculus of variations by introducing a *needle-like variation.* These were mathematical studies, but the motivation for them were the problems

---

[3] The principle of dynamic programming was rediscovered many times, under diverse names; e.g. a name popular today is *dynamic back-casting.*

of automatic control, e.g. the research of the Pontryagins group was motivated by the collaboration with the control engineer Alexander Feldbaum, see e.g. (Feldbaum 1965), who beside the problem of *time-optimal control* (how to achieve a desired final state of a dynamic system in the minimal time possible) and a related problem of the principle of discontinuity of control (later called *bang-bang control*) formulated also many other ideas, related e.g. to an optimization of a control process coupled with learning about the model of controlled plant. Such combined optimization and learning was called *dual control* by Feldbaum and he observed that in order to learn we must partly give up the optimality or efficiency of control (we must learn by our errors, in an additional feedback loop).

Further studies of optimal control included also *stochastic models* of controlled processes, already used, since the works of Wiener (1949), for communication and control, but developed in combination with the problem of optimal control by Kalman (1960) who proposed a method of determining an *optimal filter,* called also *Kalman filter.* These results were used, among others, in the design of the control systems for the first landing on the Moon, see e.g. Åström (1970). In all these studies we observe a significant synergy of technology, in particular automatic control, with mathematics: technology provides examples of practical problems resulting from its inventions, mathematics finds a generalized solution, which is in turn naturally used by technology. It is not, however, a one-sided relation, it is much rather a positive feedback loop that results both in new technological inventions and in new mathematical results.

Optimal control does not exhaust diverse contemporary trends in automatic control, adaptation, learning, intelligent controllers etc.; for example, automatic control contributed also to the origins of *hierarchical systems theory,* see e.g. (Findeisen et al. 1980), which is discussed in more detail in one of further chapters.

8.3   Robotics and Automation

While in the case of automatic control and analog computers the most important thing is their conceptual impact, in the case of robotics and automation their socio-economic contribution is equally important (even if robotics is often presented today from the perspective of artificial intelligence, but the conceptual impact of the latter field will be discussed in next chapter).

As already mentioned, first robots created in 1957–1961 were designed for the purposes of automation of industrial production, with a parallel development of the concept of CAD-CAM (computer assisted design and manufacturing). From that time the development of robotic technology was immense, and it is difficult to describe it shortly. Let's note then only several examples: mobile robots able to evade obstacles on their way; diverse solutions of robotic arms, inclusive of micro-manipulators, highly sensitive robot "fingers"; humanoid robots, together with a mimics of a comics-like face of the robot; flying robots (e.g. unmanned helicopters); medical robots (in diagnostics and surgery); and many others. On the other hand, applications of robots developed more slowly and often in other directions than originally forecasted.

Let us start with automation and robotization of industrial production. Currently, the corresponding technology is so far advanced that not technical issues but socio-economic concerns decide of the degree of automation and robotization of production, often differently in the case of different production domains, by the way. We should remember that at the top of industrial civilization in the first half of the 20th century such ideas as *fordism* and *technocracy* developed, propagated not by technicians, but by managers using technology for a more efficient organization of production and more strict supervision of workers employed at a conveyor belt.[4] This resulted in a great social resistance, coming from labour unions and humanist critics of industrial civilization. Among the latter, the greatest influence was exerted by Marcuse (1964), whose book *One-dimensional Man* became the basis of ideology of student revolutions in 1967–1968. However, if we read Marcuse attentively, we see that he strongly censures rather the socio-economic system of advanced industrial civilization, not technology in the meaning discussed in Chap. 3, even if he calls this system (imprecisely) *technology*. Moreover, Marcuse searches also for a hope for a change and sees such hope (significantly!) in the automation and robotization of production which would make the labour conditions more tolerable for workers.

Technicians were always motivated by precisely this goal, quite independently from Marcuse (to whom they were mistrustful because of his imprecise censures). Today, the automation and robotization of production have enabled the achievement of this goal, and as the labour conditions might be much better than 50 years before, the demand for equal rights for women might be fully realized (and if it is not, it results from socio-cultural reasons). But the social resistance against full automation increases, even if it is not fully consistent. The reason for this resistance is a justified fear that a speedy introduction of full automation will result in a greater unemployment, or a shift of employment towards more impermanent work in services. The reason of the inconsistency of this resistance is that a contemporary consumer prefers to buy products (say, automobiles) manufactured in highly automated factories, because such products are usually better (say, more durable). Thus, the full automation and robotization of manufacturing are unavoidable, they are a sign of informational revolution. But the related socio-economic process is long-lasting and diversified, and in some domains its pace is faster, while in other ones, slower. Full implementation of the so-called *third industrial revolution* (the construction of unmanned factories) will first of all require a solution for difficult social problems (e.g. how to provide steady work for average people, see Chap. 14) and will last at least 100 years on the global scale.

Applications of robots, however, are nowadays much wider than the automation and robotization of industrial manufacturing. An important domain of such

---

[4] The adjective *technocratic*, used very often as a negative epithet by humanist critics of technology, is thus used incorrectly, a-historically, since technocrats were great managers or *technology brokers* in the first half of the 20th century, not engineers.

applications turned out to be *exploration in conditions dangerous for people,* such as cosmic exploration, on planets, asteroids, etc., or exploration of volcano craters on Earth. Such applications are also an excellent testing range for diverse features of robots. E.g., in the year 1993, the eight-legged robot Dante, developed in the laboratory of robotics at the Carnegie Mellon University in Pittsburgh, descended deep into the crater of the Mount Erebus in Antarctica, but fell off and melted in lava. In 1994, its improved version, Dante II, descended into the crater of Mount Spurr in Alaska and succeeded in the mission of data gathering.[5] In relation with cosmic application of robots, a new concept was introduced: a *rover* which denotes a small exploratory robot. A debate about the old theme *small is beautiful* developed: the issue is whether it is more efficient to send a bigger and expensive mobile robot, or a swarm of smaller, more brisk *rovers* behaving like artificial ants?

We are yet far away from a full robotization of medical care (perhaps fortunately, the idea of such robotization includes many dangers), but diverse electronic appliances with robot-like functions are broadly used, both in surgery (starting with laser scalpel) and to even broader degree in medical diagnostics: advanced diagnostic systems have nowadays diverse features of artificial intelligence, such as support of diagnostic decisions, recognition and affirmation of patients identity, etc.

To some extent surprising, even if predicted already by Asimov, is the use of robots as people companions. This idea was implemented for the first time in Japan, where it was difficult to keep even dogs or cats in the great density of population in big agglomerations, so there was a demand for robots that could take over their functions. The construction and distribution of this type of robots, called AIBO, was initiated by Sony corporation in 1999; since that time many new versions or generations of such robots have been developed. Japanese are also convinced that larger humanoid robots (with comics-like faces, a result Japanese fascination with comics) will take over the functions of old people care in the future.

Bearing in mind all these human-friendly application of robots we must stress that also a dangerous trend developed in robotics, aimed at using robots as fighting machines, mostly as a result of using technology for profit by entrepreneurs or *technology brokers* without ethical restriction. Two human features were exploited by such entrepreneurs. The first one is human demand for entertainment: since we cannot have gladiators, we could use robots as actors of fight and contest. This idea was put into practice by the entrepreneur Marc Thorpe, who organized *Robot Wars* in the Fort Mason Center, San Francisco. I am afraid that this event will go down infamously in the history of robotics that started with the laudable *three laws of robotics* of Isaac Asimov and ends up building fighting robots that do not obey

---

[5] How does it look like in comparison with the opinion of Karl Popper, who in (otherwise excellent) paper *Three Views Concerning Human Knowledge* (1956) maintained that technology does not use critical tests and does not abandon falsified solutions? However, Popper was a physicist, not a technician, did not know what technology is actually doing and believed that technology is a simple application of the results of hard science.

such laws. The foregoing is aggravated by the second feature: military applications of any new technology. Since we use autopilots, aim assisting devices and homing missiles on combat planes or helicopters, we surely also could construct unmanned fighting vehicles, or military robots in general? This question indicates a fundamental ethical dilemma: one issue is when the decision about killing people belongs to other people, quite another issue is automation of such decision by leaving it to a robot. This general ethical question motivates also my reservations towards an unlimited use of artificial intelligence, discussed in more detail in the next chapter.

8.4 Conceptual Importance of Feedback

The concept of feedback resulted in perhaps the greatest conceptual breakthrough brought by contemporary technology, even if not yet fully understood by other cultural spheres. As already mentioned, this concept was introduced (even if patented, only reinvented) by Harold Black in the year 1928 (Black 1934), and was soon developed further in diverse works of telecommunication, together with its theory (Nyquist 1932), originally for the purposes of stabilization of the properties of telecommunication amplifiers, but soon extended to automatic control engineering, analog computers, etc. Actually, *the phenomenon of feedback is much older and is a foundation of the industrial civilization,* since James Watt in the years 1769–1783 did not invent steam engine, but only used negative feedback to stabilize already existing but unstable steam engines which were prone to explosion, making their broad social use possible. The traces of feedback principle in use can be found in ancient times; as it often happens in technology, the phenomenon of feedback was employed in diverse technical branches long before naming the concept and developing its theory.

This is, by the way, an excellent example of the general principle that technology is not a simple application of scientific theories, because many specialists after Watt and before Black, starting with Airy (1840), actually used feedback, they were "speaking prose without knowing it", without the concept and the theory of feedback. Even Black introduced the concept of feedback before a more detailed theory of this phenomenon was developed, and Watt (or engineers actually using similar concepts even before Watt, see Bennet 1979) wanted only to solve a problem, for Watt, how to stabilize rotational speed of steam engine, and did not use any theoretical foundation. On the other hand, automatic control engineering and mathematical control theory are an excellent example of a positive feedback and synergy between engineering and hard science. Mathematicians derived inspiration from technical problems, such as in the case of Lew Pontryagin who found that for a correct mathematical representation of the problems of optimal control one has to revise classical assumptions of the calculus of variations. Technicians used such mathematical solutions, but they took up also new problems and made inventions that in turn stimulated mathematicians.

A great contribution of Nyquist (1932) was the development of the theory of feedback systems, with the distinction of a *closed-loop system* (with feedback) and an *open-loop system* (after disabling feedback). This is a fundamental conceptual contribution: *in order to understand well the behaviour of a system with feedback, we have first to understand the behaviour of the system without feedback and to realize that the feedback can be broken in various points of the loop.* It is also of a fundamental importance for understanding of the ostensible character of the paradox of *vicious circle,* which was discussed in Chap. 6.

The feedback relation must be understood in its dynamic aspects, as a causal-reflexive dependence of two time streams of causes and effects.[6] The stream of effects must be, even if slightly, delayed in respect to the stream of causes, otherwise we would encounter a real paradox. The feedback relation has essentially two types. A *positive feedback* relation concerns a situation when the time stream of effects supports in feedback the time stream of causes; this results in a fast, avalanche-like development. If the development is stopped by some nonlinear effects, e.g., nonlinear saturation, then positive feedback results in a self-support for one of (at least two) possible equilibrium states. The term *positive feedback* is technical and might be misleading, since the effects of positive feedback can sometimes be *negative.* Nevertheless, the positive feedback relation is observed in many social phenomena, such as the relation between hard science and technology mentioned earlier.

A *negative feedback* relation concerns a situation where the time stream of effects counteracts in feedback the time stream of causes, which usually results in a *positive* phenomenon of stabilisation. Today, such negative feedback is commonly used in technology, starting with robotics. However, there is one essential exception: we also widely use positive feedback, in computer memory. Each computer contains millions or billions of bi-stable memory elements of diverse character (originally, bi-stable transistor micro-switches called *flip-flops,* but memory technology is also quickly developing, see next chapter), which are all based on a positive feedback or self-support.

Stabilization of human body temperature is based on a biological negative feedback, similarly to the stabilization of the movements of a robot. If such a negative feedback is too strong, it might lead to self-induced oscillations (which was noted for the first time by Airy, in a technical system used in astronomy, but it actually happens to all older people with trembling hands). In more complex, nonlinear cases, such oscillations can also lead to generation of chaotic oscillations, or even to new forms of movements or order emerging out of chaos (see next chapters).

Although it was telecommunications and control engineering that introduced and developed the concept of feedback, and *robotics cannot function without feedback,* there are also quite different interpretations of the history of this concept.

---

[6] The same dependence was later called *reflexivity relation* by Soros (2006) who did not realize that this is much older, well practically and theoretically studied *feedback relation.*

Wiener (1948) popularized the analysis of the concept of feedback in living organisms and in society, calling such analysis *cybernetics*. Jay Forrester (Forrester 1961) appropriated, from control engineering and analog computers, the idea of block diagrams of dynamic systems with feedback and applied this idea to socio-economic problems, calling it initially *industrial dynamics,* and later *systems dynamics*, although the concept of systems dynamics originated from the modelling of dynamic systems on analog computers (invented in 1931 by Vannevar Bush). This example shows us a definite tendency, indicating how large is the split between epistemai of different cultural spheres nowadays, of social and management science to appropriate systemic concepts which were developed earlier by technology or strict sciences. Many of the so-called *soft systems thinkers* (Jackson 2000; Midgley 2003) *maintain today that it was Wiener who invented feedback and Forrester who invented systems dynamics.*

Coming back to the distinction of *open-loop* and *closed-loop feedback system* we shall see how the so-called *Nyquist criterion* (1932) for generation of oscillations by a system with negative feedback (formulated before Wiener and much before Forrester) can be used to analyse various cycles (economic crises, civilization turns, etc.) in economic and social systems. The open-loop system in Fig. 8.5 has a *spectral transfer function*, the ratio of Fourier transforms of the output and input signals, expressed by $G_0(j\omega) = k\ e^{-j\omega\Delta}/j\omega$, where j is the imaginary unit, ω is the angular frequency, $\omega = 2\pi f$ where f stands for the frequency of oscillations, k is the resultant amplification coefficient, $e^{-j\omega\Delta}$ is the transfer function of the *delay element* with *time delay Δ*, while $1/j\omega$ stands for the transfer function of an *integral element* that accumulates the former results.

Let us start with a classical analysis of such a system. A typical problem in designing a feedback system is the choice of its parameters, in this case, the choice of the amplification coefficient k in the open-loop, given the delay time Δ, in such a way as to obtain the properties of the closed-loop system corresponding to its applications. For example, if the system is a servomechanism, then we must choose the parameters of the open-loop in such a way that the *step response of the closed-loop system* (the response of the system to a sudden change of the set value x_0) is stabilizing rather fast, without too much oscillations. We must remember that feedback is a casual-reflexive relation of two time-streams of causes and effects, theoretically infinite in time; it is an *infinite recourse or regress,*[7] in philosophy called a *hydra;* but the design problem consists precisely in taming that hydra by obtaining a fast stabilization of the output variable in the closed-loop system. Figure 8.6 illustrates a typical[8] step response (after a sudden change of the set

---

[7] *Recourse* is repeating an action over time, *regress,* repeating an action or reasoning backward in time or in depth of reasoning; in the automatic control theory (particularly in optimal control, where the principle of dynamic programming requires a combination of backward and forward direction of time) such concepts must be combined, both directions of time-flow must be considered.

[8] Typical, but only for a system with inertial delays; for a system with pure time delay, as in Fig. 8.5, the start of the step response would be additionally delayed by the time Δ.

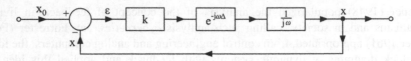

Fig. 8.5 A closed-loop feedback system with an integral element and time delay element

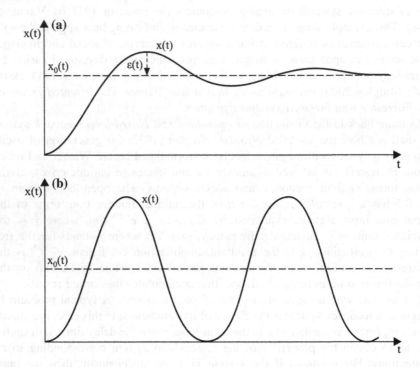

Fig. 8.6 A typical step response of a closed-loop feedback system **a** for a stable system, **b** at the stability boundary

value $x_0(t)$ at the time $t = 0$) of the closed-loop system in two variants: (a) after a correct choice of the amplification coefficient k; (b) with an excessive value of the amplification coefficient, resulting in oscillations. The correct choice is achieved usually after a repeated computer simulation of models of the system (in a kind of virtual laboratory) and observation of diverse resulting step responses; however, we can rely also on a theory that helps in such choice or makes it possible to compute some properties of the closed-loop system analytically.

Such theory and the possibility of its application is illustrated by the following example. Since Nyquist (1932) we know the formula for the *spectral transfer function of the closed-loop system*:

$$G_Z(j\omega) = G_0(j\omega)/(1 + G_0(j\omega)) \qquad (8.1)$$

Suppose that the feedback system in Fig. 8.5 represents not a technical control system, but approximates dynamic properties of a social system,[9] e.g. represents economic phenomena. In such a system we might observe an accumulation of income expressed by the integral element and a delay resulting e.g. from the delay of investments; moreover, the system might include a feedback (e.g. we make investments in order to counteract a decrease in income).

If such a system generates oscillations, shows cyclical changes, then what will be the period T or the frequency f = 1/T of such oscillations?

The Nyquist criterion of stability says that for linear systems, a closed-loop system is stable if the open-loop system is stable and the denominator of the transfer function (5.1) of the closed-loop is non-zero (more precisely, the graph of the changes of $G_0(j\omega)$ with the change of angular frequency ω bypasses the point -1 in an appropriate way). The closed loop system is on the boundary of stability and generates oscillations if the denominator of its transfer function is zero, $G_0(j\omega) = -1$.

Thus, according to Nyquist criterion, the angular frequency $\omega = 2\pi/T$ of such oscillations must be such that $|G_0(j\omega)| = 1$ and Arg $G_0(j\omega) = -\pi$ (in angular measure of radians, which corresponds to $-180°$ in classical measure). Using the complex number calculus, we conclude that Arg $G_0(j\omega) = -\omega\Delta - \pi/2$ and $|G_0(j\omega)| = k/\omega$; the requirement Arg $G_0(j\omega) = -\pi$ results in $-2\pi\Delta/T - \pi/2 = -\pi$, hence $T = 4\Delta$, and the requirement $|G_0(j\omega)| = 1$ results in $k = \pi/2\Delta$. The conclusion is: *the period of a cycle is equal to four delay times,* if the amplification is sufficiently great (in the above economic example, if the reactions to the changes are sufficiently fast and deep).

This is rather a general conclusion. It can be used to analyze diverse economic cycles if we identify the appropriate delay times; let us consider a socio-economic cycle of supply of graduates of a given specialty of university studies in relation to their shortage or excess on the labour market. If the delay time (the time of study with possible extensions) is circa $\Delta = 5$ years, then $T = 20$ years. In this case we do not even need to use the Nyquist criterion and assume the linearity of the system; we can obtain the same result by direct analysis. If in a given moment of time a shortage of specialists, say, in management, appears on the labour market and the graduates of secondary schools learn about that, then they start to go in larger cohorts to the management studies (without assuming linear proportionality, just following a mass trend limited only by the number of young people of

---

[9] Naturally only approximately and in a simplified model; I agree with Soros (2006) that social phenomena are much more indeterminate than physical or technical phenomena. However, I disagree with his opinion that there are no regularities in social phenomena, because he introduced the concept of *reflexivity relation without noting that it is actually feedback relation* for which there exists a thorough, well developed theory. Regularities in social relations exist and can be described with approximate models, see e.g. a model of cyclic changes of supply of specialists in a given discipline, described below.

according interests). However, the results of such change will appear on the labour market for the first time after 5 years; during this time the shortage of specialists will be increasing, hence it will take the next (second) 5 years of increased supply for the supply and demand to equilibrate. Therefore, after 10 years (half of the cycle) an oversupply of specialists appears on the labour market, but at the university we still have 5 yearly cohorts of large numbers of students. Therefore, the oversupply of specialists will grow for the next (third) 5 years, even if the graduates of secondary schools learn immediately about this fact and start to go for management studies in smaller cohorts (if they do not learn about it with some delay, then it is equivalent to some increase of the delay time; we have anticipated this by assuming $\Delta = 5$ years, even if university studies are usually 4 years long). After these (third) 5 years we have a substantial oversupply on the labour market, hence it will need next (fourth) 5 years of diminishing supply until the supply becomes smaller than the demand. Therefore once again: *the period of a cycle is equal to four delay times.*[10]

Nevertheless, the above example illustrates the thesis of Harry Nyquist that the delays or inertias in a feedback system shift the reciprocal relation of cause and effect: if the shift amounts to $180°$, hence half a cycle, then the system generates cycles or cyclic development (because cycles can overlap with other long term trends).

In the case of civilization development and cycles, if the civilization delays amount to 30–40 years (as those identified in recent history of information technology), the expected time of a civilization cycle amounts to 120–160 years. We see that the theory of feedback systems allows for various conclusions concerning socio-economic systems. Clearly, such conclusions are only approximate and theoretical, based on very simplified models of socio-economic reality, but nevertheless they provide useful information.

It should be stressed once again that *the concept of feedback has essentially a dynamic character, it is a circular, dynamic interaction of two time streams of causes and effects.* A cause must obviously precede its effect in time, hence effects react in a feedback to causes with a smaller or larger delay. If understood differently, as a single static relation of a cause and effect, feedback would be clearly a paradox, a vicious circle (a contradiction in the case of negative feedback, a self-supporting argument in the case of positive feedback). However, dynamic feedback systems are not only possible, but also work and support us in contemporary

---

[10] The description of the behaviour of young secondary school graduates in this example might appear too simplified, mechanistic. Indeed, a young graduate is sufficiently educated to use the description of this example in order to make a more informed decision. In order to do this, however, (s)he would have to (a) know and understand this description; (b) obtain additional information about the duration of the recent boom (or recession) on the market of specialists of a given specialty, such as management, hence when a next change of demand will occur. On the other hand, young graduates would most probably not have the information necessary to take such more informed decisions, hence the cyclic character of the supply of specialists on labour markets might be a systemic, unavoidable feature resulting from the delay in education; see Chap. 14 for further discussion.

technical civilization. Moreover, *they function on the basis of an infinite regress or recourse* (in the case of digital systems; in analog systems, the recourse is not only an infinite sequence but has a continuous temporal character). In Fig. 8.6, if there persists a nonzero control error ε*(t)*, the controller takes actions to bring the error to zero in next moments and continues such actions in an infinite recourse. Such actions express the phenomenon of *self-correction* or *self-reference; which clearly shows that the concept of feedback is a break-through.* Philosophical, epistemological consequences of this fact we discussed already in Chap. 6.

It should be also stressed once again that a lack of education in automatic control or robotics results in lack of understanding between representatives of social sciences and humanities as regards the concept of feedback. An example is Soros (2006) who, following Karl Popper, correctly stresses the fallibility of all social theories and the belief in an open society, but he also uses an argument about the universality of the *reflexivity relation,* not noting, as we commented above, that he actually speaks about *feedback relation,* a phenomenon analyzed much earlier and deeper in technology. Indeed, feedback relation can be a source of instability or cyclic or wave-like behaviour, but it does not always result in indeterminate phenomena, as suggested by Soros. The description of the boom-recession cycles in the book by Soros actually illustrates the principle known from Nyquist (1932) that a phase shift of about 180° in a system with feedback can result in instability.

References

Airy, G.B.: On the regulator of the clock-work for effecting uniform movement of equatorials. Mem. R. Astron. Soc. **11**, 249–267 (1840)

Åström, K.J.: Introduction to Stochastic Control Theory. Academic Press, New York (1970)

Bellman, R.: Dynamic Programming. Princeton University Press, New Jersey (1957)

Bennett, S.: A History of Control Engineering 1800-1930. IEE, Peter Peregrinus (1979)

Bertallanfy, L.: General Systems Theory. Gen. Syst. **1**, 1–10 (1956)

Black, H.S.: Stabilized feedback amplifiers. Bell Syst. Tech. J. Electr. Eng. **53**, 1311–1312 (1934)

Carson, J.R.: Electric Circuit Theory and the Operational Calculus. McGraw-Hill, New York (1926)

Feldbaum, A.A.: Osnovy Teorii Optimalnych Avtomaticzeskich System. Nauka, Moscow (1965)

Findeisen, W., Bailey, F.N., Brdyś, M., Malinowski, K., Tatjewski, P., Woźniak, A.: Control and Coordination in Hierarchical Systems. Wiley, Chichester (1980)

Forrester, J.W.: Industrial Dynamics. MIT Press, Cambridge (1961)

Groszkowski, J.: Generacja i stabilizacja częstotliwości (Generation and Stabilization of Frequency). WNT Warsaw (1947)

Hall, A.C.: The Analysis and Synthesis of Linear Servomechanisms. MIT Press, Cambridge, Mass (1943)

Házen, H.L.: Theory of Servo-mechanisms. Journal of Franklin Institute (1934)

Heaviside, O.: Electrical Papers. Macmillan, London (1892)

Hurwitz, A.: Über die Bedingungen, unter welchen eine Gleichung nur Würzeln mit negativen realen Teilen besitzt. Mathematische Annalen **1885**, 273–284 (1885)

Jackson, M.C.: Systems Approaches to Management. Kluwer Academic Plenum Publishers, New York (2000)

Kalman, R.E.: A new approach to linear filtering and prediction problems. ASME Trans. Part D (J. Basic Eng.) **82**, 35–45 (1960)

Kudrewicz, J.: Nieliniowe Obwody Elektryczne: Teoria I Symulacja Komputerowa (Nonlinear Electrical Systems: Theory and Computer Simulation). WNT, Warsaw (1996)

Lucertini, M., Gasca, A.M., Nicolo, F.: Technological Concepts and Mathematical Models in the Evolution of Modern Engineering Systems. Birkhauser, Basel (2004)

Marcuse, H.: One-Dimensional Man. Beacon Press, Boston (1964)

Maxwell, J.C.: On Governors. Proc. Roy. Soc. **16**(1868), 270–283 (1868)

Midgley, G.: Systems Thinking. Sage Publications, London (2003)

Minorsky, N.: Directional stability and automatically steered bodies. J. Am. Soc. Naval Eng. **34**, 280–286 (1922)

Nyquist, H.: Regeneration theory. Bell Syst. Tech. J. **11**, 126–147 (1932)

Oldenbourg, R.C., Sartorius, H.: Dynamik Selbstätiger Regelungen. München Verlag, München (1944)

Oppelt, W.: Grundgesetze der Regelung. Wolfenbüttel Verlag, Hannover (1947)

Pontryagin, L.S., Boltyansky, V.G., Gamkrelidze, R.V., Mishchenko, E.F.: Matematiczeskaja Teoria Optimalnych Processov. Nauka, Moskwa (1961). English translation The Mathematical Theory of Optimal Processes, Wiley, New York (1962)

Popper, K.R.: Three views concerning human knowledge. In: Lewis, W.H.D. (ed.) Contemporary British Philosophy, Third Series, Allen and Unwin, London (1956)

Routh, E.J.: A Treatise on the Stability of a Given State of Motion. Macmillan, London (1877)

Smith, E.S.: Automatic Control Engineering. McGraw-Hill, New York (1942)

Soros, G.: The Age of Fallibility: Consequences of the War on Terror. Public Affairs, New York. Polish translation (2006) Nowy, okropny świat: era omylności, Świat Książki, Warsaw (2006)

Vishnegradsky, I.I.: On Controllers of Direct Action. Izviestia SPB Tekhnologiczeskogo Instituta (1877)

Wiener, N.: Cybernetics or Control and Communication in the Animal and the Machine. MIT Press, Cambridge, Mass (1948)

Wiener, N.: Extrapolation Interpolation and Smoothing of Stationary Time Series with Engineering Applications. Wiley, New York (1949)

Wierzbicki, A.P.: Zagadnienia dynamiki regulatorów krokowych (Problems of dynamics of step-wise controllers). Pomiary, Automatyka, Kontrola **9**(1), 12–26 (1963)

Wierzbicki, A.P.: Modele i wrażliwość układów sterowania. WNT, Warsaw. English edition (1982) Models and Sensitivity of Control Systems, Elsevier-WNT (1977)

Chapter 9
Digital Computers, Transistors and Integrated Circuits

9.1 Introduction, Fundamental Events

Abacus, a mechanical appliance (e.g. beads on a rod or a wire) making counting easy, was known already in Babylonia around 3000 years ago, rediscovered or improved in China, and later in Europe in 12–13th century.

In the years 1614–1617, John Napier invented *logarithms* and constructed an appliance based on this invention, a prototype of *logarithmic sliding rule* that until 1970 was a basic computational support of an engineer. In the years 1624–1625, Johannes Kepler published first complete logarithmic tables.

In 1642, Blaise Pascal proposed a prototype of an *arithmometer,* a computing machine supporting addition. The initial idea was far from perfection. In the years 1672–1710, the idea was improved by Gottfried Wilhelm Leibniz, who used complex mechanical gears to construct an arithmometer that performed both addition and multiplication. Ideas of Leibnitz were not technically perfect and were for some time forgotten, and only in 1820 Charles Xavier de Colmar from Alsace rediscovered, by improving the ideas of Leibnitz, an arithmometer that was produced serially and commercially distributed. If we use this example to determine the civilization delay "from idea to industry", then this delay amounts to ca. 180 years.

The results of de Colmar reminded the world of the problem of how to construct a computing machine. As a result of market competition, the arithmometer was further improved by subsequent inventions (e.g. a gear with changeable number of cogs, such as used today in automotive gearboxes, invented around 1875). In the years 1822–1834, Benjamin Hershel Babbage developed several versions of a more universal computing machine (a "differential engine" or "analytic engine"). British sources tend to attribute the concept of a computer to Babbage, although equally well we could start from Napier, Pascal, Leibnitz, de Colmar or even from ancient Babylonians or Chinese.

More essential for the construction of digital computers were the studies of George Boole who in 1847 published *The Mathematical Analysis of Logic,* in

© Springer International Publishing Switzerland 2015
A.P. Wierzbicki, *Techneₙ: Elements of Recent History of Information Technologies with Epistemological Conclusions*, Intelligent Systems Reference Library 71, DOI 10.1007/978-3-319-09033-7_9

which he formulated mathematical principles of *binary logic*. This logic has logical values one (true) and zero (untrue) and is universally used nowadays in construction of digital computers. But for such application *a conceptual breakthrough was necessary: an interpretation of logical values as the state of closed (conducting a current) and open (non-conducting) electrical circuit*. Formally, such interpretation was provided for the first time by Claude Shannon in his doctoral dissertation (1938, over 90 years after Boole), but actually it was used intuitively much earlier by engineers, e.g. to control elevators, or in automation of telecommunication switchboards, and since 1936 in the construction of first prototypes of digital computers.

A very important invention for the development of digital computers was the construction of a *typewriter*, developed (also not for the first time) by Christopher Sholes, Samuel Soule and Carlos Glidden in the year 1847. The *QWERTY* keyboard used by them resulted from the need of such placement of key letters that hitting neighbour keys subsequently was improbable, which was aimed to avoid collisions of mechanical typing levers. Even if this reason has lost technical sense long ago (and resulted in a keyboard structure that is far from ergonomic efficiency), we still use this keyboard practically in all computers as a basic *user interface* (an input appliance connecting the user with the computer). The reason is simple: generations have learned fast writing using this keyboard and a change of this habit would require much effort, so consumers would not buy computers with a different keyboard. This is a basic example of the phenomenon of *path dependence* (Arthur 1994) in economics that contradicts the classical assertions that free market results in an optimal selection of the standards of production.[1] Typewriters had obviously a broader impact, e.g. in telecommunications, as prototypes of telexes and telefacsimiles, on the social life in general, since over 100 years, generations of writers worked using typewriters, and later also on the construction of computers, their keyboards, computer printers.

The invention of *electric punched card tabulator* patented by Herman Hollerith in 1889 was also important for future development of digital computers. It was almost immediately (1890) used for processing data from a census of population in the United States; this success led to the establishment of the IBM corporation. Also first versions of cryptographic coding machines Enigma, which came into being around 1919, were a combination of the principles of arithmometer with future tasks of digital computers.

As described in previous chapters, first electromagnetic and electronic computers were invented by Vannevar Bush in MIT in the years 1930–1931; but they were analog, not digital computers, and solved systems of ordinary differential equations. Their broad applications and their rather complex way of programming strongly influenced the development of digital computers. In parallel to the works

---

[1] Clearly, it is the market that contributes to the phenomenon of *path dependence*, that a change of the keyboard would be so difficult, but this means that free market supports rather the path dependence, not an optimization of production technology standards.

of Bush, several theoretical papers were published that today are counted among the foundations of digital computing. In 1931, Kurt Gödel published a paper proving the incompleteness of mathematics, more precisely, he has shown that the question of truth might not be decidable in a given mathematical system. In 1933, Alfred Tarski has shown that reasonable statements about truth in a given language require the use of a *metalanguage,* a language of a higher order; we have discussed this issue in former chapters. These works stimulated Alan Turing, who in 1936 published the paper *On Computable Numbers,* defining (binary) numbers that can be computed when using a scheme of universal computing machine proposed by him. In 1937, Alonzo Church called this type of machine a "Turing machine". Since these works were used later to develop the fundamental theory of digital computers, theoreticians of informatics maintain today that these works initiated the construction of digital computers.

Such an interpretation is, however, rather distant from historical facts. First electromechanical digital computers were constructed by telecommunication engineers that had no idea about the works of Gödel and Turing, but they were well acquainted with relay telecommunication switchboards. In the years 1934–1936, Konrad Zuse in Germany developed (and in 1936, before the publication of the paper of Turing, patented) a prototype of an universal electromechanical digital computer, based on relays and on an analogy to telecommunication switchboards (he could not know the works of Turing or Shannon about the interpretation of Boolean algebra, the work of Shannon dates back to 1938). Zuse did not, however, succeed in the first implementation of a functioning version of such a computer, even if the patent shows such possibility and the first German prototype called Z3 was implemented in 1941. Konrad Zuse was surpassed by *George Stibitz and Samuel Williams who in 1939 constructed the first functioning electromechanical (relay based) digital computer Complex Number Calculator at Bell Laboratories;* concurrently, IBM started to develop the digital computer Harvard Mark I under the direction of Howard Eiken. In 1940, Norbert Wiener and John Mauchly demonstrated a remote access to the Complex Number Calculator, using for this purpose an advanced (for this time) telegraphic artefact, a telex. Thus, the historical truth is that the inventions of telecommunication engineers resulted in the construction of the first digital computer, even if these inventions stimulated theoreticians, e.g., Norbert Wiener, to use and interpret them. Later on, theorists of computer science interpreted the results of Turing as the foundation of digital computer construction, which led to further improvements of computers. However, first computers actually implementing the idea of Turing machine, called ACE and DEUCE, were constructed almost 20 years after the Turing's paper.

Nevertheless, initial improvements of computers were related rather to the development of electronics than mathematical logics. In 1941, Helmut Schreyer (a co-worker of Konrad Zuse) obtains a doctoral degree in Germany for the construction of vacuum tube switches (later called flip-flops, see further sections of this chapter), an electronic equivalent of a relay, replacing electromechanical relays. A year later, independently, John Mauchly repeats the invention of

Schreyer in the United States. This leads initially to an experimental version of a fully electronic digital computer called Vacuum Tube Multiplier, constructed by IBM in 1943. At the same time, in the years 1943–1945 at the Moore School of Electrical Engineering, the University of Pennsylvania, a full version of an electronic digital computer was constructed, called ENIAC (Electronic Numerical Integrator and Computer) by its creators, engineers John Mauchly and Pres Eckert. The name ENIAC (and especially the emphasis put on *numerical integrator*) is telling: it indicates a strong impact of analog computers that supported an *integration* of differential equations. Indeed, ENIAC was expected to help in working out naval artillery tables that were developed before with analog computers. In 1944, Pres Eckert added magnetic discs (or rotating cylinders, drums) to ENIAC in order to record and read data.

After World War II, digital computers quickly found diverse non-military applications, together with first applications of *symbolic computing,* namely preparing tables of mathematical functions with their derivatives (Eiken did it in 1945). Such tables were prepared before computers by mathematicians, but required long time and contained many errors. Another broad field of applications was statistics. In 1947–1948, the United States Census Bureau commissioned the construction of UNIVAC computer to Electronic Control Company (founded by Pres Eckert and John Mauchly after they left Moore School, later called Eckert-Mauchly Computer Corporation, sold in 1950 to Remington Rand). Tested in 1949 on programs sent by the United States Census Bureau, UNIVAC was delivered to the Bureau in 1951 and became the first broadly used computer in civil service.

In 1952, Grace M. Hopper wrote the first compilation code for the UNIVAC computer, today called *compiler*. A compiler is a programme that serves for an automatic translation of a code written in one language, e.g. in a higher level programming language, into an executable code in another language, e.g. in the language of commands of a computer processor. In the same year Hopper published a book *The Education of a Computer,* the first textbook of computer programming. The results of Grace Hopper initiated the development of programming in higher level languages and a gradual separation of software from hardware. This separation became a reality around 1970, together with the development of the UNIX operating system. Before that, diverse computer programming languages were developed. *Fortran* language was created by John Backus in 1954 in IBM, *Algol* was created also by Backus in 1958. *Lisp* language, preferred in applications for artificial intelligence, was developed by John McCarthy in the years 1958–1960. An universal language for business applications called *Cobol* was developed by a large team (William Selen, Gertruda Tierney, Howard Bromberg, Howard Discount, Vernon Reeves and Jean Sammet) under ARPA, a programme promoted by the Department of Defence of the USA. The universal operating system *UNIX* was created in 1969–1970 by Kenneth Thompson and Dennis Ritchie at Bell Laboratories as a basis for connecting computers into a network. From this date on we can count the final separation of software from hardware in computer technology. As we can see, the emergence of software out of hardware was a lengthy process, and each step in this process had a technical substantiation, generally related to the complexity of problems that had to be mastered, but it was in a sense spontaneous, nobody planned it, even Grace M. Hopper, it was a self-supporting civilization development.

In the years 1950–1980, diverse versions of computers were developed in many countries. For example in 1951, the enterprise J. Lyons and Company started the production of LEO I, the first computer for business applications; IBM introduced its computer IMB 701, intended both for scientific computing and for business applications, in the end of 1952 (and called it *data processing machine,* perhaps in order to distinguish it from UNIVAC that was called *computer*). Earlier, but also in 1952, Sergiey Lebiediev constructed the first digital computer in Russia, MESM, and Heinz Billig build the first complete digital computer in Germany, G1 (at the Max Planck Institute in Göttingen, under the direction of Werner Heisenberg). A commercial production of another computer, Z-22, was started in Germany in 1958 by Konrad Zuse, the author of the first patent for a digital computer from 1936 (thus, with a delay of 22 years).

The first computer constructed in Poland in 1954 by the Group of Mathematical Machines of the Institute of Mathematics of the Polish Academy of Science was, similarly as other computers in the world, an analog computer, called ARR (it was an analyzer of differential equations). A digital computer XYZ was constructed in 1958 by a group of both engineers[2] and mathematicians under the leadership of Leon Łukaszewicz. The Group of Mathematical Machines has grown into the Institute of Mathematical Machines where further computers were constructed (ZAM-2 in 1961, later equipped with a system of symbolic addresses SAS and programming language SAKO; ZAM-41 in 1963). Concurrently, the Chair of Mathematical Machines Construction of the Warsaw University of Technology (officially launched in 1963 in result of combination of two other chairs that constructed computers even earlier) prepared several prototypes: EMC in 1960, UMC-1 in 1961 (transferred to industrial production in the factory ELWRO in Wrocław in 1962), a transistor version, UMC 10, in 1965; these developments were directed by Antoni Kiliński with the participation of Zdzisław Pawlak. Another group at the Institute of Physics of the University of Warsaw developed in 1965, under the leadership of Jacek Karpiński, a minicomputer KAR-65 with an original architecture, but built of transistors brought from England (despite embargo). The machine was faster than other computers in Poland but difficult for industrial implementation. The ELWRO factory in Wrocław constructed also, in collaboration both with the Warsaw University of Technology as well as the Wrocław University of Technology, its own versions of computers called ODRA (version 1003 produced since 1964, transistor version 1204 produced since 1967, and version 1304 produced since 1968 that supported the software of an English computer ICL-1904).

---

[2] The group of constructors of XYZ included e.g. Zdzisław Pawlak, at that time a young engineer that graduated from the Department of Telecommunications of the Warsaw University of Technology; later, Pawlak participated in the construction of EMC computer and much later (1991) he created the *rough set theory,* see further sections.

The invention of transistors was groundbreaking for the development of computers: it resulted in miniaturization and increase of reliability. In the world, the first projects of transistors were patented by Julius E. Lilienfeld in the year 1928 in Germany, but the first actually working *point-contact transistor* was constructed in the USA in Bell Telephone Laboratories in 1947 (by John Bardeen, Walter H. Brattain and William B. Shockley). The first transistor radio was produced by Texas Instruments in 1954 using *germanium bipolar transistors*; at the same time, Texas Instruments started the production of *silicon bipolar transistors,* more durable and reliable than germanium transistors. Thus, even in the case of transistors, where the development, production and applications were much accelerated by military and space exploration use, the civilization delay "from the invention to industry" exceeded 20 years; this is one of the shortest delays of this type.

A further development of transistors initially concentrated on *integrated circuits,* integrated electronic devices comprised at first of a few pieces, and today of many billions of electronic elements of diverse types, such as transistors, diodes, resistors, capacitors, etc. The first integrated circuits were constructed almost at the same time by Jack Kilby at Texas Instruments and (a few months later) by Robert Noyce at Fairchild Semiconductors in 1958. Later, integrated circuits became the basis of *microprocessors,* main parts of computers, *central processing units (CPU),* produced as a single integrated circuit, today of the large scale of integration. The first microprocessor was constructed and patented for military purposes of control of fighter airplanes in the years 1968–1970 (it was that much effective that it was kept as strictly classified and secret until the end of the 20th century). First commercial microprocessors worked on 4-bit data (processed digital numbers of only four binary positions) and were developed concurrently by Texas Instruments (Gary Boone) and Intel (Marcian F. Hoff) in 1971, but Texas Instruments preceded Intel by 2 month and patented this idea, which resulted later in long negotiations. Soon after that, 8-bit microprocessors (Intel 8080) and 16-bit microprocessors were developed; the latter were used in minicomputers PDP by Digital Equipment Corporation (DEC).

The introduction of transistors and microprocessors resulted in miniaturization of even large computers; before that, electromechanical relay computers or even vacuum tube computers were often artefacts of large size, often filling up immense rooms or halls. The first transistor versions of IBM 709 computer were produced in 1960; simultaneously, transistors were used by DEC to miniaturize PDP-1 (called *programmed data processor*) that can be counted as the first *minicomputer.*

The further development of transistor and integrated circuit technology was, and remains today, strictly related to the needs of computer technology. It included the construction of *field effect transistors* that are controlled by input voltage as opposed to *bipolar transistors* that were controlled by input current. First field effect transistors were developed by William B. Shockley in 1952 (earlier patents of Julius Lilienfeld were based on similar principles). Diverse early versions of field effect transistors included MISFET and MOSFET transistors that had an isolating layer between the controlling electrode (gate) and the semiconducting base; this results in an essential increase of the input resistance of the transistor.

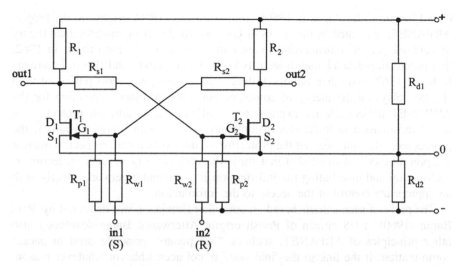

Fig. 9.1 A diagram of a flip-flop (bi-stable) circuit RS on transistors JFET N

Field effect transistors with very large input resistance resulted in the development of DRAM (Dynamic Random Access Memory) by Robert H. Dennar at IBM in 1966. In a DRAM, a single transistor (and not a pair of them as in a typical flip-flop switch presented in Fig. 9.1, described later in this book) can remember the occurrence of a voltage impulse on its gate, the controlling electrode. The principle of DRAM operation is the same as in the integrating element of analog computers described earlier in Fig. 8.3a: a small natural capacitance between the electrodes of the transistor together with large input resistance result in an integrating element, remembering former input impulses. The use of DRAMs results in doubling the density of memory elements in an integrated memory circuit.

Field effect transistors were the subject of further development until today. However, currently, several completely different solutions are investigated, quantum effect circuits with the use of quantum indeterminacy of electronic state, or carbon nanotube transistors that can be controlled by a single electron, all in order to realize the fundamental functions in a digital computer, such as logical switches, flip-flop or dynamic memory.

The military and cosmic applications initiated also the development of *computer networks,* a groundbreaking aspect of computer use. In the beginning of 1958, President of the United States Dwight Eisenhower established the Advanced Research Project Agency[3] (ARPA) at the Department of Defence of the USA, as a response to the sputnik sent around the Earth by Soviet Union in October of 1957. ARPA decided to concentrate on computer technology development around 1959–1960, starting with sponsoring the standardization of software (including the

---

[3] Originally called Advanced Research Planning Agency. The word "planning" was changed to "project" because of ideological connotations.

Cobol language developed in 1960). Soon after that ARPA originated the project ARPANET, motivated by the works of Leonard Kleinrock on mathematical theory of network packet[4] transmission (master thesis in 1961, doctoral thesis in 1962, first publication delayed due to security reasons until 1964), and also of Lawrence Roberts (1967) and other authors writing (and having their works published with similar delay) on the theory of computer networks. Military motivation for the ARPANET project, clearly expressed by ARPA, was a sufficient protection of secret information in multi-access computer systems with shared resources; the project took also into account the protection of information with respect to nuclear weapon attacks and concluded that the only effective way of such protection is multiplying and distributing the information in a computer network, clearly with an appropriate control of the access to the information.

The proof of such a thesis based on computer simulation was delivered by Paul Baran (1960), a US citizen of Polish origin. Afterwards, Baran developed also other principles of ARPANET, such as "hot potato" principle used in packet commutation: if the link to the final node is not accessible for whatever reason, send the packet to any other node.

The results of ARPANET were classified, but already in 1969 Leonard Kleinrock in UCLA implemented the first connection of computer network (originally only between a computer in UCLA and a router of telecommunication network, extended afterwards to a computer in Stanford Research Institute), a year later adding a new node of computer network in Cambridge, Massachusetts. In 1972, ARPANET had already 15 nodes and Ray Tomlinson invented email for this network (he also introduced the @ sign in email addresses). In 1974, Vinton Cerf and Robert Kahn devised TCP (Transmission Control Protocol), extended and split in 1978 into TCP/IP (Internet Protocol). In 1983 ARPANET was declassified for the first time (with a separation of strictly classified MILNET) and the protocol TCP/IP was accepted as a basic standard of transmission in computer networks; it was the start of a broad penetration of Internet. Therefore the year 1983 is the date on which social penetration of computer networks began.

It is significant that the development and dissemination of personal computers occurred parallel to (well, maybe only slightly earlier than) the development and dissemination of computer networks. The so-called minicomputers did not result in such dissemination, they were simply too expensive. The road to personal computer was rather long: in 1961–1962, Wesley A. Clark in MIT constructed desktop LINC (Laboratory Instrument Computer), relatively "inexpensive" (43,000 dollars, the DEC enterprise that started the production of this computer sold only about fifty pieces). In 1965, Honeywell tried to open market for home computers H316 (one of versions was called *Kitchen Computer* with a price less than 10,000 dollars), but these turned out also too expensive for mass use. In 1968, Hewlett Packard introduced a small calculator HP9100A; in a few years the price of

---

[4] The very concept of a *packet* denoting an organized sequence of binary signals of a given length was introduced later (1967) by Donald Davies.

calculators decreased substantially and they became a popular computing tool; therefore, around 1970 the era of logarithmic sliding rules ended ultimately.

The development of a full personal computer was continued in many places. In 1975, Bill Gates and Paul Allen developed a version of Basic language intended for personal computers. However, the first personal computer that achieved a mass market dissemination was Apple II introduced in 1977 by Stephen G. Wozniak and Stephen Jobs, and their enterprise, Apple. The connection of personal computers to computer networks occurred even before the Internet has been declassified: in 1979, the Compuserve company offered email service to owners of personal computers. Together, the dissemination of computer networks and personal computers started the *informational revolution,* hence its start should be dated for the years 1977–1983, approximately for the year 1980.

9.2 Social and Conceptual Importance of Computers: The Limits of Artificial Intelligence

The social and conceptual importance of the informational revolution, with the dissemination of personal computing and network communication, is tremendous, and exceeds the importance of computer technology, computers and microprocessors significantly. But even the social importance of computer technology itself is very large and it deserves separate comments.

Microprocessors and computer technology enabled the implementation of automation, robotization and computerization of all heavy work. As a result, today all bigger machines, e.g., increasingly more cars, but also excavators, bulldozers, agricultural and mining combines, etc., include subsidiary control systems not only based on the feedback principle, but also implementing this principle with a microprocessor. Due to mass production, simple microprocessors are so much inexpensive that a popular slogan today is the concept of *ambient intelligence* (*AmI,* called also the *Internet of Things*). AmI means a saturation of the ambient environment of human life with networks of communicating microprocessors. To illustrate this idea with some exaggeration, imagine a wall paint primed with lots of miniature microprocessors used for testing the presence of a person in the room, controlling the temperature and humidity inside as well as the identity and authorisation, measuring health parameters, etc. of that person. Ambient intelligence belongs to the future, perhaps not very distant; today we observe such basic social results as freeing people from most heavy work and thus creating conditions for actually equal rights of women and men.

Thus, not the computers as such, but more generally computer technology or, even broader, informational technology, has tremendous economic and social importance nowadays. For example, the cost of electronic equipment in aeroplanes, microprocessors, sensors, radar equipment, control systems for engines and aerofoils, etc., exceeds currently often 70 % of the total costs of the airplane. In automobiles, not so much informational and electronic technology is used, but their share grows fast. This concerns almost all artefacts of everyday use, such as washing machines and refrigerators, photographic and film cameras, not to speak of all contemporary tools of electronic communication, radios, stationary and mobile telephones, tablets, which by their nature are electronic devices including specialized microprocessors.

The final social effects of the mass use of personal computers (or tablets) and of dissemination of electronic communication are difficult to foretell even today, because they encompass both immense chances and serious threats that are emerging more and more expressively. An example is the impact of informational technology on market mechanisms. A positive impact of informational technology on market mechanisms is well known and it has been invoked many times. The global web of Internet provides a fast information about price differences in diverse parts of the globe and helps in organization and logistic aspects of transport of goods, supporting globalization of commerce in that way.

On the other hand, it was information technology that constituted a partial cause of the last financial and economic crisis because of a specific degeneration and corruption of the financial market mechanism. The impact is twofold. On the one side, information technology enables mass use of some statistical techniques, and an American mathematician of Chinese origin David X. Lee (see e.g. Salmon 2009) provided a formula accelerating the computation of correlation coefficients. Financial institutions used this formula for constructing (ostensibly) most secure investment portfolios of uncorrelated assets. On the other hand, financial institutions used contemporary communication technology and Internet for unfettered advertisements of such ostensibly secure portfolios (offered as derivatives, options, etc.).

This method and these advertisements inflated the bubble of investments in complex financial instruments, but only specialists in mathematical methods of risk management knew where was the real danger. The danger results from a false assumption that correlation coefficients are constant in time, that stochastic processes on financial markets are stationary. This assumption is clearly false: if a crisis comes, the prices of all assets jointly fall down, all assets become strongly correlated. Thus, the false advertisements about alleged safety of complex financial instruments have blown up the investment bubble that had to burst, and it is not important from where the impulse that pierced the bubble came. Thus, the opinion of neoliberal economists, that it were botched interventions of the US government that resulted in the crisis, must be judged as an attempt to defend the neoliberal paradigm maintaining that free market should be left alone. Since any impulse could result in bursting the bubble, it was the mechanism that inflated the bubble which was guilty, the market mechanism degenerated by irresponsible advertisement and informational asymmetry: the buyers did not have sufficient information and knowledge about the methods of providing alleged security of investment portfolios, hence they could not make an informed decision in risky situation. In short, in new informational conditions, market mechanism can be degenerated or even corrupted by profit lust (because also mechanisms can be corrupted).

Thus, when assessing socio-economic impacts of new technologies, we should always reckon with new chances, but also unforetold dangers. This concerns also, or even especially, the idea of *artificial intelligence.* The idea that it is possible to construct intelligent machines emerged actually with the construction of robots, and later it influenced computer construction (the first laboratory of artificial intelligence was created by John McCarthy and Marvin Minsky at MIT in 1959, two years after the construction of the first robot in 1957). Since that time the concept of artificial intelligence has been strongly developed in the world, but particularly in the United States. The general goal is to construct a machine that is equally, or even more, intelligent than humans. This goal has many diverse interpretations and parts, and it relates to the development of logics, pattern and image recognition, learning of machines and computer programs, perceptrons and neural-like networks,[5] recognition of hand-written letters, checking and correcting spelling in texts, decision support and research on methods of human decision making, on the functioning of human brain, finally, to *cognitivism.* The latter is a direction of research and thinking relating a part of studies on artificial intelligence and on functioning of human brain to a philosophical current based on the conviction that *both human brain and the entire world can be treated as immense computer,* see e.g. Gardner (1985). Such direction of thinking was supported by various research attempting to reduce thinking to language and logical operations, which also inspired many useful specialist research directions, such as *pattern recognition* (automatic classification of patterns by computers), *recognition and artificial synthesis of speech, artificial translation,* etc.

Reduction of thinking to language and logical operations has long tradition, starting with classical Greek philosophy, but this tendency strengthened and intensified in the end of the 19th century, with the works of (Frege 1893), (Russell and Whitehead 1910–1913); (Wittgenstein 1922). From that time most of the 20th century philosophy concentrated on lingual aspects of cognition; fundamental studies of formal foundations of language included the concepts of *language of thought* (Fodor 1994) or of *universal grammar* (Chomsky 1986) and at the same time led to post-structural theory of language (Derrida 1974) with its stress on the subjectivity of statements. On the other hand, critical attitudes to the reduction of thinking to language and towards cognitivism developed too, e.g. in (Dreyfus 1972; Dreyfus and Dreyfus 1986; Searle 1992). They included the conviction, confirmed afterwards by more rigorous studies of human brain functions, that *we often think and above all decide in a preverbal manner, hence neither human brain nor the entire world can be reduced in an analogy to a digital computer,* at least in its current versions. This was already discussed in Chap. 5 of this book, concerning the evolutionary theory of intuition; but a parallel fact at the turn of the 20th and 21st century was the crisis of the concept of artificial intelligence: *despite spectacular specific achievements in partial artificial intelligence, we did not succeed in*

---

[5] Specialists in this field use the concept of neural networks, but actually these are only neural-like networks, much less complex than real biological neural networks.

constructing a computer or another machine that would possess creative abilities such as displayed by living organism and humans in particular.

Because of the above, many new directions emerged: new micro-theories of knowledge creation stressing the role of an interaction between *explicit knowledge,* which is rational and expressed in words, and *tacit knowledge,* preverbal, intuitive and emotional one, in the processes of knowledge creation; new analyses of the role and importance of intuition, new explanations of the phenomenon of emergence of new properties and effects on higher levels of complexity. These problems will be discussed in more detail in other chapters, but I would like to stress here a fundamental example of emergence: the civilization development during last 50 years that led to spontaneous emergence of *software* as a concept and object of creative activity, which is irreducible to *hardware,* computer equipment. Software cannot function without hardware, but it is not reducible to hardware in the sense that its functioning cannot be explained by hardware; thus, it is an *irreducible complexity that emerged in the evolution of civilization.*

The following question arises against this background: could computers become truly intelligent? That depends on the definition of intelligence, which changed several times in the history of artificial intelligence, as it will be discussed in one of following chapters. However, the new directions of research suggest that intuition and emotions of computers can emerge along with their sufficient complexity, as new computer properties; but this needs time and it will not be fully rational intelligence, expressed in words. Moreover, it is related to the *danger of computer domination over humans,* because how shall we control computers when they become truly intelligent?

9.3 Electronic Flip-Flops: The Law of Moore

It was already mentioned that each contemporary computer or microprocessor contains millions or even billions of *electronic flip-flops.* It is useful to understand well how such a circuit operates, because it is based on the concept of positive feedback and exemplifies one of the reasons why the concept of vicious circle should be replaced by the concept of self-supporting positive feedback.

The principle of operation of an electronic flip-flop or a *bi-stable electronic circuit RS* using two field effect transistors (of the type JFET N[6]) is illustrated in Fig. 9.1. If a positive voltage impulse occurs, even for a short time, at the input in_1 (denoted also as S from *set*), then it results in a current conducted by the transistor T_1, thus in a decrease of the voltage on its *drain* D_1 (because of the increase of the voltage measured on the resistor R_1). This voltage decrease is transmitted by the

---

[6] Junction Field Effect Transistor, with a channel of n type, a rather commonly used type of a junction field effect transistor. However, the principle of a flip-flop illustrated in Fig. 9.1 does not depend on the type of transistors used; it is actually the same as used in a vacuum tube flip-flop by Helmut Schreyer in 1941.

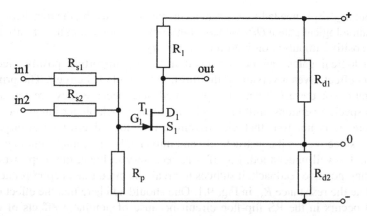

Fig. 9.2 A diagram of a logical gate *NAND* (or *NOR*, depending on the choice of resistor dividers R_{s1}, R_{s2}, R_p) using a transistor JFET N

resistor divisor R_{s1}, R_{p2} on the *gate* G_2 of the transistor T_2 and results in its clogging, in its current not being conducted, which results in a voltage increase on its *drain* D_2 (because of the decrease of the voltage measured on the resistor R_2).

Here occurs the decisive effect of the positive, self-supporting feedback: the increase of voltage on the drain D_2, transferred by the resistor divisor R_{s2}, R_{p1} on the *gate* G_1 of the transistor T_1, result in a current conducted by this transistor *even after the disappearance of the controlling impulse on the input in_1*; hence the flip-flop *remembers* that it had once a positive impulse on its input in_1. In order to reset this memory, to return to the original state without the current conducted by the transistor T_1, it is necessary to provide a positive voltage impulse on the input in_2 (denoted also as *R* from *reset*); this second input might be also used by a clock coordinating the work of a bigger set of flip-flops.

The operation of memory flop-flops can be combined with logical gates *NAND* (the negation of *and*) or *NOR* (the negation of *or*).[7] The diagram of such a circuit with the use of single field effect transistor (of the type JFET N) is presented in Fig. 9.2. Only one diagram is presented there: whether the effect will be *NAND* or *NOR*, it depends entirely on the resistor dividers R_{s1}, R_{s2}, R_p. If these resistances are chosen in such a way that positive voltage impulses on both inputs in_1, in_2 are needed to result in a current conducted by the transistor T, then we obtain a logical gate *NAND*: a negative voltage impulse arises on the output *out* (because of the voltage measured on the resistor *R*) which corresponds to the negation. If the resistances R_{s1}, R_{s2}, R_p are chosen in such a way that a positive voltage impulse on

---

[7] In 1984, Fujio Matsuoka at Toshiba invented the use of logical gates NAND or NOR as a dynamic memory preserving its state after switching off the power supply (due to a small capacitance between the drain and the gate of a field effect transistor which creates a kind of integrating circuit; the memory is transitory, impermanent, but can preserve information for over 10 years). Such transitory memory is commonly used today in *pendrives*, plugged into USB ports of computers, see also Chap. 11.

any input in_1, in_2 is needed to result in a current conducted by the transistor T, then we obtain a logical gate *NOR:* we have high voltage on the output *out* only if there are no positive impulses on both inputs in_1, in_2.

Two logical gates can be combined in a self-supporting positive feedback which in effect gives a version of the flip-flop bi-stable RS circuit. The principles of operation of these elements are presented here because they are repeated, in various specific versions, million or billion times in each bigger integrated circuit. Therefore, it is not true that vicious circle is a logical error or unsurpassable paradox, since we use it constructively and universally in digital technology. Only, it should be well understood. E.g. if it is necessary to break the loop of the self-supporting positive feedback, it suffices to break the circuit in an appropriate place, e.g. delete the resistance R_{s2} in Fig. 9.1. One should add here that the effect of self-support occurs in the RS flip-flop circuit because of nonlinear effects of current saturation in its transistors; if there were no such effects, then current in a circuit with positive feedback would grow avalanche-like, until the transistor would burn. Therefore, if we encounter self-supporting positive feedback in socio-economic practice, we should first of all assess whether this is an avalanche-like process or it displays saturation and self-support effects. After that we should analyze the mechanism of positive feedback in order to determine its critical elements that would break the feedback if removed or modified.

The historical development of the number of such elements, e.g., logical gates, that can be contained in a single integrated circuit of average size, e.g. on a square inch, is quite a different question. Already over 45 years ago, Moore (1965) formulated an empirical law resulting from observation of the development of this parameter. Today it is called *Moore law* and says that the number of elements contained in an average integrated circuit doubles every 18–24 month. Later it turned out to be a true observation, sustained for over 45 years, but in the following version: the number of elements grows about ten times each 5 years, or a hundred times in a decade. The reason is a continued development of the electronic technology of integrated circuits, with new types of transistors introduced, etc. Even if we can suspect that the possibilities of silicon technology would be exhausted soon, the Moore law will be probably valid at least through next several decades, because of new transistor technologies mentioned earlier, such as graphene, etc. This means that the number of transistors on a single integrated circuit, and thus also the computational capacity of computers, could grow yet 10^6–10^{10} times.

If this really occurs, we should seriously reckon, around the years 2040–2050, with the possibility that a true intelligence of computers could really emerge, together with emotional and intuitive aspects. This results from the assessment presented in Chap. 5 in relation to the evolutionary theory of intuition, that pre-verbal processing in human brains is *at least 10^4 times more powerful* than logical, verbal processing. If we assume that today computers are approximately equal to humans in terms of logical, verbal processing, then we have yet at least 20 years of relative quiescence and of happy but not fully substantiated conviction that humans are essentially better than machines. After that, the related problems will hit us, depending on what means *at least 10^4 times* in the above assessment. If it means *perhaps 10^8 times,* then we shall start to worry in the years 2040–2050, but perhaps it is better to start to worry earlier.

9.4 Software and Hardware: Computational Complexity

Extremely fast development of hardware capabilities of computers often results in a naive conviction that it would solve everything: why to learn complicated mathematics, develop software and computational methods, when computers in 10 years will be hundred times faster anyway? Such conviction is groundless, and even dangerous, because of many reasons explained below. First, however, I shall quote a personal anecdote how I dealt in a dispute with such convictions. Several times in life (e.g. in Japan) I encountered the constructors of supercomputers; I used to bet with them that in a day I will write a program (correct and realistic, that is, concerning an actual computational problem) for their computer that will saturate it. I did not lose such a bet; below I explain how to write such a program.

   The above stems from my knowledge of *computational complexity* of problems to be solved on a computer. The theory of computational complexity is a substantive part of computer science, rather complex in itself, so I present it here in a maximally simplified version. Computational problems to be solved by a computer are characterized by their type, e.g. the problem of *routing* in a network consists in determination of the shortest or the cheapest path connecting two points in the network, while the problem of *travelling salesman* in the logistic theory is similar to the routing problem but additionally requires several nodes of the network to be visited on the selected path. Another important aspect of specification of a computational problem is its *dimension,* defined either as the number of input data to be processed when solving the problem, or the number of variables in a mathematical characterization of the problem.

   The general problem of computational complexity is to determine what computational effort (e.g. how many elementary operations) is necessary to solve a problem of a given type depending on the dimension of the problem. Such estimates are not precise, but of the "at least" or "at most" type; they characterize the general type of the dependence of computational effort on the dimension of the problem. A fundamental result of the computational complexity theory says that, except for some problems which are particularly simple, *the dependence of computational effort on the problem dimension is nonlinear.* Moreover, such nonlinear dependence has a relatively mild *polynomial* character only for rather simple problems, while for more complex ones it is *non-polynomial,* that is, *exponential* or *combinatorial,* and grows very fast with problem dimension.

   Consequently, it is very easy to devise a program that will saturate even most powerful computer. It suffices to choose a problem of more complex type, with an exponential dependence of computational complexity on the problem dimension, and add an outer loop in the program that will gradually increase this dimension.

To be fair, we should add a stopping test, but of the type "stop when the problem dimension exceeds 10^8 or computation time exceeds 1,000 h".[8] Various conclusions result from this fact, more practical for computational science, more general for epistemology.

Each representative of applied computational science, a technician, biologist, meteorologist, physicist etc. who uses computers to solve complex simulations or optimizations in research or design, knows well from practical experience that s/he cannot use a fully accurate model of the problem incorporating her/his full knowledge in the specific field, but rather an approximate discretization of time and space and other simplifications, because otherwise the necessary computation will take too much time. Therefore, computational science is an art of compromise between a possibly greatest accuracy and a reasonable computing time. While this compromise changes with the increase of computational capabilities of computers, it remains a compromise.

From the above it follows that there are no universal and at the same time-efficient algorithms: for each area of computational science and each type of computational problem we must work intensively on specific algorithms that would be effective for a given class of problems. This concerns also parallel or cloud computations that employ not only one processor or computer, but a number of processors in a supercomputer, or large number of computers in a network, because each parallelization of computations requires an additional effort for their coordination, and it is very difficult to obtain an acceleration of computations proportional to the number of processors used.

A classical example here is the history of genetic and evolutionary algorithms that rely on the use of primitive evolutionary models ("best adapted survives") to devise computational algorithms. They were regarded as a prototype of an universal algorithm and can be used in various applications, of which those for optimization, and especially multi-objective optimization, are particularly useful. Actually, they turned out to be very effective in the case of difficult problems, and particularly in global or multiple criteria optimization, but for problems of small dimension. For higher dimensions the computational complexity intervened, because evolutionary algorithms were not constructed with this complexity taken into account (actually, they are characterised by an exponential type of dependence of computational effort on the problem dimension). Thus, the argument that we need not learn complex mathematics and difficult algorithms, that it is sufficient to use simple principles of evolutionary adaptation in an universal algorithm, turned out to be false. The belief in an absolute power of computers and universal algorithms can be dangerous, because it justifies ignorance.

---

[8] A program without a stopping test would be not fair in respect to computer constructors, but as we test the saturation of the computer here, we can use large numbers. Stopping after 1,000 h is obviously not realistic, computer administrator would stop the program much earlier and admit that the computer was saturated. A real test consists in solving a problem of a high dimension, say, 10^8, in a reasonable time.

And the Moore law, the fact that computational abilities grow exponentially with time, would not help if we similarly increase the volume of recorded data (that actually grows even faster than according to the Moore law): the latter corresponds to the dimension of the problems solved, and the computational effort grows exponentially with this dimension. Thus, the compromise between a possibly greatest accuracy and a reasonable computing time will remain with us.

In addition, this means that the problem of computational complexity has deep epistemic consequences. From the foregoing it follows that cognitive capacities of humans are limited not only by our subjective properties, but also by the imperfection of tools, including the most advanced computers and their software, that we use in cognitive processes. It appears that *humans as the subjects of cognition always transcended, among other things due to redundancy of human minds or brains, tools available at a given time, always tried to improve the tools.* This is one of the reasons to assume that creation of tools, together with cognitive curiosity and interpersonal communication, is a fundamental feature defining humanity.

References

Arthur, W.B.: Increasing Returns and Path Dependence in the Economy. Michigan University Press, Ann Arbor (1994)

Baran, P.: Reliable digital communications systems using unreliable network repeater nodes. RAND Corporation papers, document P-995. http://www.rand.org/pubs/papers/P1995.html (1960). Accessed 29 Mar 2011

Chomsky, N.: Knowledge of Language: Its Nature, Origin and Use. Praeger Special Studies, New York (1986)

Derrida, J.: Of Grammatology. John Hopkins University Press, Baltimore (1974)

Dreyfus, H.: What Computers Can't Do: the Limits of Artificial Intellgence. Harper & Row, New York (1972)

Dreyfus, H., Dreyfus, S.: Mind over Machine: The Role of Human Intuition and Expertise in the Era of Computers. Free Press (1986)

Fodor, J.A.: The elm and the expert: mentalese and its semantics. MIT Press, Boston (1994)

Frege, L.F.G.: Grundgesetze der Arithmetik, Begriffsschriftlich Abgeleitet. Verlag Hermann Pohle, Jena (1893)

Gardner, H.: The Minds New Science: A History of the Cognitive Revolution. Basic Books, New York (1985)

Moore, G.A.: Cramming More Components onto Integrated Circuits. Electronics Magazine **38**(8), 4–10 (1965)

Russel, B., Whitehead, A.N.: Principia Mathematica. Cambridge University Press, Cambridge (1910)

Salmon, F.: Recipe for disaster: The formula that killed wall street. Wired magazine, Tech Biz: IT (2009). Accessed 17 Mar 2009

Searle, J.R.: The Rediscovery of Mind. MIT Press, Boston (1992)

Wittgenstein, L.: Tractatus Logico-Philosophicus. Cambridge University Press, Cambridge (1922)

Chapter 10
Systems Theory, Theory of Chaos, Emergence

10.1 Systems Theory: Introduction, Basic Events

This chapter has a specific, dual character. Firstly, it presents only selected elements of the history of systems theory and systems technology, or more generally, systems research. Secondly, it focuses on one aspect of systems research, namely on chaos theory and the phenomenon of emergence.

We should start with a definition of the concept of a *system*. In everyday language it is often used as a synonym of a scientific theory (such as Copernican system, etc.). Such meaning was popularized by August Comte (1830–1842) who called his social theories a system. More precisely, however, present-day systems theory and systems research use the concept of a *system* in the sense of *a set of elements together with relations between them,* see e.g. (Marchal 1975). Thus, *an organized whole* (as it was proposed in the definition of a system by Bogdanov 1910) is not treated as a system until we define the parts of the whole and relations that organize it. Similarly, the set of the Sun and its planets is not a system (even if it is commonly called the solar system) until we take into account at least the relation of gravitational forces determining its behaviour. Obviously, there are more relations in a planetary system, starting with radiation of the Sun warming its planets. However, we cannot take into account and usually do not know all possible relations and influences (moreover, even if we knew them precisely, we could not conduct their computational analysis in a reasonable time, as discussed in Chap. 9). Therefore, the actually analyzed system is always a simplified model of reality. Systems research postulates possibly thorough, interdisciplinary (e.g. in the case of a solar system, not only astronomic, but also biological, including the questions relating to the conditions of life on various planets), and whenever possible overall analysis of relations between elements of a system and their consequences.

© Springer International Publishing Switzerland 2015   175
A.P. Wierzbicki, *Technen: Elements of Recent History of Information Technologies with Epistemological Conclusions*, Intelligent Systems Reference Library 71,
DOI 10.1007/978-3-319-09033-7_10

The history of systems research can be dated back to Heraclitus (since systems research should also stress dynamics and variability, the principle that *everything flows*) and to Aristotle, but contemporary systemic research started with a cross-connection of three originally independent disciplines:

(1) The first of them is the *theory of automatic control,* already discussed in Chap. 8, established in 1940–1944 by the works of Smith (1942), Hall (1943), Oldenbourg and Sartorius (1944), even if actually much older (starting with Watt, Airy, Maxwell and many others). It was concentrated on the interdisciplinary analysis of the dynamics of technical control systems, and later generalized to arbitrary control systems, also those occurring in nature, and called, in such general case, *cybernetics* (Wiener 1948).

(2) The second one was a parallel or slightly younger discipline called *operations* (or operational) *research.* There is a quarrel today about the original start of this discipline between British and American sources (see e.g. Dahan and Pestre 2004), while mathematical modelling of operations research problems has actually American and Russian roots. Tjaling Koopmans developed a mathematical model of supply for naval ships on an extensive area already at the beginning of the Pacific fights in the World War II (Koopmans 1942). Around the same time, George Dantzig invented an algorithm computing optimal solutions for linear mathematical models such as proposed by Koopmans. This method, called *simplex method of linear programming* (see Dantzig 1963) became the foundation for the development of operations research and the fundamental computational algorithm in this field, still broadly used today. The simplex method has an explicitly algorithmic character; George Dantzig used it originally for paper computations, but eventually it became one of the first applications of digital computers, and it is not clear whether the word *programming* was used for the first time in relation to the linear programming method or to programming of computers in general. Nowadays, solving linear programming problems still constitutes a large part of the workload of large computers, because codes can be usually reformulated as a large linear programming problems solving of which means breaking the codes. As it often happens, Koopmans and Dantzig were not the first to consider the problem of optimization of linear mathematical models; before them, Leonid Kantorovich from Russia formulated this issue in relation to economic planning (Kantorowicz 1939). Independently from this dispute about the origins of operations research, the discipline rapidly developed towards an interdisciplinary analysis of mathematical models of diverse systems from the perspective of optimization of the operation of these systems. However, it concentrated mostly on static models, as opposed to automatic control focused on dynamic ones. Today, operations research has tremendous practical and socio-economic importance, it is used in various logistic systems, such as maritime supply and transportation, air transportation (e.g. systems of airline connections, even ticket reservations, would not function today without operations research support), see e.g. (Hillier and Liebermann 2002).

(3) *Systems analysis* is a discipline introduced in military and air industry of the USA around 1950, among others by RAND corporation, see also Quade (1985). It uses the tools of the two disciplines mentioned above, but concentrates on socio-political aspects and on decision making, and also on holistic and intuitive aspects of interdisciplinary analysis by including some aspects of social and behavioural sciences. Because of the concentration on decision making, the systems analysis used also the game theory (starting with the fundamental work of von Neumann and Morgenstern 1944) and led to further development of the statistical decision theory (see Keeney and Raiffa 1976). The foregoing led to development of the mathematical decision theory with its diverse aspects: multiple criteria decision-making (see e.g. Gal et al. 1999; Wierzbicki et al. 2000), applications of fuzzy set logics (see Zadeh 1965; Kacprzyk 2001) or rough set logics (Pawlak 1991; Słowiński 1995). These approaches, even if they actually started with the systems analysis, are treated today as parts of operations research or more generally *hard systems research*. However, systems research has also another, sociological branch that stresses approaches not using mathematical modelling, see e.g. (Midgley 2003).

The above resulted from a concurrent development of the *general systems theory* for economic applications (e.g., Boulding 1956) according to the proposal of Bertallanfy (1951, 1956) who stressed the generality of systems research methods, the importance of interdisciplinary and holistic approaches and the effect of *synergy* (*the whole is bigger than the sum of its parts*). It is telling that Bertallanfy substantiated the necessity of development of the *general systems theory*, pointing out the increasing specialization of disciplinary sciences, the dearth of interdisciplinary approaches and the possibility of using the same mathematical models in diverse disciplines, while the need of holistic approaches resulted, according to Bertallanfy, from the frequent occurrence of teleological problems.

This last issue became a reason for a controversy. Strict sciences, accustomed to atomistic reduction (explanation of properties of a system by the properties of its parts), maintained that holism and synergy have no sense, since they must assume an intuitive perception of the whole, while the systems theory is allegedly[1] analytic in its nature (see e.g. Bunge 1977). At the same time, social sciences introduced the concept of *systems thinking* that highlighted the importance of systems complexity and holistic approaches, but questioned the arguments of Bertallanfy concerning the generality of mathematical modelling: they maintained (groundlessly, as it will be shown below) that the use of mathematical modelling in systems research cannot address systems complexity and is an expression of technocratic tendencies. According to the book *One-Dimensional Man* by Marcuse (1964) that had a tremendous impact on the social sciences at that time, technocratic thinking enslaved humanity; and the conclusion of *system thinkers* was that

---

[1] I have shown earlier in Chap. 5 on the rational evolutionary theory of intuition that all knowledge is based on intuitive elements, hence also the systems theory must use intuition, especially in relation to holistic approaches and the phenomenon of synergy.

they should exclude mathematical modelling. However, such a conclusion is inconsistent not only with classic arguments of Bertallanfy, but also with another fundamental assumption of *systems thinking:* the demand of *pluralism,* a postulate to accept diverse perspectives. Applying the principle of pluralism consequently, we should first of all check what mathematical modelling has to say about systems complexity (see e.g. Gutenbaum 2003), and it would turn out then that the problems of the theory of catastrophes, the theory of chaos, the emergence of order out of chaos, and the problems of emergence in general are based on mathematical modelling, as it will be shown below.

This controversy resulted in a conflict between *soft* and *hard systems analysis.* There are many aspects of this conflict (it suffices to search the Internet with the key phrase *hard and soft systems*), but I shall describe here only three issues.

The first issue is an epistemological dilemma emerging with the development of soft systems thinking, concerning the problem of *reality construction.* With the development of the general systems theory it became clear that systems research contributes, among others, to the construction of reality, see e.g. (Foerster 1973). However, the interpretation of this fact by engineers and postmodern sociologists became quite varied. Today, engineers use systemic mathematical modelling when constructing any new equipment and say that it is obvious: when we construct a new systemic model we create a new, virtual aspect of reality, but later we submit it to real testing and experimental falsification. Postmodern sociologists that question the very concept of objectivity and do not understand the falsificationist episteme of technology proper try to attach absolute attributes to the virtual reality constructed in computers.

The second issue are the attempts of social scientists working in the field of soft systems research to prove that social sciences created the systems theory independently of technology, see e.g. (Midgley 2003). In such perspective, the systems theory is supposed to develop from the theory of organizations called *tectology* by Bogdanov (1910) for whom a system was an *organized whole.* Then it is supposed that it was Norbert Wiener who invented feedback and that Jay Forrester invented systems dynamics. All this is maintained in order not to recognize any role of technology in development of the systems theory,[2] according to the hermeneutic perspective formed under the influence of Marcuse. The objections concerning Midgley's biased presentation are fobbed off by him with a statement that it is *winners who write the history* (Midgley 2003, p. xix) which is an argument not acceptable for the hermeneutic perspective of technology striving for objectivity. The belief of Midgley that the soft systems analysis has won results from its actual domination between sociologists, with absolute neglect or even conscious negligence of the fact that the hard systems analysis became an essential tool in all engineering construction, since computer simulation of models of various artefacts in *virtual laboratories* is today (see e.g. Makowski et al. 2007) an indispensable

---

[2] It should be recalled here that the first scientific journal which used the word *system* in its title was a technical telecommunications journal, *Bell System Technical Journal,* issued since 1923.

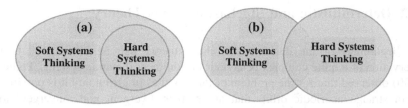

Fig. 10.1 **a** The relation of *soft systems thinking* and *hard systems thinking* according to Checkland (1978); **b** The same relation resulting from the diversification of episteme of different cultural spheres

early stage of any complex construction, preceding fairly more expensive real tests of such constructions. At present, we cannot construct, e.g. an airplane without computer simulation and optimization of mathematical models of its elements, wings, fuselage, even entire airplane; the same concerns construction of a car, home, bridge, etc. Therefore, it is not the soft systems analysis that has won, only the epistemai of social sciences and technology became very and disturbingly separated.

In relation to this victorious attitude of the soft systems analysis, we should consider yet another example, *Soft Systems Methodology (SSM)*, see e.g. (Checkland 1978, 1981). SSM stresses pluralism, admitting diverse perspectives called *Weltanschauungen, problem owners,* and more recently, *open discourse* representing these various perspectives. In fact, if we use the different perspective of hard mathematical modelling, SSM (if limited to its systemic core) must be recognized as an excellent approach, consistent with experiences resulting from the art of modelling of technical systems and described even earlier (see e.g. Wierzbicki 1977).

More doubts arise when we include the paradigmatic motivation of SSM. Checkland (1978, 1982) clearly stresses his motivation and belief in *enslaving, degrading and functionalist role of technological thinking and mathematical modelling.* This belief is clearly based on the slogans of Marcuse and leads Checkland to cultural imperialism: in his papers, he presents the relation between hard and soft systems thinking as depicted in Fig. 10.1a, while the principles of pluralism and open discourse would indicate rather the relation as depicted on Fig. 10.1b.

The fact that *hard systems thinking* has much to say on the issues of complexity, synergy and emergence, results from the history of quite different problems closely related to mathematical modelling of dynamic systems: catastrophe theory and deterministic, or more recently, stochastic chaos theory. The catastrophe theory explains the phenomenon of sudden change of the trajectory of an ostensibly continuous dynamic system in cases when the system has many equilibria or bifurcation elements (see Thom 1975; Gutenbaum 2003). The theory of deterministic and stochastic chaos will be discussed in more detail in the next section.

10.2 Deterministic and Stochastic Chaos Theory

The beginnings of the deterministic chaos theory can be dated back to the observation of a quite complex behaviour of nonlinear dynamic systems. Poincaré (1890) observed that for the problem of three bodies in astronomy it is possible to obtain strange aperiodic orbits that are neither divergent nor convergent to a periodic orbit or an equilibrium. Hadamard (1898) provided another example of a similar phenomenon. Hopf (1942) connected such behaviour with the phenomenon of bifurcation in nonlinear systems (which was later called catastrophe theory). In standard works on the chaos theory, written usually from the perspective of strict sciences, the contributions of technical sciences are often not highlighted. However, complex behaviour of nonlinear systems was a typical subject of interest of technical science, starting with (Van der Pol 1920; Van der Pol and Van der Mark 1927), through (Cartwright and Littlewood 1945) and many others (see Lucertini et al. 2004). In Poland, pioneering works in this field were carried out by Groszkowski (1947), and later, Jacek Kudrewicz (see the description in Kudrewicz 1996). Personally, I utilized atypical behaviour and aperiodic oscillations of nonlinear dynamic systems for achieving a desired even if unexpected properties of automatic controllers, first in a patent from 1961, then in a doctoral thesis based on that patent (see Wierzbicki 1963).

However in my opinion, two events are decisive for the development of the concept and theory of deterministic chaos. The first of them is usually not noticed in the works on the chaos theory, even by the excellent popularizing work of (Gleick 1987). It is the concept of a *pseudo-random number generator*. The first non-military broad application of a digital computer occurred in statistics (UNI-VAC in 1947), and soon it was necessary to simulate random processes with the use of a fully deterministic artefact, because such is a digital computer. As it is documented by a short paper of von Neumann (1951), an idea of using recursion (discrete time dynamics) of highly nonlinear transformations in order to produce a sequence of numbers with properties similar to random numbers[3] emerged already in that time. The simplest algorithmic prescription is as follows. Take a binary number with a large number of bits. Square it; the number of bits will increase twofold. Cut one quarter of bits at the beginning and one quarter of bits at the end of the number, thus reducing it to the original amount of bits. Repeat such a transformation arbitrarily long. It turns out that we obtain a periodic sequence in that way (after a very large number of repetitions, depending on the amount of bits used in the original number, the sequence will return to the origin), but inside a period it behaves like a random sequence. Since then, many other methods of generation of pseudo-random numbers have been developed, together with simulation of a given probability distribution, e.g. normal, or uniform on a given interval. A pseudo-random number generator is nothing else as a generator of deterministic chaos, but it was devised to solve an important practical problem

---

[3] Even if von Neumann, who was a probabilist, censured this idea as "sinful".

without theoretical basis, because the analysis of deterministic chaotic phenomena and the very name *deterministic chaos* starts with a very important but later work of Lorenz (1963); this is again an example how the praxis precedes the theory.

The second event decisive for development of the deterministic chaos theory is the work of Lorenz[4] mentioned above. Lorenz worked on the use of nonlinear and dynamic mathematical models for weather prediction. He observed that such meteorological models display "curious" behaviour: they are not only very sensitive to any change of initial conditions (he invented the metaphor *a swing of butterfly wings in Beijing can cause a hurricane in Florida*), but also, in certain conditions, they display aperiodic oscillations. The trajectories of such dynamic models in the so-called phase space (the state space of the model) move in a neighbourhood of a closed, periodic trajectory, but they are neither convergent to that trajectory nor they diverge away from it (in a normal behaviour of a dynamic system one of these two cases should occur). This phenomenon or rather a periodic trajectory with such neighbourhood was called by Lorenz a *strange attractor,* and the general behaviour of such (after all) deterministic system he called *chaotic behaviour.*

Almost concurrently,[5] Benoît Mandelbrot encountered *fractal phenomena* (see e.g. Mandelbrot 1963, 1982; Mandelbrot and Hudson 2005) in diverse sets of data, say, describing the prices on a speculative financial market, but also describing the geographical coordinates of the shore of Great Britain. Fractals mean self-similarity of data considered in different scales of analysis (similarly as a small part of a fern leaf is similar to the entire leaf). However, the adjective *fractal* has a mathematical sense and derives from observation that the so-called topological dimension (see e.g. Kuratowski 1965) of this type of data has a fractional value: the line describing the change of prices in time or the coordinates of a shore on a map is neither one-dimensional (a straight line) nor two-dimensional (a plane); it has a topological dimension between 1 and 2. A more complex, rugged shore has a higher topological dimension, e.g. the dimension of the shore of Great Britain is 1.3 while the dimension of the shore of Norway is 1.52. The phenomenon of self-similarity can be treated as an example of an order emerging out of deterministic chaos.

All this resulted in an enormous interest in the phenomena of deterministic chaos: we should mention here the work of Feigenbaum (1978) and an excellent,

---

[4] Before him, many others achieved similar results, using analog computers—myself in 1960 in Darmstadt, Yoshisuke Ueda in 1961 in Japan (see Abraham and Ueda 2001)—but we did not publish them sufficiently soon and we did not dare to call them deterministic chaos.

[5] Some authors—see e.g. (Kotyński 2000)—ascribe the authorship of deterministic chaos theory to Mandelbrot. However, in his very important work (1963) Benoît Mandelbrot described fractal phenomena without calling them deterministic chaos; he provided such an interpretation later, while Edward Lorenz (1963) was the first to use the term *deterministic chaos.* In a similar fashion I could say that my work on technical applications of chaotic behaviour preceded both the work of Lorenz and of Mandelbrot, but I must fairly admit that it would be only a delayed interpretation: in 1960, I did not use the concept of deterministic chaos and strange attractors—which I observed in experiments with an analog computer—I treated them only as a method to generate nice pictures with a computer. See also (Gleick 1987).

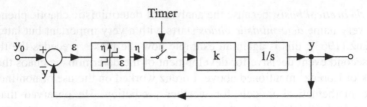

Fig. 10.2 Schema of a simple digital servomechanism

even if slightly one-sided, review of the problems and development of the deterministic chaos theory in Gleick (1987). A sign of the one-sidedness mentioned above is not only the fact that technological contributions to the chaos theory are omitted, but also the fact that the stochastic chaos theory which will be described subsequently in a further part of this section, is absolutely neglected there. But first, we should consider in more detail an excellent example of chaotic behaviour in a deterministic system, resulting from the technological research of Jan Konstanty Kurman (1975) on the dynamics of digital servomechanisms.

A simple digital servomechanism is used for positioning of various mechanical objects, e.g. the arm of a robot, on the basis of a digital measurement of location and with the use of a microprocessor controlling an appropriate engine that moves the object. Its simplified schema is presented in Fig. 10.2.

A step-wise nonlinear characteristics represents the process of transformation of an analog measurement into a digital one,[6] and the timer, the cadence of microprocessor work. The microprocessor must convert the measured error of positioning ε into a desired movement of the engine and does it, in the simplest case, by multiplying it by an amplification coefficient k. An integrating element with the transfer function[7] $1/s$ expresses the simplest model of the dynamics of the engine and the controlled object.

If the conversion of the positioning error into the desired movement of the engine was correct, which corresponds to the amplification coefficient $k \approx 1$, then the engine can achieve the desired position in one step (in the case of the simplest dynamics of the controlled object; with more complex models of controlled object dynamics the behaviour is also more complex, see Kurman 1975). In Fig. 10.3 such a course of systems behaviour is presented in part a. However, if the conversion is not precise and compensates the error excessively, which corresponds to the amplification coefficient $k \approx 2$, the servomechanism will possibly overshoot up to the next quantum of measurement.

---

[6] Actually, such a characteristics occurs in real systems both at the input and in the feedback loop, thus the system presented in Fig. 10.2 is a simplification, but equivalent to the real one concerning the properties of the closed-loop system.

[7] Recall that the transfer function of a linear dynamic element is the ratio of Laplace (or Fourier, in the case of spectral transfer function) transforms of its output and input signals, with zero initial conditions. If the output signal is an integral of the input signal, the transfer function is $1/s$.

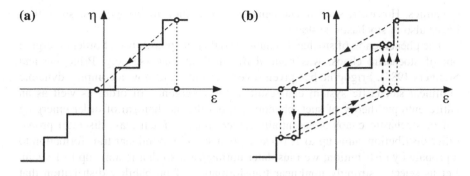

Fig. 10.3 Trajectories (shown by *dashed lines*) of the equilibration of an original error in a simple digital servomechanism: **a** in the case of correct assessment of the movement needed, $k \approx 1$; **b** in the case of excessive error compensation, $k \approx 2$

In such a situation, depending on micro displacements of initial conditions, invisible in macroscopic scale (contained within one quantum of measurement), aperiodic oscillations of the amplitude of one or two measurement quanta, shown in Fig. 10.3b, emerge in the servomechanism.

It should be added that a similar phenomenon occurs in iterative computational processes in digital computers, when a computational process is convergent in macro-scale and approaches a quantum of digital representation of real numbers, then we can often observe chaotic oscillations at the level of one or several quanta of accuracy.[8]

Thus, a deterministic digital servomechanism can generate chaotic oscillations, similarly as deterministic computations at the limit of digital representation of real numbers. More strange might appear the fact that from such chaotic behaviour, in certain conditions, a new order might emerge.

However, for a technician who uses the properties of nonlinear dynamic systems to obtain desired behaviour of a complex system, emergence of order out of deterministic chaos is nothing strange. In my own work on the dynamics of stepwise controllers with nonlinear feedback (Wierzbicki 1963), the behaviour of that controller on the detailed micro level corresponded to aperiodic oscillations; it was chaotic, but on the averaged macro level I obtained the desired behaviour of the controller in that manner, even if these properties did not result in a simple way from the properties of elements used in the construction of the controller. Thus, order emerged out of chaos. The phenomenon of order emerging out of deterministic chaos, or of new, qualitatively different properties of nonlinear dynamic systems with feedback was noted more recently by Maturana (1980) who developed a theory of *autopoiesis*, treated as a general property of nonlinear systems

---

[8] Experienced specialists in computational engineering know this and select appropriate criteria stopping the computational process, because an inadequacy of several quanta of digital representation is usually immaterial.

dynamics. However, such phenomena are not limited to *autopoiesis,* since they occur also in stochastic systems.

The phenomenon of stochastic chaos, or rather the possibility of order emerging out of stochastic chaos, was noticed by Ilia Prigogine (see e.g. Prigogine and Stengers 1984). Prigogine has given several examples of how a complex dynamic nonlinear stochastic system can result in the emergence of order as well as an entire entropic theory of such phenomena, but the mechanism of order emerging out of stochastic chaos is essentially rather simple.[9] Each (say, discrete) probability distribution sums up to 1. If we apply a strongly nonlinear transformation to a probability distribution, we must later normalize it so that it sums up to 1 again. Let us select a strongly nonlinear transformation of probability distribution that results in an increase of the largest probability (if several events have the same largest probability, we increase the probability of only one of them). Then, after normalization of the distribution, the probabilities of all other events must decrease. Let us repeat such a transformation indefinitely. Then, with appropriate assumptions about the speed of increase of the largest probability,[10] it will converge to 1, while all other probabilities will converge to zero, thus order will emerge out of chaos.

10.3 Complexity and Emergence

We see, therefore, that the emergence of order out of chaos is a fully rational phenomenon, fully substantiated mathematically, it occurs in sufficiently complex systems, e.g. nonlinear systems with recourse or with feedback, and it is, in a sense, a product of this complexity. We should add that this new order can have, as opposed to the simplest example with nonlinear recourse of probability distribution, essentially new, unpredicted or unexpected properties that do not result simply from the properties of elements of the system in which the emergence of new order occurs. This phenomenon was already discussed in earlier chapters and was called *emergence principle.* Here I recall its formulation, originally given in Wierzbicki and Nakamori (2006, 2007):

The Emergence Principle: *new properties of systems emerge with the increase of their degree of complexity; these properties are qualitatively different from the properties of parts of the systems and irreducible to them.*

It might seem that the emergence principle results form the general systems theory and is a natural addendum to the general systemic concepts of holism and synergy. Indeed, it is an addendum, but farther-reaching. The emergence principle

---

[9] We discovered this mechanism together with Yuri Ermoliev, an eminent probabilist from Kiev, when we listened to a lecture by Ilia Prigogine and tried to simplify his rather complex arguments.

[10] Example: it is sufficient to assume that the largest probability always increases by an amount that is greater than a given part of its distance to 1.

stresses that *the new properties emerging with the growth of systems complexity might be qualitatively different from and irreducible to the properties of parts of the system.* This is not a conclusion stemming from holism and synergy: *synergy and holism say that the whole is bigger than the sum of its parts or different than this sum, but they do not stress irreducibility. Thus, in a classical general systems thinking, the whole might be bigger, but is still reducible to its parts.*

The emergence principle is not a metaphysical or religious belief: it results not only from mathematics, but also from historical experience, including technical aspects, especially of informational technology. The historical process of spontaneous emergence of software from hardware was described in preceding chapters. Software cannot function without hardware, but it is irreducible to hardware; if we essentially change the hardware, e.g. introduce quantum computers, we will uphold software principles, maybe introducing some necessary modifications, even if only because the existing software tools resulted from many decades of work of many programming specialists. The phenomenon of emergence of software out of hardware was a spontaneous, unplanned effect of civilization evolution, simply, it was easier to further develop programming when it became separated from hardware.

Another example of, this time partly planned, emergence is the rise of the Internet as a complex of not only technical solutions but also commercial, service, network social aspects (it will be discussed in next chapter). In the history of informational science there are several other examples of either spontaneous or planned, goal-oriented emergence in order to conquer the complexity of technical systems.

An example of a development aimed to conquer complexity is development of the hierarchical systems theory, in particular in relation to multi-layered systems, where it is assumed (see e.g. Findeisen et al. 1980), in order to master the complexity of such systems, that subsequent, higher layers feature special tasks and functions, independent form the functions of lower layers, although assuming their correct functioning.

Another example is the layered structure of protocols of telecommunication and computer networks. Today, the Internet or generally any computer network is one of the most complex technical systems in the world (and even more complex if we take into account also its social functions). As it will be described in more detail in the next chapter, creators of the Internet deliberately separated, while predicting future complexity and trying to master it, the functions of that system into four, and afterwards even into seven fully independent layers. A higher layer simply assumes that the functions of a lower layer, e.g., of the layer of physical transmission of signals, will be, if possible, correctly executed, and the tasks of a higher layer are different and independent from the lower layer. Such a separation is only a model; in practice, functions of some layers relate to each other or are even combined.

Although the emergence principle does not require metaphysical substantiation, and it is a fully rational consequence of the theory and technology of chaos as well as historical development of technology and civilisation evolution, it might nevertheless have some metaphysical or epistemological consequences. For example, the emergence principle has certain impact on the metaphysics of the Absolute, since it negates the arguments of creationists who maintained that irreducible complexity could not spontaneously emerge in result of evolution, hence it was a result of an intelligent design of the universe. To the contrary, the civilisation evolution during last 50 years gives examples of spontaneous emergence of irreducible complexity, starting with the emergence of software from hardware.

In the perspective of epistemology, *the emergence principle is opposed to reductionism: not all can be reduced to atomistic properties of the parts of a system*, which is still a paradigmatic belief of many physicists. It should be stressed that the majority of strict and natural sciences, more paradigmatic than technical science, continues to favour reductionism. For example, physicists believe that quantum computations will essentially change computing, while, as already mentioned, they will essentially change hardware, whereas software develops to a large extent independently from hardware, in line with irreducible principles.

The conceptual difficulty related to the acceptance of the emergence principle is fundamental, since from the time of Aristotle (see e.g. Blandzi 2009), to understand something is usually treated as an ability to provide causal explanation, even if we use a broad meaning of the concept of a cause; however, such interpretation does not encompass essentially new phenomena, where the "cause" is a concatenation of circumstances in a complex dynamic process. Therefore, what was the "cause" of emergence of software out of hardware? Historically, the cause was constituted by the difficulty in managing the complexity of programming and the attempts of first computer programmers to nonetheless conquer the difficulty. Does this cause sufficiently explain all diverse properties of software? Not at all, we should rather go deeper into the specific historical process of the development of this new field, software, together with its emergent, irreducible properties. We could find in that way causal explanations of these properties, but they will be *ex post* explanations; their emergence we must simply accept.

References

Abraham, R.H., Ueda, Y. (eds.): The Chaos Avant-Garde: Memoirs of the Early Days of Chaos Theory. World Scientific Publishing Co, Singapore (2001)
August Comte: La mèthode positive en seize leçons (przekład Polski: Metoda pozytywna w 16 wykładach, PWN 1961, Kraków) (1830–1842)
Bertallanfy, L.: General systems theory. A new approach to unity of science. Hum. Biol. **23**, 303–361 (1951)
Bertallanfy, L.: General systems theory. Gen. Syst. **1**, 1–10 (1956)
Blandzi, S.: Arystotelesowska filozofia pierwsza a metafizyka (Aristotelian primal philosophy and metaphysics). In: Motycka, A. (red) Nauka a metafizyka (Science and Metaphysics), Wydawnictwo IFIS PAN, Warsaw (2009)
Bogdanov, A.: The historical necessity and the scientific feasibility of tektology. In: Dudley, P. (ed.) Boganovs Tektology, Centre of Systems Studies, Hull (1996, orgin. 1910)

Boulding, K.: General systems theory—the skeleton of science. Manage. Sci. **2**, 197–208 (1956)

Bunge, M.: General systems and holism. Gen. Syst. **22**, 87–90 (1977)

Cartwright, M.L., Littlewood, J.E.: On non-linear differential equations of the second order, I: The equation $y'' + k(1 - y^2)y' + y = b\lambda k\cos(\lambda t + a)$, k large. J. Lond. Math. Soc. **20**, 180–189 (1945)

Checkland, P.: The Origins and Nature of 'Hard' Systems Thinking. J. App. Syst. Anal. **5**, 99–110 (1978)

Checkland, P.: Systems Thinking, Systems Practice. John Wiley, Chichester (1982)

Dahan, A., Pestre, D.: Transferring formal and mathematical tools from war management to political, technological, and social intervention (1940–1960). In: Lucertini, et al. (eds.) Technological Concepts and Mathematical Models in the Evolution of Modern Engineering Systems, op.cit (2004)

Dantzig, G.B.: Linear Programming and Extensions. Princeton University Press, Princeton (1963)

Feigenbaum, M.: Quantitative universality for a class of nonlinear transformations. J. Stat. Phys. **19**(1), 25–52 (1978)

Findeisen, W., Bailey, F.N., Brdyś, M., Malinowski, K., Tatjewski, P., Woźniak, A.: Control and Coordination in Hierarchical Systems. Wiley, Chichester (1980)

Foerster, H.: On constructing a reality. In: Preiser, E. (ed.) Environmental Systems Research, Dowden. Hutchingson & Ross, Stroudberg (1973)

Gal, T., Stewart, J.T., Hanne, T.: Multicriteria Decision Making: Advances in MCDM Models, Algorithms, Theory and Applications. Kluwer, Boston-Dordrecht-London (1999)

Gleick, J.: Chaos: Making a New Science. Viking Penguin, New York (1987)

Groszkowski, J.: Generacja i stabilizacja częstotliwości (Generation and Stabilization of Frequency). WNT Warsaw (1947)

Gutenbaum, J.: Modelowanie Matematyczne Systemów (Mathematical Modelling of Systems). Akademicka Oficyna Wydawnicza EXIT, Warsaw (2003)

Hadamard, J.: Les surfaces à courbures opposées et leurs lignes géodesiques. Journal de Mathématiques Pures et Appliquées **4**, 27–73 (1898)

Hall, A.C.: The Analysis and Synthesis of Linear Servomechanisms. MIT Press, Cambridge, Mass (1943)

Hillier, F.S., Lieberman, G.J.: Introduction to Operations Research. McGraw-Hill, New York (2002)

Hopf, E.: Abzweigung einer periodischen Lösung von einer stationären Lösung eines differential System. Berichte Math.-Phys. Kl. Sächs. Acad. Wiss. Leipzig **94**, 1–22 (1942)

Kacprzyk, J.: Wieloetapowe sterowanie rozmyte (Multistage Fuzzy Control) WNT, Warsaw (2001)

Kantorowicz, L.V.: Matematiczeskije Metody Organizacji I Planirowanija Proizwodstwa. S.U. Press, Leningrad (1939)

Keeney, R., Raiffa, H.: Decisions with Multiple Objectives: Preferences and Value Tradeoffs. J. Wiley and Sons, New York (1976)

Koopmans, T.J.: Exchange Ratios between Cargoes on Various Routes; Memorandum for the Combined Shipping Adjustment Board. W Scientific Papers of T.J. Koopmans, Springer, New York 1970 (1942)

Kotyński, J.: Teoria chaosu a metody oceny ryzyka i analizy rynków kapitałowych (Chaos Theory and Methods of Assessing the Risk and Analysing Capital Markets. Zarządzanie Ryzykiem **1**, 72–90 (2000)

Kudrewicz, J.: Nieliniowe Obwody Elektryczne: Teoria I Symulacja Komputerowa (Nonlinear Electrical Systems: Theory and Computer Simulation). WNT, Warsaw (1996)

Kuratowski, K.: Wstęp Do Teorii Mnogości I Topologii (Introduction to Set Theory and Topology). PWN, Warsaw (1965)

Kurman, J.K.: Teoria Regulacji: Podstawy, Analiza, Projektowanie (Control Theory: Foundations, Analysis, Design). WNT, Warsaw (1975)

Lorenz, E.: Deterministic nonperiodic flow. J. Atmos. Sci. **20**, 130–141 (1963)

Lucertini, M., Gasca, A.M., Nicolo, F.: Technological Concepts and Mathematical Models in the Evolution of Modern Engineering Systems. Birkhauser, Basel (2004)

Makowski, M., Wierzbicki, A.P.: Virtual laboratories. In: Wierzbicki, A.P., Nakamori, Y. (eds.) Creative Environments, op.cit., pp 233–254 (2007)

Mandelbrot, B.B.: The variation of certain speculative prices. J. Bus. 36, 394–419 (1963)

Mandelbrot, B.B.: The Fractal Geometry of Nature. W.H. Freeman & Co, New York (1982)

Mandelbrot, B.B., Hudson, R.L.: The (Mis)Behaviour of Markets a Fractal View of Risk Ruin and Reward. Profile Books, London (2005)

Marchal, J.H.: On the concept of a system. Philos. Sci. 42, 448–468 (1975)

Marcuse, H.: One-Dimensional Man. Beacon Press, Boston (1964)

Maturana, H.: Biology of cognition. In: Maturana, H., Varela, F. (eds.) Autopoiesis and Cognition. Reidel, Dordrecht (1980)

Midgley, G.: Systems Thinking. Sage Publications, London (2003)

Oldenbourg, R.C., Sartorius, H.: Dynamik Selbstätiger Regelungen. München Verlag, München (1944)

Pawlak, Z.: Rough Sets—Theoretical Aspects of Reasoning About Data. Kluwer, Dordrecht (1991)

Poincaré, J.H.: Sur le problème des trois corps et les équations de la dynamique. Divergence des séries de M. Lindstedt. Acta Mathematica 13, 1–270 (1890)

Prigogine, I., Stengers, I.: Order Out of Chaos. Bantam, New York (1984)

Quade, E.S.: Analiza systemowa: możliwości i ograniczenia (Systems Analysis: Possibilities and Limitations). In: W. Findeisen (ed.) Analiza systemowa—podstawy i metodologia (Systems Analysis—Foundations and Methodology). PWN Warsaw (1985)

Słowiński, R.: Rough Set Approach to Decision Analysis. AI Expert 10, 18–25 (1995)

Smith, E.S.: Automatic Control Engineering. McGraw-Hill, New York (1942)

Thom, R.: Structural Stability and Morphogenesis. Benjamin, Massachusetts (1975)

Van der Pol B.: A theory of the amplitude of free and forced triode vibrations. Tijdschrift Ned. Radiogenoot 1, 3–31 (1920)

Van der Pol, B., van der Mark, J.: Frequency demultiplication. Nature 120, 363–364 (1927)

von Neumann, J.: Various techniques used in connection with random digits. In: Householder, A.S., Forsythe, G.E., Germond, H.H. (eds.) Monte Carlo Method, National Bureau of Standards Applied Mathematics Series, 12, pp. 36–38. U.S. Government Printing Office, Washington, D.C (1951)

von Neumann, J., Morgenstern, I.: Theory of Games and Economic Behavior. Princeton University Press, Princeton (1944)

Wiener, N.: Cybernetics or Control and Communication in the Animal and the Machine. MIT Press, Cambridge, Mass (1948)

Wierzbicki, A.P.: Zagadnienia dynamiki regulatorów krokowych (Problems of Dynamics of Step-wise Controllers). Pomiary, Automatyka, Kontrola 9(1), 12–26 (1963)

Wierzbicki, A.P.: Megatrends of information society and the emergence of knowledge science. In: Proceedings of the International Conference on Virtual Environments for Advanced Modeling, JAIST, Tatsunokuchi (2000)

Wierzbicki, A.P., Nakamori, Y.: Creative Space: Models of Creative Processes for the Knowledge Civilization Age. Springer Verlag, Berlin-Heidelberg (2006)

Wierzbicki, A.P., Nakamori, Y. (eds.): Creative Environments: Issues of Creativity Support for the Knowledge Civilization Age. Springer Verlag, Berlin-Heidelberg (2007)

Wierzbicki, A.P.: Modele i wrażliwość układów sterowania. WNT, Warsaw. English edition (1982) Models and Sensitivity of Control Systems, Elsevier-WNT (1977)

Zadeh, L.: Fuzzy Sets. Inf. Control 8, 338–353 (1965)

Chapter 11
Informational Revolution: Personal Computers and the Internet

11.1 Introduction: Important Caesurae and Events

Three main megatrends of the informational revolution were discussed in Chap. 2. Since more specific aspects of this revolution are discussed in the followings sections, here I would like to concentrate on caesurae and important events.

As described in the previous chapters, computer networks has started to be constructed already in the sixties and seventies of the twentieth century, but within a strictly classified project *ARPANET* in the USA. A ground-breaking point for their broader utilisation was the declassification of (a part of) this project in the year 1983. The *protocol TCP/IP*, devised in 1974, which will be shortly described in a further part of this chapter, was fundamental for the operation of these networks. Email was devised earlier, in 1972; but in the case of both of them, their full socio-economic use was enabled for the first time in 1983, together with the declassification of computer networks which were then called *Internet*.

Computer networks would not have their today's importance, if they had not coincided with marketing of the first popular *personal computer* Apple II by Apple Computers in 1977. To be more precise, the history is (as usual) more complicated. One of the previous chapters describes the history of computer miniaturisation and unsuccessful attempts to market small computers. Since Steve Jobs and Steve Wozniak, creators of Apple II, started with trying to sell a set of computer elements for individual assembly, called Apple I, in 1976, another firm surpassed them and introduced personal computer Commodore PET on the market in the beginning of 1977. Several months later, Apple II was introduced as a ready product, achieved a bigger success and dominated on the market for at least several years, until IBM decided to overcome its addiction to big computers for big business and introduced an IBM PC (denoted as IBM 5250) in 1981. Personally, I brought an early version of

© Springer International Publishing Switzerland 2015
A.P. Wierzbicki, *Techne_n: Elements of Recent History of Information Technologies
with Epistemological Conclusions*, Intelligent Systems Reference Library 71,
DOI 10.1007/978-3-319-09033-7_11

Apple II to Poland,[1] used it for many years at home and handed it over to the Museum of Technology in Warsaw. Personal computers started to be used broadly for many purposes, including research and computations, although text composition and edition, email, computer games, enlarged today by the Internet and all richness of network services, were their dominant functions.

The beginning of industrial civilisation is dated by historians to the year 1760, the approximate time of James Watt's improvement of steam engine (existing almost hundred years earlier, starting with Necoman invention) that made its broad socio-economic use possible. Therefore, *as a caesura of the beginning of informational revolution and the related new epoch I propose approximately the year 1980*, between 1977 and 1983, not the years 1931 or 1936, when analog and digital computers were, correspondingly, invented.

Today, about 30 years after 1980, the socio-economic effects of the growing informational revolution are already well visible and advanced, even in Poland. However, we can assume that today, about twofold acceleration of events occurs when compared to the times of Watt. Thus, the socio-economic penetration of informational techniques in the recent Poland should be compared with the socio-economic penetration of industrial technology (also in Poland) 60 years after Watt, in 1820. It was the time of Stanisław Staszic, a Polish pioneer of industrial technology; similarly to the opposition he encountered, the development of information society in Poland is for various reasons delayed, which will be discussed in the next chapters.

After 1980, an essential acceleration of changes took place on the world scale. It can be argued that also the dissolution of the so-called communist system (or Eastern bloc) was related to the informational revolution. Firstly, the dematerialization of work resulting from increasing automation, robotization and computerization of production brought about an invalidation of the slogan of the leading role of proletariat; and proletariat did not want to lose that role, which led to such movements as Solidarity in Poland. Secondly, president Ronald Reagan consciously used the pressure of high technology, including computer techniques, to exert a pressure on the countries of the Eastern bloc, which he directly stated in his speeches. Thirdly, political leaders of the Eastern bloc, both in the USSR and in Poland, were also aware that high computer technology requires a market and democratic system for the spread of information society, which has finally led to a relatively peaceful systemic transition.[2]

---

[1] First in 1985, even if I bought it and used in Austria earlier; after computer networks had been declassified, embargo on high technology products became less stringent, thus I could bring this computer to Poland.

[2] In that time, I was using both the example of Apple II computer brought by me to Poland and the arguments from the Tofflers book *The Third Wave* (1980); in this book, Tofflers actually predicted the fall of the communist system (although worded this prediction cautiously) to convince the political leaders of Poland about the necessity of a democratic turn. The necessity of a market turn was already known to them, but could not be implemented because of the embargo; the arguments of Tofflers were more convincing than mine, because nobody is a prophet in his own country.

The concept of *information society* was introduced in 1980 by the Japanese researcher Masuda (1980).[3] After a dozen years, Martin Bangemann presented to the European Commission a strategy of the development of information society, called Bangemann report (Bangemann 1994). The European Commission used this strategy as a basis for its developmental actions for many years, contributing to the spread of the informational revolution in Europe.[4]

11.2 Personal Computers

In 1977, 48 thousand personal computers were sold globally (mostly in the United States); in 2001, 125 millions personal computers were sold, and their aggregate number since 1977 amounted to one billion.[5] In the year 2007, 263 millions personal computers were sold and it was expected that until 2008 their aggregate number will reach two billions. During thirty years, their prices decreased almost tenfold (on average, to about 550 dollars), while their memory capacities and processor speeds increased several thousand times, thus the ratio of the offered quality to the price improved astoundingly.

Personal computers became a dominant tool of work and entertainment for people almost all around the world, at least in developed countries, but they gradually penetrate also developing countries. At work, still a majority of them (about 80 %) are *desktop computers* or *workstations,*[6] usually connected to a local computer network. Privately or at home, we more often use *laptops,* slowly replaced nowadays by *tablets.* However, because of a radio access to the Internet, even those mobile personal computers can be used as workstations, we can work also in travel, while maintaining connections to colleagues from the workplace. On the other hand, desktop computers have spread in households also as film browsers, called *home theatre PCs.* This trend of taking over the functions of television by a personal computer connected to the Internet represent a challenge for the producers of television sets and media distributors, and new versions of

---

[3] As usual, the authorship of this concept is contested. The formulation *information society* was used earlier (1963) by another Japanese, Tadao Umesao. See also (Zacher 2007; Wydro 2008).

[4] I participated in these developments personally, among others as a member of ISTAG (Information Society Technology Advisory Group) of the European Commission.

[5] See http://news.cnet.com/2100-1040-940713.html.

[6] The meaning of these concepts changes with time. A *working station* before 1980 was a terminal connecting an employee of a large company to the main computer of this company, called *mainframe.* After 1981, IBM PC were often used as working stations, and *desktop* meant that they were typically situated on a desk. To the contrary, a *laptop* described a mobile personal computer, with many other versions, such as *notebook, netbook, palmtop;* today, *tablets* are evidently the most popular mobile personal computers. At home, a *desktop* evolved to a *home theatre PC* with a large screen to watch films or play games.

television sets have also some functions of a personal computer or an Internet browser.

There is no need to describe computer architecture, as its detailed description can be found e.g. in Wikipedia (to paraphrase old saying: *what is a computer, anybody sees today*). From the perspective of a user, a review of typical applications of personal computers is more important. And here we encounter a difficulty: it is impossible to present typical applications shortly, because there are simply too many of them.

Today, computers are used in education, in various ways, starting with preparation of materials for lectures and exercises with the use of the information available on the Internet, up to electronic education with multimedia textbooks. They are used in creative work, starting with composition, edition and publishing of texts, to musical compositions or architectural design. Computers are used in medicine, not only to handle databases of patients, but also in diagnostics, in biomedical engineering to interpret diverse signals in computerised tomography, and even to read brain waves of paralysed people who can control a wheelchair or pronounce synthesised speech in that way. Computers are used in business management and organisation, starting with small enterprises, with supporting organization of commerce, accountancy of small shops, through accounting of supermarkets, up to electronic (networked) commerce on the Internet. In banking, computers are used universally.

This list is certainly not complete, since computers and the Internet are used today practically in all domains of life, and it is the essence of the informational revolution. For any such domain of applications, we can expect that a specific preparation of computer users is needed. Even in such simple application as the use of computers to support commerce, a special preparation of users is required: in using text editors, spreadsheets, examples of specialized computer software (e.g. in finance and accounting), preparation of documents related to sales, calculation of net, pre-tax and post-tax prices, finding information in various web sources, etc.

Even if we use computers at home, we must acquire specific expertise depending on a given field of use: we might need specific hardware, we must learn how to use this hardware and which software might be helpful. This concerns even the use of a personal computer in its classical application: as a text editor (for email, letters, research papers); but imagine a more contemporary application as an organizer of a home movie collection. We might require a specific hardware (large screen, integration of computer and television set), specific software (starting with a simple data base of movies that we bought on DVD discs, to a data base of all films stored in sufficiently enlarged computer memory), and for fans even dedicated software for editing one's own photos and films,[7] etc. It is important that the user of such collection must specialize in this field, learn the possibilities of

---

[7] After taking into account intellectual property rights, which is discussed in a separate point, we can imagine even a creative activity consisting in editing and creating new versions of classical films, naturally with references to the source material.

utilizing and enlarging their own hardware and software; it is not a simple issue. The same might concern every domain of home computer use, such as organization of shopping or entertainment or making financial investments via the Internet.[8]

For this reason, in times of the information revolution it is reasonable to postulate that all university curricula, not only technical ones, but also arts and humanities, social, medical studies, etc., included several technical courses, such as robotics and biomedical engineering (in order to make people oriented in the contemporary world), but above all informatics with applications connected with a given faculty, concentrated on the hardware and software typically used for the purposes of such studies. Even specialists in, say, Polish literature should learn advanced text editors and DTP software including, for example, a good understanding of the differences between LaTeX and MsWord. This requirement is clearly revolutionary, contrary to the accepted paradigms of humanities that required teaching of some humanistic subjects, e.g. philosophy, at technical studies, but did not found it proper to respond with reciprocity.

This requirement is the more substantiated that we can expect a further integration of computers, or microprocessors and devices similar to personal computers, with other artefacts of everyday use. An example is modern automatic washing machine which is actually a specialized robot controlled by a microprocessor; but we can expect a barrage of microprocessors in the ambient environment of human life, a development of the so-called *ambient intelligence,* or rather intelligent home, intelligent car, *Internet of things,* etc. We should clearly understand both chances and threats (e.g. of an excessive invigilation of people by technical systems) related to such a direction of development. And this direction is inevitable, since microprocessors enable integration of many artefacts: the above mentioned integration of home computer with large screen television, integration of *smartphone* with personal notebook, photo camera and mobile television with a small screen, etc. We should expect further development of such integration (e.g. of a car with a computer, mobile telephone and GPS satellite localization) that will open the way for *ambient intelligence.* People must be sufficiently educated to know which steps on this way are safe and which are dangerous, and require new safeguards (legal, technical, etc.).

11.3 Computer Networks, WWW, Pendrives

A short history of the ARPA project, ARPANET, the emergence of email and TCP/IP protocol was already presented in Chap. 9. Since the declassification of ARPANET in 1983, an eventful development of computer networks has occurred,

---

[8] One of examples of the actual improvement of the position of women, enabled by the computer use, is the success of a married Japanese woman (in Japan, married women usually stay at home) in financial investments via the Internet.

together with a specific class of services offered in these networks called World Wide Web (WWW).

First we should explain, however, why TCP/IP protocol was selected as the basis of the development of computer networks. Clearly, historical reasons were decisive (which supports the theses about *path dependence*, the dependence of socio-economic developments on the development path, see Arthur (1994)). The ARPA project was aimed at the protection of military telecommunication from a possible nuclear bomb attack. Therefore, ARPANET concentrated on packet transmission of information which was dispersed in a sense throughout the net-work, with an automatic search for a connecting route independently for each packet. Also, ARPANET emphasized the equal authorisations of network users, because after destruction of the command centre, each other user should be able to take up that function. An essential role in accepting these principles was played by a computer simulation of such a network with possible destruction of a part of it, performed by Baran (1960). In a sense, this remained in contradiction with the habits of classical telecommunication that used centralized management of net-work resources; the new principles were more democratic and egalitarian. Such situation had diverse consequences, and I shall mention here only two of them. Firstly, big telecommunication operators resisted to accept TCP/IP as a standard even a dozen or more years after 1982, when the protocol was accepted as a standard for computer networks; only nowadays, it is a standard for all electronic communication. Secondly, a specific, egalitarian subculture of Internet users emerged, stressing the equality of user rights and the freedom of access to information. After the commercialisation of the Internet, it has led to a conflict between the so-called *intellectual property rights* and the so-called *network piracy* that will be described in more detail in one of the next sections, and to further conflicts between the freedom of access to the Internet and diverse attempts of limiting this freedom; see also (Winston 1998).

However conceptually, the most important are *multilayered principles* of TCP/IP. The creators of TCP/IP, Robert E. Kahn and Vinton Cerf, who since 1973 has worked on this protocol, understood well that the complexity of the problem cannot be mastered unless they subdivide the issues related to functioning of computer networks into several layers of different type, with different functions and objectives. Originally, they selected four layers (counting it bottom-up):

- *The layer of network access,* called also physical layer, including the operations of modems and network cards realizing actual connections to the network;
- *Internet layer* (or Internet protocol layer (IP)) that transforms datagrams with Internet IP addresses and determines routing, a tentative selection of a route to the destination computer in the network;
- *Transport layer* is responsible for the robustness of data transmission, sends appropriate information to selected applications using appropriate ports;
- *Application layer* includes a collection of protocols related to applications, software systems in this uppermost layer that execute the actual services for the user, such as the software for email service, for network search engines, etc.

In the next years, inspired by the example of TCP/IP, the International Standard Organisation (ISO) worked on the standards of models of telecommunication networks (Open Systems Interconnection (OSI)), in order to enable connections between diverse networks. In 1984, ISO extended the multilayered model of TCP/IP to seven OSI layers:

- Physical layer;
- Link layer;
- Network layer;
- Transport layer;
- Session layer;
- Presentation layer;
- Application layer.

Without describing the functions of these layers in detail (see e.g. Jankowski et al. (2003), or on the Internet under the key phrase *OSI 7 Layers Reference Model For Network Communication*), we should stress the groundbreaking importance of such an approach. It rejects the atomistic attitude of strict sciences by assuming that a programmer for an upper layer should forget the details of operation of lower layers, while assuming only that they should function approximately correctly, and concentrate on the goals and functions of the layer in concern. This is an example of a planned use of the emergence principle, discussed in Chap. 10, but we must remember that this principle was formulated later, as praxis often precedes theory, especially in technology.

The next great turning point was the emergence of World Wide Web (WWW), based on the concept of *hypertext*, i.e. text composed of independent parts called *nodes*, interconnected by *hyperlinks* (such places that a reference to them, a click, results in a transfer to another node). The idea comes from Vannevar Bush, the constructor of the first analog computer, who in the paper *As We May Think* (1945) suggested such organization of information in order to achieve its more efficient employment by people and machines. Douglas Englebart, the constructor of a computer mouse, also worked on this idea, but the term *hypertext* was proposed by Theodor H. Nelson in 1965, while the practical application of these concepts[9] occurred in the works of Timothy Berners-Lee, prepared with his co-workers from CERN (a centre of research on particle physics in Switzerland) who in the years 1990–1992 implemented the hypertext idea in the HTTP protocol created especially for this purpose and the HTML language. On this basis, in 1992, CERN[10] activated the first WWW server, and in 1993, the first graphical search engine

---

[9] As usual, we encounter a dispute here as to who was the author of the idea. Timothy Berners-Lee maintains that he created WWW independently, even if the basic ideas were actually much older. Theodor Nelson criticises the HTML language saying that it uses unacceptable simplifications that he wanted to avoid. Vannevar Bush did not use the terms *hypertext* or *hyperlink*, even if he described a corresponding organization of information.

[10] If we use the example of hypertext to count the delay between the concept and its broad social use, the delay amounts to 47 years.

Mosaica became available, and the easiness of access to information from the entire world by one click secured a very fast development of WWW network.

Starting with 1992, when the number of Internet servers exceeded one million, a very fast development of events belonging to the informational revolution can be observed. In 1993, Timothy O'Reilly created Global Network Navigator, the first commercial WWW portal. In 1994, the number of WWW portals and pages exceeded 10 thousand, and within 2 years, until 1996, 100 thousand. In 1994, Bell Laboratories demonstrated the first mobile access to the Internet and, reciprocally, the first Internet radio, InterOp, started transmissions in Las Vegas. In 1996, the number of emails sent exceeded the number of traditional post letters (Norman 2005). In May of 1997, *Deep Blue*, a supercomputer constructed by IBM, won in tournament conditions with the current world champion in chess, Garry Kasparov.[11] In 1998, Google Corporation was founded, a firm specializing in the search for content of the Internet. In 1999, the Supreme Court of USA declared Internet addresses (the names of domains) to be property objects. In 2000, the number of WWW pages and portals exceeded 25 million. In 2001 (in English language) *Wikipedia, the Free Encyclopedia* started its functioning as a project ensuring free access to knowledge, based on a common efforts and end editing of all users, with all deficiencies of such solution (primarily, an insufficient control of the objectivity of the entries), but also with a great advantage of free access from any point on the globe where the Internet extends.

In the years 1998–2000, Trek, IBM, SanDisk (former M-Systems) and several others companies developed first *pendrives* (sometimes imprecisely called USB plugs) that revolutionized personal storage of data, texts, video items, software programs, etc., replacing floppy disks and DVDs. Pendrive plug is an integrated electronic memory comprised of large numbers of NAND gates (described in Chap. 9), exceeding currently hundreds of Gigabits and contained in a plug of the size of a small pen. USB is the name of a standard of connections, common for many appliances (computers, telephones, cameras, video recorders, etc.). The invention of a dynamic memory, transitory but preserving its state long after switching off the power supply, based on small capacitances between the electrodes of transistors in a NOR or NAND gate, is attributed[12] to Fujio Matsuoka from Toshiba (in 1984); but Toshiba did not market this invention. A dozen years later a group of young researchers from Singapore tried to introduce pendrive plugs on the market, but the crucial factor was market competition that developed with some additional delay: when Trek from Singapore marketed a pendrive plug called *Thumb Drive* in 2000, IBM and other companies quickly responded and a fierce competition started between American and Far East companies. Even in the

---

[11] We should note here that in (1958), Simon and Newell predicted that in 10 (!) years a computer will become a chess world champion, but it actually happened 40 years later. On the other hand, in 1986, Dreyfus brothers from Berkeley University suggested (in their excellent book *Mind over Machine,* compare Chap. 5) that such an event is not probable.

[12] The idea is actually much older, since the same principle has been used in integrating elements in analog computers since 1931, see Chap. 8.

case of an invention responding to an universal and obvious social need, the pure delay time (counted from 1984; from 1931 it would be much longer) extended to 16 years. The total time of civilization delay, taking into account the dynamics of social penetration of pendrives (see Chap. 4), that is, the time needed until an average user in the world will have a pendrive, will probably amount to at least 40 years.

In the years 2000–2001, a "burst of Internet bubble" took place, that is a collapse of stock market investments in excessively priced shares of Internet-related companies. This was only an relatively small economic crisis (the great financial and economic crisis of 2007–2009 was much bigger) that resulted from irrational behaviour of stock market players who did not evaluate the actual techno-economic opportunities of companies, and were satisfied with the word "Internet".

The pace of development in the next decade was similarly fast, starting with the publication of the first results of the Human Genome Project, utilizing advanced computer technology to create a complete map of human genome, in 2001. Afterwards, the time has come for *grid computing,* a type of processing with the use of many computers connected into a network to solve advanced research problems, and then for *cloud computing,* using network computing of unspecified localization, etc. An important development was constituted by new approaches to distance learning via the Internet. In 2000, MIT launched the *OpenCourseWare* project of free access to multimedia courses offered on the web. In 2006, this has led to the idea of *Open Access,* free availability of the results of research projects financed with federal funds in the USA. Today, the applications of computer technology and Internet access influence all domains of life, as indicated in former sections. However, especially important is their impact on mobile telephony.

11.4   Mobile Telephony

The trend towards constructing a fully mobile telephone was already described in Chap. 7: it is an old trend, radiotelephony can be counted from the beginnings of radio, and cellular telephony from 1943, but mass social penetration of mobile telephony started in the beginnings of 1990s.[13] Even in the 1970s and 1980s, mobile telephones were called "brick telephones" or "car telephones", since they weighted over one kilogram and required a strong case to be carried. Such mobile telephones were included to the so-called *zero generation* or to the *first generation* initiated in 1973 by Motorola, but with a telephone whose weight quickly exhausted the user.

---

[13] The delay between the original concept and mass social penetration, depending on how we count it, amounts in this case to 50–90 years.

Really miniaturized mobile telephones appeared in the *second generation* (2G), originated in 1991 by Radiolija, a company based in Finland and using the GSM standard (Global System of Mobile Telecommunications).[14] Radiolinja was also a pioneer of various mobile services. For example, it offered the choice of melodies indicating an incoming mobile call to the users. But Radiolinja was soon overtaken by another Finnish company using GSM standard, Nokia (as from 1992). Other companies, such as Ericsson, Motorola, Phillips, Siemens, were delayed in comparison to Nokia, even if e.g. Motorola held important patents for diverse aspects of mobile telephony. The reason for the delay was mostly related to standards: various companies wanted to keep their own standards in order to preserve their monopolistic or oligopolistic positions on the market, while the acceptance of a common GSM standard accelerated the development and opened the market. As a result, Finland, and then Sweden, became a leader in the development of mobile telephony, even if the first cellular networks of *first generation*[15] were created earlier, in 1979 in Tokyo and in 1983 in Chicago.

Because of the great market success of mobile telephony of the *second generation*, the slogan of *third generation* was advanced already in the middle of 1990-ties: third generation means integration of the functions of a telephone with many other services and appliances, such as a photo camera, mobile television, radio access to the Internet, personal notebook, elements of artificial intelligence called generally a *smartphone,* in various physical forms such as *tablets* (mobile phones integrated with personal computers), etc. While the development in these directions has been remarkable, standards of the third generation have not been well defined, they are left to market competition. This has led to several competing, sometimes not fully compatible standards, later called networks of *2.5* or *2.75 generation.* The first of such networks was DoCoMo of NTT from Tokyo, created in 2001 with the use of WDCMA standard, and in 2002, also SK Telecom and KTF networks in South Korea, and the Monet network in the USA, based on different standards, as well as two networks in Europe, in Italy and Great Britain, using WDCMA. Until 2007, networks of the "third" generation supported 9 % of mobile telephony in the world, while their share in Japan or South Korea was much larger. Fourth generation networks are a subject of research. But this process would be much faster and more efficient, if a global common standard were accepted. The markets of mobile telephony confirm the theoretical thesis of Brian Arthur (1994) that standards are not selected optimally by the market, they are *path dependent* (historically determined)

---

[14] The GSM standard was eventually, if slowly accepted in Europe, while in the USA a fight developed between the companies preferring two similar CDMA and TDMA standards, which delayed the development of 2G mobile telephony.

[15] The generations of mobile telephony differ mostly in the way they use radio spectrum and in the services they offer, but also in their standards and technical details that are not discussed here. In 1G, the analog standards NMT, AMPS were applied; 2G used mostly GSM with diverse bands of radio spectrum; 3G uses, e.g., IMT-2000, UMTS; 4G is mostly under development. There are also differences in the ways of separation of channels, such as Wireless Code Division Multiple Access (WCDMA), or Time Division Multiple Access (TDMA), TD-CDMA, FDMA.

and the process of market selection of standards is costly and lengthy. The resistance against a global common standard results from striving for monopoly or at least oligopoly on this very profitable market: a common standard opens the market, making it easier for smaller companies to enter.

11.5 Internet and Network Services

The Internet evolves: it was different before WWW, and today researchers speak about the *third semantic generation* of the Internet (even if this is rather a promotional slogan than actuality, see next section), the *fourth generation* is expected to encompass sensor networks with agent software features, etc. However, this section concentrates on the services existing on the Internet today.

The Internet, taken together with the global WWW network and mobile access to the web, became the largest set of information and services in the history of mankind. Its users can communicate, by email, *Voice over IP* services, videophony (e.g. *Skype*), video conferencing, etc.; they can cooperate, do research either individually (starting with searching through this tremendous set of information) or in groups, using diverse methods of communication. And this is only a small part of services offered by the Internet; others include e.g. electronic commerce, electronic distance learning, etc. Network services of the Internet will be described in more detail later. Here, it should be stressed that because of the tremendous socio-economic importance of these services the understanding of the word *Internet* might today concern at least three dimensions:

(1) *Technical dimension of the Internet* is constituted by complex teleinformatic networks based on TCP/IP protocol, together with data resources indexed in line with hypertext principles;
(2) *Service dimension of the Internet* is the set of resources and services offered on the web;
(3) *Social dimension of the Internet* is constituted by the community that uses and develops these services further, often for free (as commented above), or also for remittance on commercial principles.

In the future, other dimensions of the Internet might also become important, such as the educational dimension. However at this point, mainly service dimension shall be discussed in order to illustrate the vastness of this dimension, in which we can distinguish the following:

Services consisting in access to the Internet. Access to the Internet can be of diverse type, starting with a classical access through a commuted link with a modem and connection to a telephone line, through a fixed access in the SDI standard with relatively low parameters of the access speed, then broadband access in the xDSL standard (x digital subscriber line, where x denotes a specific type, e.g. ADSL with a standing for asymmetry, used among others as part of *Neostrada* in Poland), and radio broadband access. For all these services, maximal access speeds in kilobits or megabits per second (for an asymmetric access they are different for the web traffic to the user and from the user) and, obviously, prices, are the basic parameters.

Search services. There are many search engines available on the Internet, such as Google. We should realize, however, that they are, on the one hand, very efficient and co-define the usefulness of the Internet (because what the vastness of information is worth if we cannot find what we search for?), but on the other hand they are far from perfection from the point of view of the user. The needs of a user are diverse: they are different for a researcher who checks what people have written on a given theme, different for a learning user, different for a consumer, different for a producer. On the other hand, the construction of search engines is focused on commerce, they are a source of income coming from advertisement and information on the preferences of an average or a specific user, useful e.g. for electronic commerce organizations. Under the pressure of various users and the threat of competition, the developers of search engines gradually (even if slowly, because it is also an oligopolistic market with limited competition) respond to the needs of the users and make it possible to make more complex requests. The issue of search techniques and interaction with the searching user are related to the problems of ontological engineering and semantic networks, commented in the next section.

Telematic video-text services. According to the standards of ITU, the International Telecommunication Union, this name (not very adequately selected[16]) denotes services requiring transmission of a fixed text or picture. There are many variants of such services: the proper video-text (teletext), telefax, email, many aspects of electronic commerce including tele-purchase, electronic transmission of documents (electronic data interchange (EDI), with the transmission of invoices, agreements etc. with a secure electronic signature), distance education services (e-learning, even if in this type of services, multimedia audiovisual elements are increasingly important and it should be rather treated separately), etc. Each of these services has a rich internal structure, for instance, electronic commerce requires a specific support for electronic transactions, needs a distinction of B2C (business to customer) and B2B (business to business), needs a service support for Internet shops, etc. Tele-purchase services support an individual customer in selecting between various electronic commerce offers.

Multimedia services. These services require a joint transmission of at least two media of communication, usually voice and image. This services include: the abovementioned video-phony (stationary or mobile with radio access), video-conferences of various types (e.g. audio-graphic conference where the voice communication is enriched with pictures such as in MS PowerPoint presentation), video-text enriched by sound, etc. Television services are not included here, they form a separate group.

---

[16] *Telematics* has a slightly different and broader meaning in the field of automatic control, which is closely related to telecommunication, but nevertheless different. In automatic control, telematics means a distant control which requires a broader range of *real time data* transmission techniques than only transmission of a fixed text or picture.

Television services. They form a separate group because of high transmission speed requirements (or requirements related to the breadth of the transmission band). They can be *distributive*, such as satellite television, or *interactive*, with a support for a search of an appropriate channel or content. These services are becoming included in mobile telephony of the third generation, with a further integration of a mobile phone and a television set (as well as other functions, photo camera or personal notebook with email, etc.).

Distance education services. These services became very diversified internally, with the distinction of *distance learning* based on access to remote educational texts or even multimedia lectures, *distance sessions* enabling communication between listeners and the teacher (here multimedia communication is desirable, since such sessions can also support consultations or distance examining), and also a participation of students in a *virtual laboratory*. There are also virtual environments for group work, virtual groups or classes, as well as services of a virtual deans office with the registration of schoolwork, listeners, teachers, presentation of requirements for participation and grading, support for creation of plans and schedules of schoolwork, libraries of teaching materials, registration of partial and final grades, statistical analysis of results, etc.; the possibilities are very rich.

Even if the selected network services are described above very shortly, and there are many others, such as games and gaming services, it should be clear after this short review that the richness and plurality of Internet services will grow in the future. It is important, however, that they should be developed not only with a view to *artificial intelligence* contained in them, but first and foremost with the focus on the human user and her/his sovereignty, they should be *human centred*.

11.6   Artificial Intelligence, Cognitivism

As mentioned in former chapters, the concept of *artificial intelligence* emerged in relation to robotics, but its slogan soon achieved its own dynamics.

From the first seminar on artificial intelligence in 1956 in Dartmouth College, a strong group of dedicated researchers formed, despite a considerable scepticism on the part of other fields of technology and science, even in information sciences. It promoted artificial intelligence in various ways, usually anticipating a very fast development in this field. I already mentioned excessive expectations or opposite opinions concerning computerized chess playing. Similarly, in 1970, Marvin Minsky (Minsky 1970) maintained that in 3–8 years a computer of intelligence equal to an average human would be constructed. All these erroneous predictions can be explained by erroneous definitions of intelligence assumed in early works on artificial intelligence. Simply, *human intelligence is a phenomenon much more complex than it originally appeared.*

The history of such misunderstandings should be started with a fundamental question: *does human mind or brain, starting with a single neuron, work on the basis of classical binary logic?* The researchers of artificial intelligence assumed, following the work of McCollough and Pitts (1943), and they still believe in it in majority, that a positive answer to this question is correct. Since neurons communicate by electric impulses and we can distinguish two states of a neuron (sleep and excitation), then we can assume that a neuron is a biological equivalent of a flip-flop device described in Chap. 9. Since neurons receive signals from many other neurons, we should use flip-flop devices of many input signals and construct a model of a network of such devices. Such reasoning resulted in a concept of a *perceptron* and *artificial neural networks;* I prefer to call the latter *neuron-like networks,* since the initial assumption of the simplicity of a human neuron turned out to be false in view of more detailed research. A perceptron uses many input signals, transforms them (using e.g. a weighted sum, and while modifying these weighting coefficients, it can also learn), and later uses an appropriate function (usually the so-called sigmoidal function) to define when the logical state of the model of a neuron should be switched. The first implementation of such neuron-like network, SNARC, was achieved by Minsky and Edmonds in the year 1951. But it was also Marvin Minsky (1986) who provided a fundamental critique of the idea of a perceptron.

The research on the functioning of a human neuron shows, on the other hand, that it is not a static logical system, but a complex dynamic nonlinear system that, when in the state of excitation, generates a sequence of output signals and achieves the state of excitation not immediately, but after an accumulation of input signals. A model of such a system is thus much more complex than a perceptron or sigmoidal function; it is not reducible to binary logic. A full simulation of the operation of one neuron in real time would require the computational power of approximately today's typical personal computer. Since there are several billions of such computers on the globe, even if we connect them all in a network (with a full knowledge of how diverse are ways in which neurons work in human brain), we would be still lacking several orders of magnitude (circa 10^9 computers, around 10^{11}–10^{12} neurons in human body) to simulate in such a network the neural system of a man. Thus, neuron-similar networks, still used and developed, are only an exceedingly simplified model of our neural system and the human brain is not reducible to a digital computer with strictly logical structure.

Therefore, much depends on the definition of artificial intelligence. The earliest proposal of such definition is the so-called *Turing test* (Turing 1950). This test assumes contact with a machine through an artificial communication channel (Turing assumed communication via a telex) and declares the machine to be intelligent if a human on the other end of the communication channel cannot distinguish its responses from those of another human. However, this test is artificial because it assumes artificiality of the communication channel: even if it would be an audiovisual, virtual communication, a true test would require a virtual face of the machine and face-to-face communication in order to include the body language. Moreover, Turing test was soon shown to be unreliable, since in the end of 1960s Joseph Weizenbaum (see e.g. Weizenbaum 1976) constructed a program, ELIZA, that could converse on the level of not too intelligent teenager by repeating parts of the questions while using some grammatical rules to convert sentences.

Thus, the history of development of artificial intelligence is a history of learning by mistakes. It was also related with the so-called *summers and winters of artificial intelligence,* a cyclical succession of periods of popularity and sound financing of the studies on artificial intelligence followed by periods of disappointment with artificial intelligence resulting in diminishing finances for the research.

The first *summer of artificial intelligence* came in the period 1958–1970, when the researchers of this domain were financed without limitations, promised much, and delivered not so much: perceptrons, development of foundations of computer logics with a special Prolog language, beginnings of automatic learning. In 1970, governmental agencies of the USA and Great Britain concluded that financing of artificial intelligence research was not effective, which limited research in this field to a small group of enthusiasts. A *winter of artificial intelligence* continued until 1980, when great corporations became convinced about the possibilities of *expert systems* (developed already by Edward Feigenbaum) in specific domains of application and started to finance this variant of research on artificial intelligence. This was actually a change of definition of artificial intelligence, since an expert system means a summary of human expert knowledge in a specific field with the use of logical rules (afterwards, also other types of models were included, see further comments). Concurrently, in Japan, a great governmental program was launched aiming to construct computers of a next generation and based on the assumption that their equipment would contain elements of artificial intelligence. The development of neuron-like networks was continued, together with new methods of their learning; this has lead to the development of *cognitivism,* discussed below. The first big research on semantic networks and ontological engineering was also started, including a great CYC project aimed at summarizing all knowledge represented in the Internet with these tools.

But again, promises were much greater than results: the Japanese program brought a limited outcome (e.g. in the field of combining syllables of *hiragana* alphabet to represent *kanji* symbolic letters) and expert systems turned out to be unreliable, requiring continuous participation of human experts to interpret new atypical data. In result, the *summer of artificial intelligence* lasted only in the years 1980–1987, and was followed by a *winter of artificial intelligence* in the years 1987–1993 resulting from the disappointment of large corporations with the effectiveness of expert systems. However, computer capabilities developed according to Moore law (see Chap. 9) and the informational revolution has started; thus artificial intelligence began to bring more substantial effects after 1993, including intelligent support for text edition, beginnings of automatic translation of texts in different languages, using *intelligent programming agents* in diverse functions, the success of *Deep Blue* computer in chess playing in 1997, etc. But the problem, what is natural and what artificial intelligence, remained unsolved.

Early definitions of artificial intelligence reduced it to the ability of verbal communication, because it seemed at that time that all human knowledge reduces to words, that mathematics can be reduced to logics, that classical logics is sufficient to describe the real world (also today many researchers are convinced about that). When it comes to words, it is sufficient to construct a machine that thinks logically to declare that it is intelligent. But each computer from its nature thinks logically, when transforming numbers and words. Thus, perhaps each computer is intelligent? Obviously, such a definition of intelligence is not sufficient.

Nonetheless, intelligence understood as logics in processing words dominated in at least one, humanistic-philosophical aspect of artificial intelligence, the so-called *cognitivism*. Many philosophers and researchers concentrated on the research of language structures, starting with *structuralism* and *post-structuralism* (see e.g. Derrida 1974; Chomsky 1986), through the concepts of the *language of thought* (e.g. Fodor 1994) or *universal grammar* (Chomsky 1986). Some philosophers, e.g. (Dreyfus 1972; Dreyfus and Dreyfus 1986) or (Searle 1992), criticised such approaches, but nevertheless a specific thinking current developed, called *cognitivism*, based on a (conscious or subconscious) comparison of human brain and mind, or entire world (or even God) to a great computer, see e.g. Pylyshyn (1984), Gardner (1985), Jacyna-Onyszkiewicz (2009).

As part of its further development it turned out, however, that cognitivism in such understanding is too shallow: specialized research on brain functions shows that brain works differently than a digital computer, at least in an analog-digital fashion, as follows e.g. from the above comments on the functioning of a single neuron. Similarly as neuron-like networks with their primitive model of a neuron are a greatly simplified models, the general comparison of a brain to a digital computer is highly unreliable. This does not mean that cognitivism has failed completely: it stimulated *cognitive research*, meaning empirical studies of diverse perceptual and cognitive aspects of brain, but the results of these studies negate the classical assumptions of cognitivism.

Another but related aspect is the search for better (even if only partial) definitions of intelligence. One of such definitions is the statement that *an intelligent being must have at least the ability to learn*. This is a correct, even if not exhaustive definition of intelligence, but again, we can provide various definitions of learning. An evolutionary definition generally accepted today is that learning is a modification of a model of environment helpful in survival. As mentioned in Chap. 8, one of the early definitions of learning as an aspect of intelligence was given by Alexander A. Feldbaum (Feldbaum 1965) who treated learning as an ability of an automatic controller in a feedback system to utilize errors (deviations from optimal functioning) to correct the model of the controlled plant (a model of reality or its relevant part in the environment of the learning subject). Further definitions of learning reduced it sometimes to logical formulae: in a large data set used for learning we should find interesting logical formulae, and their application to new data will confirm or contradict the effectiveness of learning. Yet another, slightly different definition of learning is related to the concept of *learning organization* (Senge 1990), where a holistic, systemic nature of learning is stressed.

The definition of learning through an adaptation of logical formulae is in fact very close to cognitivism and classical artificial intelligence, and it is the main reason of successes of the latter. This definition is related to the issue of dividing a large data set into smaller subsets or classes related to distinct logical formulae; as tools of such division, computational optimization or neuron-like networks are used. On such basis, tools for translating texts from one language into another emerged (still far from perfection), or for checking spelling in text editors (often frustrating users of such editors, if the automation of such tools is excessive), and finally, very useful tools for finding hidden and tacit knowledge in large data sets (*data mining*) or large text repositories (*text mining*). The latter applications required a broader set of tools including diverse software for statistical reasoning, optimization, multiple criteria decision making etc. (see e.g. Granat and Wierzbicki 2009). In actual applications it turned out that in order to answer the question what logical formulae (or statistical or other decision models) we should characterize as really useful and valuable for the user, *the crucial factor in data mining is the participation of the user of a computer system and the tacit, intuitive and emotional knowledge of this user.*

Similar conclusions concern *text mining* in large text repositories. Although in order to browse the Internet, highly advanced *search engines* are necessary and there is a highly advanced theory of *semantic networks* and *ontological engineering*, they nevertheless serve in a sense their own goals. For example, search engines supply free information to the Internet users because they reciprocally collect information about the Internet users and use this information for various commercial purposes, starting with focused advertising. Thus, they are by no means constructed to serve the Internet users in a best possible way. Semantic networks and ontological engineering concentrate on determining the structure of the most often used concepts and relations between them (in large text repositories or even in entire Internet). These structures and relations are technically called *ontologies* (they are of course virtual digital beings, thus such technical vocabulary is not quite contradictory to the concept of ontology as a theory of being). However, the problem of usability of such structures and relations for a specific user with her/his own specific tacit knowledge emerges here again: an automatically formed ontology might be useless for a specific user and purpose.

Ontological engineering is focused on an automatic, or partly automatic, allowing for an interaction with experts and users (which usually gives better results), construction of an ontology based on large text repositories. Semantic networks in the CYC project mentioned above set an ambitious and laudable, even if rather utopian goal to create a general ontology and semantic interpretation of all the information contained on the Internet. The question is, however, whether such a general ontology will be useful for a specific user.[17]

All this leads to a conclusion that the classical concept of artificial intelligence should be replaced today by the concept of *knowledge engineering*, broadly understood, including not only creation and usage of tools of classical artificial intelligence, but also other tools whose aim is to bring out *tacit knowledge hidden in large data sets and text repositories*. I am using the term *tacit knowledge* here

---

[17] If we apply tools of automatic ontological engineering to a large repository of texts from telecommunications, we will obtain a statement that *a network consists of nodes and connections between them*, true and obvious for any telecommunication specialists, but how can it help a specific user?

consciously in the sense of *preverbal knowledge* since even in a large text repository we might find thoughts not expressed directly in words and requiring an interpretation according to an internal, preverbal knowledge of an user (see Chaps. 5 and 12). This is also related to the aspect of *human centred computing*, human user focus.

11.7 Human Centred Computing, Human User Focus; Knowledge Engineering

After *Deep Blue* won with the chess world master in 1997, a reflection followed: what's the big deal? The concentration of attention of computer scientists on artificial intelligence was aimed at creating a situation where a computer would be more intelligent than a human, and it was achieved after many, over 40, years of work, but only for a logically reducible game such as chess. Does it imply that such intelligent computers can better serve humans? Do we want at all the computers to become more independent and creative than us? Is the definition of artificial intelligence correct, can computers be creative?

Such doubts coincided in time with the first signs of *computer domination over humans*. This is not yet an explicit domination, but the domination of conviction of computer programmers that human users of computers should behave only as imagined and coded by them, enforcing certain behaviour of computer users. The simplest examples in text editors include an automatic change to a capital letter after a dot or an automatic correction of the spelling of a word that the text editor dislikes without asking the user for an approval. Clearly, the user can correct back such interventions, if they are not needed, but with a large amount of text such a correction might be overlooked which can lead to serious errors. There are also text editors (such as LaTeX) that preserve full sovereignty of the user, but they are more difficult in use, and programmers creating the more popular ones concluded (because of the popularity) that they can violate that sovereignty.

In the beginning of twenty-first century, such controversies resulted in a new concept of *human centred computing* or *human user focus*. This concept can be diversely understood; e.g. http://sites.google.com/site/egonvdb/ lists three aspects of this concept:

(1) A computer application should provide an optimal support for the user, independent from her/his advancement in computer technology. In other words, the technology must be adapted to the cognitive abilities of the user and the interaction between the computer application and the user should be optimized;
(2) Artificial intelligence technologies should gain by learning from people;
(3) An understanding of the needs of users has an importance going beyond the issue of improving computer technology.

However, aspects (2) and (3) appear to be an unnecessary since obvious justification before programmers fascinated with artificial intelligence, while a decisive aspect of the concept of *human centred computing* seems to be aspect (1), even in a stronger and broader form. This aspect should be understood as characteristic of a relation between computer and its human user, with many dimensions of this relation taken into account. The concept of *user friendly software*, corresponding in fact to the requirements of aspect (1), is much older than *human centred computing*. It is about something more; personally, I tried to express this as a requirement of *users sovereignty* on the basis of long experience in computerized decision support and broadly understood knowledge engineering: *the final decision and final interpretation of knowledge belongs to the user and a computer should not violate this principle.*

It is perhaps worth explaining how that conflict between artificial intelligence and the sovereignty of a human user arises in the case of computerized decision support. The multiple criteria decision analysis allows for many *efficient* solutions, called also *Pareto-optimal.*[18] The goal of computer modelling and the software for decision support is to ensure that the decisions proposed to the user are selected only between efficient decisions. However, the choice between efficient decisions should belong to the user or decision-maker.

Meanwhile, the classical decision theory, closely related to the game theory and to the market theory, postulates that each decision maker (player, consumer, etc., here the user) expresses her/his preferences through a *utility function* (or *value function*)[19] that aggregates criteria and transforms them into a scalar measure. Therefore, the classical approaches to computerized decision support, see e.g. Keeney and Raiffa (1976), assume that the decision maker should answer many questions in order to identify her/his preferences or utility function and afterwards the computer will optimize this function and announce the optimal decision to the user. Soon it was realized that such an approach has certain drawbacks, because it corresponds essentially to a full automation of decisions after an initial identification of the utility function, and what happens, when the decision maker does not like the supposedly optimal decision? The proponents of such an approach answered that it could not be helped, the user should understand that the only rational decision is one proposed by the computer.

However, less paradigmatically inclined researchers of decision support admitted the possibility of decision corrections, and proposed various methods of *interactive decision support.* Hence historically, the problem of computer-man interaction (from the work of Geoffrion et al. 1972) was first investigated in research on computerized decision support. This research included to a large extent also some aspects of artificial intelligence and decision automation, see e.g.

---

[18] Which means that no single criterion value could be improved without worsening the value of some other criterion.

[19] This is a specialized and not broadly used distinction: a value function means an aggregation of diverse criteria without taking uncertainty into account, a utility function performs such aggregation while taking it into account.

Gal et al. (1999), but as tools of decision support, while .the essential problem became the issue of sovereignty of the decision maker in the selection of any efficient decision. This issue was resolved and such sovereignty obtained through a special method of user interaction, called *reference point approach,* proposed in Wierzbicki (1980) even if its broad applications were developed later, see (Wierzbicki et al. 2000).

Another area of computer applications is finding knowledge in large data sets; it has a shorter history, but it creates a similar conflict. Because of rapidly increasing amount of information and digital data gathered either on the Internet, or by telecommunication operators and other entrepreneurs, or at universities and other research institutions, an important problem became the issue of finding, in large data sets, not only information interesting for us, but also more, namely useful relations between such information, or in other words *tacit knowledge hidden in large data sets.*

I am using here the term *tacit knowledge* on purpose, even if it originally applies to knowledge difficult to express in words, preverbal, hidden in human mind, see (Polanyi 1966; Nonaka and Takeuchi 1995; Wierzbicki and Nakamori 2006, 2007). This is because preverbal knowledge, not yet expressed in words, is contained also in large data sets or even in large text sets, and it is precisely a difficult task to bring it out from these sets, which is expressed by terms *data mining* or *text mining.* We could use as well the term *knowledge mining* with the goal of expressing this knowledge in words, logical formulae or any other form of model. The conflict discussed here concerns the way in which this knowledge shall be expressed: should it be done automatically using artificial intelligence methods, or should we rather assume an active, perhaps decisive role of a human user?

Some sources use the concept of *knowledge science,* even if this would imply a strong connection to epistemology. A related field, both to epistemology and to knowledge extraction from large sets of data and texts, is *knowledge management* that historically emerged from computer science (see e.g. Wierzbicki and Nakamori 2007) but is today treated as a part of management sciences. For all these premises, it is reasonable to speak today about *knowledge engineering,* encompassing the construction and utilization of computerized tools of knowledge extraction and transformation. Knowledge engineering can be subdivided into several parts:

1. Narrowly understood engineering of artificial intelligence and automatic learning, tacitly assuming the domination of computers over users.
2. Engineering of knowledge extraction from large data sets, *data mining,* using such theoretical background as diverse parts of logics, statistics, multiple criteria decision theory, etc. In applications, it is important to provide a verbal interpretation of resulting models of knowledge according to the requirements of the user that often have the character of tacit knowledge, difficult to express in words. Therefore, this part, in contrast to part I, strongly stresses the issue of methods of user interaction, the sovereignty of the user, the role of intuition and tacit knowledge.

3. Engineering of text processing and knowledge extraction from texts. Basic fields for this part are: ontological engineering (construction of taxonomies enhanced by various relations between concepts based on large sets of texts), semantic networks and *Semantic Web,* Internet search engines, etc. Important here is the possibility of an interaction with the user in order to obtain her/his interpretation of the selected verbal knowledge, again based on her/his tacit knowledge.

Therefore, the concept of *human centred computing,* even if new at the beginning of the twenty-first century, expresses, nevertheless, conflicts and problems that have grown and were research issues at least since 1972. This concept, although its main dimension is the issue of computer-man interaction, has many other aspects. It is related to the issue of human intuition and its role in cognition, discussed in Chap. 5. Another aspect is *social informatics,* concentrating on the impact of computer and information technology on the change of social practices and customs, or on social practices of using such techniques. Yet another aspect is constituted by computerized support of creativity that will be discussed in Chap. 12.

11.8 Intellectual Property Rights, Network Piracy, and the Conflict Concerning Privatization of Intellectual Heritage of Humanity

Since the informational revolution leads to knowledge-based economy or even to knowledge civilization, no wonder that the issue of property rights to knowledge becomes a subject of a strong dispute that can be even treated as the fundamental conflict of the new era. This conflict has many dimensions.

The concepts of "intellectual property right" and "network piracy" are based on the following argument (I quote after Lawrence Lessig 2004, p. 18) "Creative work has value; whenever I use, or take, or build upon the creative work of others, I am taking from them something of value. Whenever I take something of value from someone else, I should have their permission. Taking something of value from someone else without permission is wrong. It is a form of piracy."

Lessig strongly criticizes this argument, calling it after Rochelle Dreyfuss an "if value, then a property right" theory, false in its essence, and I fully agree with this critique. If somebody develops my own research ideas, e.g., concerning the method of the reference point approach to multiple criteria optimization and decision theory, then I am happy, even if I expect a reasonably correct quotation of my own work in this field. But I am not against it even in the case of a lack of quotation, if the new work presents an essential development of my method. And it would seem for me nonsensical to require other authors to ask my permission to work on my method. This example clearly shows that the argument "if value, then a property right" does not apply to the development of science, even if some economists, see e.g. Cellary (2011), believe that in knowledge economy, knowledge becomes a commodity.

Lessig rightly suggests that law should distinguish at least two situations of utilization of a creative work. One is when the work is copied, and in this case we should obviously observe good practices of intellectual property. These good practices sometimes require a correct quotation of the source, in other cases (e.g. larger texts) obtaining permission of the author, but such permission is not a universal requirement.[20] Another one is the use of creative value of the work for further transformation and development, and in this case it is in social interest and interest of humanity to maximally limit possible restrictions of such use of a creative work.

The conflict concerning intellectual property rights and the question what might be accounted as piracy in this field obviously intensifies with the development of informational network techniques, providing various and easy ways of copying of works accessible on the Internet. It results in the concept of "network piracy". But many users of the Internet have quite different interests and customs in this respect than desired by large corporations making business in distribution of music, films and other creative works. Moreover, many authors maintain, see e.g. Vaagan and Koehler (2005), that there are ethically motivated *laws of access* that should dominate over strengthening of intellectual property rights.

The legal initiative to strengthen intellectual property rights and to fight against network piracy comes from large corporations and law firms serving them, not from creative workers. Similarly, the interpretation of the extent of such rights e.g. in the United States pursues the interests of large corporations and related law firms, not the interests of the entire American society and its creativity. Beside the legal initiatives, large corporations finance also the development of new technical solutions that should supposedly protect intellectual property rights. Such technical solutions are distributed under a hypocritical name of *digital rights management, DRM,* and serve to limit the rights of access to and utilization of network resources. There are many such solutions that serve to limit the utilization of films, television, documents, etc.; they might have the form of metadata, watermarks, etc. DRM solutions encountered strong criticism also between computer scientists. This criticism is best expressed by the opinion of Richard Stallman who believes that DRM should be called *digital restriction management,* that it is a malicious addition to a code, a feature designed to disadvantage the user of that code, hence an attribute that cannot be tolerated.[21] Another opinion, expressed by music composers, stresses that good artists gain (on reputation etc.) from free distribution

---

[20] For example, I did not ask the permission of Lawrance Lessig to quote and discuss his opinions, but I do not think that he would have something against it.

[21] See, e.g., http://www.gnu.org/philosophy/right-to-read.html.

of their works, can get higher remuneration at concerts etc., hence they are against using DRM. Many other arguments are used: that DRM violates the laws of free market competition, also the laws of a consumer, that records using DRM will not be readable in the future and thus historically lost.

All conflicts described above are only parts of a general conflict about right of access to and property of knowledge. There are also other dimensions of this conflict, such as the oligopolistic fight for standards and patents on all high technology markets. For example, pharmaceutical concerns set their prices without any relation to classical market rules and marginal costs of production, while they try to convince consumers that their prices result from high costs of research borne by these concerns, and nobody can directly[22] check such statements, since the knowledge possessed by a corporation is fully secret and large corporations prohibit their employees even from publishing the results of their research; they also excessively extend the periods of validity of their patents. Thus, we observe an intensification of *corporate privatization of knowledge,* contrary to the tradition of the industrial civilization era, where the periods of validity of patents were short and even the knowledge created for the needs of market companies was soon accessible to all people as *intellectual heritage of humanity.* This privatization is accompanied by the phenomenon of pollution of intellectual heritage, similar to the pollution of natural environment: large corporations publish only such part of their knowledge that serves their interests, increase the *informational asymmetry.* Who knows well and objectively today, what is the efficiency and collateral effects of e.g. flu vaccines?

A related general question is as follows: *are there any rational arguments for privatization of knowledge that is a public good?* As it was shown both by Lawrance Lessig and by other authors, e.g. (Wierzbicki and Nakamori 2007), *knowledge is not a degradable good: it does not diminish, but often increases, by intensive common use.* Therefore, with respect to knowledge we do not observe a *tragedy of commons,* that is, an excessive degradation of a common good through its intensive common use. It is precisely tragedy of commons that is a rational argument for the privatization of a common good; but in the case of the common use of knowledge the situation is opposite: a society gains, not loses, if most knowledge remains a public good.

---

[22] Indirectly, it is possible to check whether these prices correspond to prices typical for an oligopolistic market; it is necessary to compute, using the game theory, what would be the price on oligopolistic market and how much it can differ from marginal production cost. We must know the market share coefficient of a company, κ, and the elasticity of demand with respect to price, ε, to obtain $p = m_c/(1-\kappa/\varepsilon)$, where p is an oligopolistic competitive price, and m_c is marginal production cost (that is, the competitive price on an ideal free market), see e.g. Kameoka and Wierzbicki (2005). If the actually observed price deviates widely from such estimation, this suggests that there is an explicit or tacit cartel price agreement and anti-monopolistic proceeding should be implemented.

Admittedly, the easiness of copying digital information has also some negative consequences; for example, it stimulates rather negative changes of customs. If young people are not taught at schools that they should not use obvious plagiarism when preparing their homework, then the easiness of copying will have obviously negative impacts. Therefore, it is necessary to discuss openly at schools the conflicts concerning the property of knowledge, but also to use existing tools, software for determining the percentage content of sources directly quoted from the Internet.

However, the academic community perceives this general conflict and responds counter-offensively: it promotes a free access to the knowledge developed by using public money from state donations. It has diverse specific forms, but generally relates to creation of web platforms offering teaching materials of best universities and results of research financed with public money. We cannot tell today, how this controversy between public, private corporate and private individual property of knowledge will end, but in the interests of survival of humanity and its sustainable development it is necessary to preserve possibly largest share of public knowledge, constituting an intellectual heritage of humanity, well tested and protected against pollution.

11.9 Binary Logic and Logical Pluralism

At the end of this Chapter, I would like to present a discussion of the impact of the informational revolution on such an important cognitive tool of humanity which is provided by logics. I have already criticised the conviction that human intelligence or computer intelligence can be reduced to classical logics; nevertheless, logics is an important tool, because it helps to check the correctness of various verbal or mathematical inferences. However, *from the fact that logics is a virtual tool it does not follow that it is an absolute, ideal tool.*

For a long time logics was treated as a part of philosophy, but from the time of Boole (1847) it has become a part of mathematics. And as in all mathematical systems, the veracity or rather adequacy of logics depends on the adequacy of its assumptions in respect to a specific field of applications. The assumptions made by Boole are classical, two logical values, *true* or *false,* hence *there is no third way* (therefore, we speak about *classical logics,* or, equivalently, about *binary logics*) and there is a static, time-free interpretation of logical operations and values. These assumptions turned out to be an excellent abstraction for an approximate description of computer hardware. However, such a description is only approximate, because these assumptions are not fully adequate even to describe computer hardware: the execution of logical operations in hardware takes some time, during which a third logical value, *indetermined,* dominates. Since computers execute sequences of commands, it is necessary to downbeat tact in *discrete time,* that is, execute new logical commands after a certain interval of time, sufficient to eliminate the indeterminacy of former operations, hence we speak about *discrete time*,[23] counted with each next tact.

The greatest contributions to the formalization of multi-valued logics, starting with a critique of binary logics and creation of foundations for triple-valued logics, were made Jan Łukasiewicz (e.g. 1910, 1911). In his book *On the Contradiction Principle by Aristotle*, very influential for the development of Polish schools of logics, mathematics and philosophy, he defended the *(non) contradiction principle*, saying[24] that no thing can at the same time have a property and not have this property, but he treated this principle critically. On the one hand, he gave some formal proofs of this principle, but he checked also its *factual veracity*. He observed that there are many historical cases when mental constructions turned out to be self-contradictory, and in the area of empirical facts he stressed that that the very fact of motion, changing reality, is a factual contradiction, besides, all classical logics is static, does not include dynamics of events.

On the other hand, Jan Łukasiewicz consistently criticised another axiom of binary logics, the *principle of the excluded middle* saying that of two contrary logical statements about properties of a thing one must be true. According to Łukasiewicz, a basic example contradicting this principle were statements about the future that have always the undetermined value *may be*. Therefore, he decided to introduce triple valued logics.

Much later, during the informational revolution, it turned out that the assumptions of Boole are highly inadequate in the field of systems theory and software programming. This was noted by Zdzisław Pawlak (1991): if we treat a large data set as an *information system* and consider the veracity of a given logical relation between the elements of this system, then for some (pairs of) elements this relation might be *true*, for others, *false*, but there are usually many (pairs of) elements for which the relation is *undetermined*. This observation became the foundation of the *rough set theory* of Pawlak, actually, a triple valued logic, but not resulting from abstract axiomatic assumptions, only from practical needs of the analysis of large data sets (see Pawlak 1991; Słowiński 1995).

Binary logics with the values *true, false* is nevertheless traditionally used in philosophy even for problems for which it is clearly inadequate (e.g. for problems of knowledge creation, which have obviously a dynamic character), as already

---

[23] The concept of discrete time has led to the question whether the physical time is truly continuous, or, similarly as mass or energy, it has a discrete, quantum character. The assumption of discrete time might give a more consistent models of quantum physics, since nonlinear systems with discrete time easily generate chaotic behaviour (as already discussed in Chap. 10) even if they are deterministic; hence the assumption about indeterminism of the universe could be treated as a result of nonlinearity and of discrete time, not as an ad hoc assumption. Recall the statement of Albert Einstein: *God does not play dice with the universe*.

[24] This principle was traditionally called *contradiction principle*, even if it actually is a *noncontradiction principle* (and the latter name is used contemporarily). I have quoted above the so-called *ontological version* of this principle; Łukasiewicz distinguished also its *logical version* and a psychological version, the belief in the veracity of the noncontradiction principle.

discussed in Chap. 6. Here again a fundamental fact should be stressed: *if there is a third logical value, then an indirect proofs by reduction ad absurdum (that is, by finding a contradiction) lose their validity,* as noted already by Brouwer (1922). Therefore, the principle of logical noncontradiction discussed above must nevertheless be treated with outmost caution: a contradiction can indicate an emergence of a new quality, opening of "a third way", as noted already by Hegel.

The history of multivalued logics was mentioned already in the earlier chapters. Fundamental works of Łukasiewicz (1910, 1911) were read mainly by mathematicians.[25] For technical applications, Zadeh (1965) had to rediscover multivalued logics and called it *fuzzy set theory.* That such logics is needed for an adequate description of imprecise statements, should be obvious. For example, the statement "number 7 is a large number in the interval of numbers 0–10" is only to some extent true, rather less true than a similar statement about number 9. The applications of the fuzzy set theory are very broad today; the Japanese even constructed microprocessors using this theory and applied them for controlling home appliances such as washing machines.

Less universal is understanding of the need of *temporal logics* taking into account dynamic relations between logical values or rather variables. I stressed this need in the former chapters in particular in relation to the concept of *feedback,* in a sense opposite to the concept of *vicious circle.* There is no doubt that temporal logics, and especially the logics of feedback, changes essentially our way of understanding the world.

The general conclusion is clear. Contrary to what is usually taught in schools or even at universities, there is no absolute, universally valid logics, ensuring full correctness of reasoning, particularly if we use logics for proofs based on the reduction ad absurdum, because it usually turns out that apparent absurd can be differently explained in more adequate logics. This does not mean that logics should not be taught, to the contrary, in times after the informational revolution diverse logics and *logical pluralism* should be taught. And together with logical pluralism, examples of adequacy or inadequacy of diverse logics for different application areas should be presented.

References

Arthur, W.B.: Increasing Returns and Path Dependence in the Economy. Michigan University Press, Ann Arbor (1994)
Bangemann, M.: Europe and the Global Information Society. Recommendations to the European Council, European Council, Brussels (1994)

---

[25] It is telling that at the end of his life Bertrand Russell complained about the readership of his fundamental work *Principia Mathematica* (Russel 1910–1913), saying that with a true understanding this work was read by only perhaps seven people in the world, while three of them were Poles.

Baran, P.: Reliable digital communications systems using unreliable network repeater nodes. RAND Corporation papers, document p. 995. http://www.rand.org/pubs/papers/P1995.html (1960). Accessed 29 March 2011

Boole, G.: The Mathematical Analysis of Logic. Encyclopædia Britannica. 2009. Encyclopædia Britannica Online. 29 Sept 2009 (1847)

Brouwer, L.E.J.: On the significance of the principle of excluded middle in mathematics, especially in function theory. With two addenda and corrigenda. In: Jean van Heijenoort (1967) A Source Book in Mathematical Logic, 1879–1931, pp. 334–345. Harvard University Press, Cambridge, (1922)

Bush, V.: As we may think. The Atlantic Monthly 176, 101–108 (1945)

Cellary, W.: Zasoby wiedzy dobrem ekonomicznym w społeczeństwie wiedzy (Knowledge resources as economic good in knowledge society). Przyszłość: Świat, Europa, Polska, nr. 1/ 2011 (2011)

Chomsky, N.: Knowledge of Language: Its Nature, Origin and Use. Praeger Special Studies, New York (1986)

Derrida, J.: Of Grammatology. John Hopkins University Press, Baltimore (1974)

Dreyfus, H.: What Computers Can't Do: the Limits of Artificial Intellgence. Harper and Row, New York (1972)

Feldbaum, A.A.: Osnovy Teorii Optimalnych Avtomaticzeskich System. Nauka, Moscow (1965)

Fodor, J.A.: The elm and the expert: mentalese and its semantics. MIT Press, Boston (1994)

Gal, T., Stewart, J.T., Hanne, T.: Multicriteria Decision Making: Advances in MCDM Models, Algorithms, Theory and Applications. Kluwer, Boston-Dordrecht-London (1999)

Gardner, H.: The Minds New Science: a History of the Cognitive Revolution. Basic Books, New York (1985)

Geoffrion, A.M., Dyer, J.S., Feinberg, A.: An interactive approach for multicriteria optimization with an application to the operation of an academic department. Manage. Sci. 19, 357–368 (1972)

Granat, J., Wierzbicki, A.P.: Inżynieria wiedzy—nowy obszar badawczy Instytutu Łączności (Knowledge Engineering—a New Research Field of National Institute of Telecommunications). Telekomunikacja i Techniki Informacyjne 3–4, 108–116 (2009)

Jacyna-Onyszkiewicz, Z.: Fizyka kwantowa a metafizyka, str. 149–161 w: Nauka a metafizyka, pod red. A. Gotyckiej, Wydawnictwo IF i S PAN, Warszawa 2009 (Quantum Physics and Metaphysics, pp. 149–161 In: A. Gotycka (ed.) Science and Metaphysics) Warsaw (2009)

Jankowski, P., Łoniewski, D., Wójcik, G., Wierzbicki, A.P.(eds.): Sieci komputerowe. Podręcznik multimedialny (Computer Networks—a Multimedia Textbook). OKNO—Ośrodek Kształcenia Na Odległość Politechniki Warszawskiej (The Centre of Distant Education of Warsaw University of Technology), Warsaw (2003-2007)

Kameoka, A., Wierzbicki, A.P.: A Vision of New Era of Knowledge Civilization. Ith World Congress of IFSR, Kobe (2005)

Keeney, R., Raiffa, H.: Decisions with Multiple Objectives: Preferences and Value Tradeoffs. J. Wiley and Sons, New York (1976)

Lessig, L.: Free Culture: the Nature and Future of Creativity. Penguin Books, London. Polish translation (2005) Wolna kultura, Wydawnictwa Szkolne i Pedagogiczne, Warsaw (2004)

Łukasiewicz, J.: O wartościach logicznych (On Logical Values). Ruch Filozoficzny 1, 50–59 (1911)

Łukasiewicz, J.: O zasadzie sprzeczności u Arystotelesa (On the Principle of Contradiction by Aristotle).Reprint (1987) by PWN, Warsaw (1910, 1987)

Masuda, J.: The information society as post-industrial society. Institute for the Information Society, Tokyo (American edition 1981, World Future Society, Washington, DC) (1980)

McCollough, W.S., Pitts, W.H.: A logical calculus of the ideas immanent in nervous activity. Bull. Math. Biophys. 5, 115–133 (1943)

Minsky, M.: Form and Content in Computer Science. J. Assoc. Comp. Machinery 17(2) (1970)

Minsky, M.: The society of mind. Simon and Schuster, New York (1986)

Nonaka I., Takeuchi H.: The Knowledge-Creating Company. How Japanese Companies Create the Dynamics of Innovation. Oxford University Press, New York (Polish translation: *Kreowanie wiedzy w organizacji,* Poltext 2000) (1995)

Pawlak, Z.: Rough Sets—Theoretical Aspects of Reasoning About Data. Kluwer, Dordrecht (1991)

Polanyi, M.: The Tacit Dimension. Routledge and Kegan, London (1966)

Pylyshyn, Z.: Computation and Cognition. Towards a Foundation of Cognitive Science. Harvard University Press, Cambridge (1984)

Russel, B., Whitehead, A.N.: Principia Mathematica. Cambridge University Press, Cambridge (1910–1913)

Searle, J.R.: The Rediscovery of Mind. MIT Press, Boston (1992)

Senge, P.M.: The Fifth Discipline: the Art and Practice of Learning Organization. Random House, London (1990)

Simon, H.A., Newell, A.: Heuristic problem solving: the next advancements in operations research. Oper. Res. **6**, 1 (1958)

Słowiński, R.: Rough Set Approach to Decision Analysis. AI Expert **10**, 18–25 (1995)

Toffler, A., Toffler, H.: The Third Wave. William Morrow, New York (1980)

Turing, A.: Computing Machinery and Intelligence. Mind **40**(236) (1950)

Vaagan, R., Koehler, W.: Intellectual property rights vs. public access rights: ethical aspects of the DeCSS decryption program. Inf. Res. **10**(3), 230 (2005)

Weizenbaum, J.: Computer Power and Human Reason. From Judgment to Calculations. W. H. Freeman, San Francisco (1976)

Wierzbicki, A.P.: The use of reference objectives in multiobjective optimization. In: Fandel, G., Gal, T. (eds.) Multiple Criteria Decision Making: Theory and Applications, pp. 468–486. Springer, Berlin-Heidelberg (1980)

Wierzbicki, A.P., Nakamori, Y.: Creative Space: Models of Creative Processes for the Knowledge Civilization Age. Springer, Berlin-Heidelberg (2006)

Wierzbicki, A.P., Nakamori, Y. (eds.): Creative Environments: Issues of Creativity Support for the Knowledge Civilization Age. Springer, Berlin-Heidelberg (2007)

Wierzbicki, A.P., Makowski, M., Wessels, J.: Model-Based Decision Support Methodology with Environmental Applications. Kluwer, Dordrecht (2000)

Winston, B.: Media Technology and Society: a History from the Telegraph to Internet. Routledge, Chapham and Hall, London, New York (1998)

Wydro, K.B.: Informacja—charakterystyki, podstawowe techniki, wykorzystanie (Information—Characteristics, Fundamental Techniques, Utilization). Wydawnictwo Naukowe "Obserwacje", Warsaw (2008)

Norman, J.M.: In: From Gutenberg to the Internet: A Sourcebook on the History of Information Technology. historyofscience.com, Novato, CA

Zacher, L.W.: Transformacje Społeczeństw: Od Informacji Do Wiedzy (Transformations of Societies: From Information to Knowledge). C.H. Beck, Warsaw (2007)

Zadeh, L.: Fuzzy Sets. Inf. Control **8**, 338–353 (1965)

Part III
Epistemological Conclusions

Chapter 12
Creative Space and Micro-Models of Knowledge Creation

12.1 New Models of Knowledge Creation at the Turn of the 20th and 21st Century

The epistemology of the 20th century concentrated mostly on problems of knowledge justification, verification or falsification, putting aside the issues of knowledge creation, even if at the end of the 20th century a few works emerged stressing the necessity of understanding issues related with the creation of knowledge, see e.g. Searle (1992), Motycka (1998). This problem area became more actual and urgent because of the growing civilization change, a turn from the industrial civilization to a new society and civilization based on information and knowledge. As discussed in the former chapters, we accept the year 1980 as the caesura of the beginnings of this turn, an approximate date of the combination of two earlier inventions: computers and telecommunication networks, and the dissemination of computer networks that brought global access to information and knowledge potentially to every household on the Earth. But however we should appraise the caesura and its importance, there is no doubt that the last 20 years of the 20th century brought a tremendous growth of the role of knowledge as the fundamental productive resource in the most profitable fields: services and industrial production of computers, software, telecommunications, pharmaceutics, biomedical engineering, etc. This in turn resulted in a fast growth of demand for a better understanding of the processes of knowledge creation, not in the sense of a long historical perspective (such philosophical models of knowledge creation in the long term sense I shall call *macro-theories*, such as the Kuhnian concept of paradigmatic knowledge development, or Popperian falsification theory, or

---

This chapter is a modified version of my paper (Wierzbicki 2011) and summarizes two books and many papers concerning the processes of knowledge creation.

© Springer International Publishing Switzerland 2015   219
A.P. Wierzbicki, *Technen: Elements of Recent History of Information Technologies with Epistemological Conclusions*, Intelligent Systems Reference Library 71, DOI 10.1007/978-3-319-09033-7_12

Lakatosian scientific programmes, etc.), but in the sense of prescriptions how to create knowledge for the current needs, called here *micro-models* of knowledge creation.

Therefore, many theories or rather models of processes of knowledge creation emerged in the last 20 years of the 20th century and in the beginning of the 21st century, in various conditions and with diverse assumptions. Most of these models do not come from philosophy proper and epistemology, but from other fields of knowledge oriented towards specific problems, such as systems analysis and in particular computerized decision support, or management theory and in particular knowledge management theory.

12.2 Organizational Processes of Knowledge Creation

Historically, the first of such methods emerged much earlier (Osborn 1957, which is an additional argument concerning the issue of civilization delays) as the method of *brainstorming*, even if it was formalized and described as *a DCCV spiral of brainstorming knowledge creation* much later (Kunifuji et al. 2004, 2007).

Brainstorming has many definitions, and its very name implies an intensive inspiration and generation of new ideas by a group, a kind of group creative transition that was called in Chap. 5 *enlightenment (abduction, illumination, aha, eureka)*. However, after the book *Applied Imagination* (Osborn 1957) the word "brainstorming" obtained a specific meaning: *brainstorming is a group process of creating new ideas with postponing the appraisal of their value.* Later it was observed that the brainstorming method can be also used individually, since its essence is creation and recording of new ideas while postponing their appraisal and selection, even if in a group process it is clearly possible not only to generate more ideas, but also to stimulate such generation through a specific positive feedback between members of the group. This most important phase of brainstorming was called a *divergent phase, divergent thinking* or *divergent idea production* (which evokes my reservations, since I think that ideas are not produced). On this ground, the following principles of brainstorming in divergent phase were formulated:

(1) The goal of brainstorming in its divergent phase is to create a large number of ideas, not necessarily the best ones.
(2) The appraisal of the quality of ideas (in the sense of good or bad ideas, implementable or not, etc.) should be suspended.
(3) Unusual ideas are especially desirable.
(4) Using or further developing already proposed ideas is also desirable.

Brainstorming has many advantages, but also drawbacks, see e.g. Kunifuji et al. (2007). Its main drawback relates to its inconsistency: after the divergent phase it must be switched to the second convergent phase consisting in screening and choosing ideas; and such switching psychologically collides with the attitude "let all flowers bloom" of the first phase. In other words: who should be responsible for the selection of ideas: all group or only the organizer of the process? To whom the ideas generated in the process belong? Despite these drawbacks, brainstorming became one of the most often used methods of problem solving or creating useful ideas in industrial and other organizations. However, it has much less importance in the processes of academic knowledge creation, see later comments. Nevertheless, it is the oldest and most broadly used process of *organizational knowledge creation*, with an *intercultural* character, preceding later models of organizational knowledge creation described below, such as the *SECI spiral* of Far East type (Nonaka and Takeuchi 1995) or *OPEC spiral* of an Anglo-Saxon Western type (Gasson 2004). The first applications of brainstorming occurred in NASA in relation to planning of outer-space research.

There were many attempts to define a general model of brainstorming, see (Kunifuji et al. 2007), nevertheless the essential phases of this process are the following, see also a graphical representation of this model in Fig. 12.1:

(a) *Divergent thinking (Divergence)*, such as in the divergent phase described above;
(b) *Convergent thinking (Convergence)*, appraisal and selection of ideas;
(c) *Crystallization of ideas (Crystallization)*, their more specific development (particularly of analytic character, since the earlier phases are highly intuitive);
(d) *Verification of ideas (Verification)* that might consist in *learning by doing*, or in an application of a quite different method of knowledge creation, e.g. the academic method of *debating* described below.

The interpretation of the model in Fig. 12.1 as a spiral results from the fact that a repetition of a brainstorming process can only increase the number and improve

Fig. 12.1   DCCV spiral of
brainstorming (Kunifuji et al.
2004, 2007)

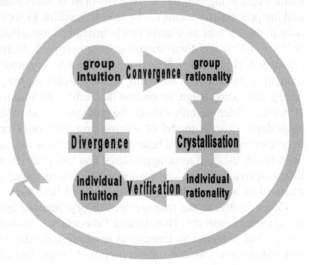

the quality of generated ideas: knowledge is not lost when it is again or intensively used. In this model, not only the transitions *Divergence-Convergence-Crystalli-sation-Verification* are important, but also their interpretation as the change of focus of attention occurring between *nodes* of the model. These nodes include: individual intuition and group intuition (*Divergence* consists in proposing individual ideas to the group), group rationality (*Convergence* is a group rationalization of intuitive ideas) and individual rationality (*Crystallisation* is an individually rational selection and specification of an idea), while *Verification* is the enhancement of individual intuition in practice, with attempts of implementation of ideas.

Another method of creation of knowledge in order to solve current problems, younger than brainstorming but the oldest one between methods emerging in the last decade of the 20th century, is *Shinayakana Systems Approach* (Nakamori and Sawaragi 1992), emerging from the interactive computerized decision support. *Shinayakana* is a Japanese concept simultaneously expressing the elasticity of a willow twig and the hardness of a sword; here it is used to signify a synthesis of the so-called *soft* and *hard systems analysis*. The history of paradigmatic discourse between these two branches of systems analysis is rather long, and only partly presented in Chap. 10. This discourse, however, has one positive, fundamental denominator: the hard systems analysis relies on using hard computerized mathematical models representing knowledge about a given problem, while the soft systems analysis is correct in saying that human behaviour cannot be well represented by mathematical models. As discussed in earlier chapters, this has resulted in the development of the *interactive decision support*, in which I took part personally together with my Japanese friends (see e.g. Wierzbicki 1983; Nakayama and Sawaragi 1984; Wierzbicki 2000). In such an approach, available objective knowledge about a given problem is represented by mathematical models, while the preferences of the user or decision maker are not strictly modelled, leaving her/him a large freedom of choice and control of the decision making process (for example, by asking her/him to specify decision requirements only in the form of aspiration levels instead of modelling her/his preferences by a utility function and then imposing the "optimal" decision on her/him). It is the conviction that human behaviour transcends the possibilities of mathematical modelling, even if such modelling is very useful for the purposes of representation of objective knowledge, motivated the authors of *Shinayakana Systems Approach*. However, under the influence of the soft systems analysis, they did not propose an algorithmic process model of solving problems or of creation of knowledge, but only a set of principles. These principles include the use of intuition, preserving open mind, using diverse approaches and perspectives when analyzing the problem, adaptive approach and readiness to learn by mistakes, elasticity of a twig and hardness of a sword, thus, using the tools of both hard and soft systems analysis.

Concurrently, in the management theory, another approach developed by Japanese authors emerged: Nonaka and Takeuchi's *Knowledge Creating Organization* (1995). Not bothered by limitations resulting from the discourse between hard and soft systems analysis, the authors proposed for the first time an algorithmic process

Fig. 12.2 SECI spiral of
knowledge creation in
organizations (Nonaka and
Takeuchi 1995)

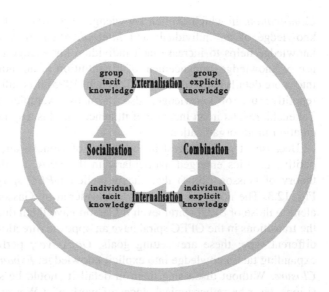

model of knowledge creation. Even if this theory concentrates on organizational knowledge creation, resulting in small improvements in knowledge useful for market-oriented organizations, it has a revolutionary importance. For the first time, it very strongly highlighted the *role of a group in processes of knowledge creation* and postulated a rational use of irrational (or rather a-rational for a Japanese) *tacit knowledge*, a novel extension of understanding of the concept of *tacit knowing* introduced by Polanyi (1966).

This theory, which is very popular in the management science today, is expressed usually by the so-called *SECI spiral,* see Fig. 12.2. This spiral also consists of four transitions between four nodes, situated on two axes. One axis is called *epistemological dimension* including *explicit knowledge* and *tacit knowledge;* another axis I prefer to call *social dimension*[1] including an *individual* and a *group*.

Subsequent transitions in this model are: *Socialization,* in which tacit knowledge of an individual is transformed into tacit knowledge of a group (this is decisive for a Far East character of this model: in Japan, employees of a firm often stay after work in order to drink beer and other beverages while informally discussing various issues related to work); *Externalization,* in which the tacit knowledge of a group is codified, transformed into explicit knowledge;

---

[1] Nonaka and Takeuchi use the concept of *ontological dimension,* but tacit and explicit knowledge might also be treated as ontological elements of discourse. Similarly, I prefer to use the concept of a *transition,* while Nonaka and Takeuchi use the concept of *knowledge conversion* instead. However, conversion suggests consumption of a transformed resource, while knowledge is not consumed with its use; hence I use here the concept of *transition,* which suggests only a change of the focus of attention.

Combination, in which explicit knowledge of a group is transformed into explicit knowledge of an individual; and *Internalization,* in which explicit individual knowledge helps to increase tacit individual knowledge (e.g. through a practical use of knowledge, enhancement of the intuition of an individual). Without going into more details (see Nonaka and Takeuchi 1995) it is sufficient to stress that each repetitive use of knowledge, also in many examples given by Nonaka and Takeuchi, results in an increase of it, hence *SECI* is a spiral leading to knowledge creation in an organization.

Because of great interest in the theory of Nonaka and Takeuchi, many competitive theories emerged, particularly in the USA. I will present here only the theory of Gasson (2004) that can be called *OPEC spiral* and is illustrated in Fig. 12.3. The nodes of the network considered by Gasson are practically equivalent to those of SECI spiral, even if Gasson gave them different names. However, the transitions in the OPEC spiral have an opposite direction and they are of quite a different type; these are: setting goals, *Objectives;* performing tasks, *Process;* expanding tacit knowledge into explicit knowledge, *Expansion;* and summarizing, *Closure.* Without discussing them in detail it should be stressed that these transitions describe rather typical stages of work of a Western organization, starting with goal setting (while the *Socialization* of Nonaka and Takeuchi, as stressed above, has a Japanese cultural background). However, another aspect is also characteristic of *OPEC spiral.*

Not only philosophy, but also other sciences in the 20th century were influenced (often tacitly) by Wittgenstein (1922) and his recommendation *"wovon man nicht sprechen kann, darüber muss man schweigen"* (you should not discuss metaphysical questions). This recommendation perhaps influenced also Gasson who in her spiral does not stress the issues of knowledge creation inside an organization strongly enough, even if usable knowledge can be obviously increased by organized problem solving, and accepts a typically Anglo-Saxon assumption: if there is not enough knowledge, hire external experts.

Fig. 12.3 OPEC spiral of knowledge creating in Western organization (Gasson 2004)

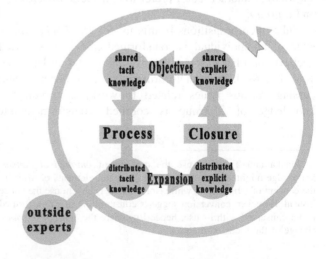

Much earlier than Gasson, in fact concurrently to Nonaka and Takeuchi, two independent theories of knowledge creation were proposed in Poland. The first one was an evolutionary rational theory of intuition published by me (Wierzbicki 1997), described in detail in Chap. 5; this theory was developed under the influence of contacts and discussions with my Japanese friends, Sawaragi and Nakamori, and their *Shinayakana Systems Approach* which recommends using intuition but does not analyze it. Practically at the same time and independently, Motycka (1998) used the Jung's concept of social unconsciousness to advance a theory of knowledge creation in times of crisis in a basic scientific field, or during a scientific revolution, which was based on the example of formation of quantum theory. This is actually a macro-theory of knowledge creation in scientific revolutions, but we shall present it below also in the form of a spiral in creative space.

Even if only the latter theory is strictly philosophical, while micro-models described above are contributions of other sciences, nonetheless taken together they indicate a kind of revolution in the last decades of the 20th and beginnings of the 21st century, resulting from informational revolution and the need to better understand the processes in which knowledge is created. There were several other related theories, some of them, such as I^5 *System* of Nakamori (2000), will be described later. The purpose of this chapter is to present a synthesis of such models or theories using the concept of *creative space*.

12.3 Creative Space

The word *irrational* is often interpreted as *emotional*. However, one of the conclusions from the evolutionary rational theory of intuition from Chap. 5 is that the old dichotomies, rational versus irrational, objective versus subjective, are too rough to describe the processes of knowledge creation in times of informational revolution and knowledge civilization. *There is a third way: between emotions and rationality there is a third, important layer of intuition.*

Thus, we should consider three layers of personality: *emotions, intuition,* and *rationality*. Similarly, beside an individual and a group we should also consider the highest layer of entire humanity, most important in the context of knowledge creation, because public knowledge, science and culture are a heritage of entire population of Earth, and actually no new knowledge is created without using a part of this heritage. Thus, if we consider three social layers of *individual, group,* and *humanity with its heritage,* and three layers of personality, then instead of four nodes as in Figs. 12.1 or 12.2 we obtain nine nodes or ontological elements: *individual rationality, individual intuition, individual emotions, group rationality, group intuition, group emotions, rational heritage (of humanity), intuitive heritage, emotional heritage.* This nine nodes, connected by diverse transitions as shown in Fig. 12.4, constitute *creative space* in its two fundamental dimensions: *epistemological* and *social* one. The creative space is treated as a *network-like model of description of creative processes* occurring as diverse transitions between

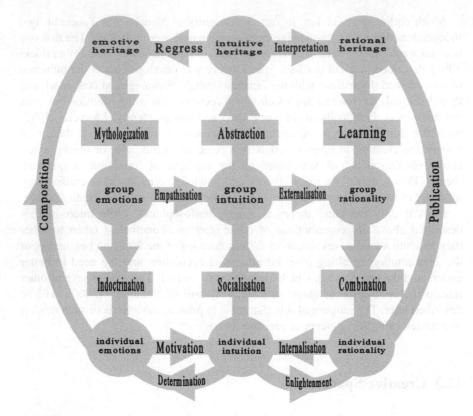

Fig. 12.4 Creative space in its fundamental dimensions: epistemological (rationality, intuition, emotions) and social (individual, group, heritage of humanity)

the nodes of the network, or entire sequences and cycles of such transitions, while any node might be treated as a starting point.

While the *individual rationality* node in Fig. 12.4 is almost equivalent to the *explicit individual knowledge* node in Fig. 12.2, the *tacit individual knowledge* node in Fig. 12.2 was subdivided in Fig. 12.4 into *individual intuition* and *individual emotions*.

Strictly speaking, a shift of elements of explicit knowledge occurs here: some emotional elements, related to arts and culture, might be treated as explicit knowledge, but they are not counted to rationality, only to emotions.[2]

---

[2] This shift is to some extent a question of convention and I am aware that such classification is very difficult, due to the specific relation of culture and arts to other resources of human knowledge. Nevertheless, we can maintain that culture and arts, which have a tremendous influence on creative activities of people in other fields, such as strict sciences or technology, execute their impact mostly by shaping of emotions. See also (Motycka 1998) on this issue.

A similar subdivision and shift occur at the second social layer, into *group rationality, group intuition, group emotions* nodes, and on the third layer, that of intellectual heritage of humanity. The latter was not considered by Nonaka and Takeuchi explicitly, but it is especially important in times of globalization and informational revolution and plays an enormous role in the processes of knowledge creation. Without our intellectual heritage we would not have processes of knowledge creation, also in organizations.

The node of *rational heritage* contains all experience and rational thinking results of science in its broadest sense (including not only *strict sciences*[3] and *technology* or *technical sciences,* but also *social sciences* as sociology and economy, *human sciences* as medicine, as well as *other sciences* such as mathematics and philosophy). This node is in a sense similar to the *third world* (or rather *world 3*) of Karl Popper, but it includes only rational aspects: while *arts and humanities* treated as sciences I include to the rational heritage, the *creations of arts and humanities* (such as music or films) I include to the emotional heritage.

The rational (and a part of emotional) heritage is recorded mostly in the form of books. The informational revolution changes that; the change from printed paper to digital electronic records might have similar, if not greater impact, as the invention of print by Gutenberg (or rather the reinvention and improvement, because print was known before Gutenberg in China). The importance of this ongoing change is related to the integration of the record of letters and photos, speech and images. Soon, the capacity of electronic memories will be so large that a simultaneous record of a book with a film featuring a cycle of related lectures will be as inexpensive as a newspaper. This will essentially change the understanding of the record of human intellectual heritage: imagine how interesting it would be to have the possibility to listen the lectures of Maxwell, Einstein, Heisenberg, Banach, Kotarbiński, Groszkowski. This will also essentially change the possibilities of *distance electronic education.*

Emotional heritage consists of artistic compositions, music, paintings, literature, all fiction created in the history of human culture, including an especially important, new form of record, films that in the time of globalization became one of the main aspects of intergenerational transfer of emotional heritage. Thus, a large part of Popperian world 3, if we understand it broadly as all explicit intellectual and cultural heritage of humanity, has actually an emotional character. But emotional heritage has also important tacit elements. For example, the globalization of films contributes to the globalization of myths of humanity, to strengthening of an aspect that was called *social unconsciousness* by Jung (already in 1953, see e.g. Jung 1971), which was used by Motycka (1998) to develop a

---

[3] Strict sciences are often called in English simply *sciences,* sometimes *hard sciences.* If we take the word *science* in the broad sense, we must add adjectives in order to represent the conviction that other sciences are also science.

theory of creation of basic knowledge in times of a crisis of a scientific discipline. There is no doubt that emotions play a tremendous role in the majority of creative processes.

A similarly great role is played by intuition and we have also an *intuitive heritage of humanity*. Recall that Kant (1781) defined a priori *synthetic judgements* as our concepts and assessments concerning space and time that appear to us as obviously true. Kant developed further the arguments of Plato, who in the dialog of Socrates with Menon has shown how a rather difficult mathematical problem (even if not quite difficult if treated as a technical problem, see comments to Chap. 5), of constructing a square with an area equal to the half of the area of an original square, can be intuitively solved by a young man without adequate education. Kant extended these arguments over the concepts in Euclidean space and Newtonian time, and later also to logics or even ethics. Today we know that these concepts and judgements that appeared to Kant as obviously true are not at all obvious and neither universally true: the space can be non-Euclidean, the time can be relative as shown by Einstein, or it may develop in two parallel scales, etc. Therefore, these concepts and judgements are not universally true, although appear to us as true; how is it possible?

It can be explained as an intuitive heritage. We learn spatial relations when playing as children with blocks or with Lego, and these relations become the foundation of our mathematical intuition that is reinforced by lessons of mathematics in school. Thus, the paradigm of teaching mathematics in school is a part of intuitive heritage of humanity. This intuitive perception of the world is not necessarily true, because we live and perceive the world in a *meso-cosmic scale,* and often we do not have a chance to see it in micro-cosmic or in macro-cosmic scale, see e.g. Vollmer (1984). However, this meso-cosmic perception provides us with a strong intuitive understanding of space and time, reinforced by the tradition of teaching mathematics at school. Kant believed that this intuitive understanding is given to us a priori, but I think that even if a part of this intuition is inherited, most of it is acquired by learning. An interesting research project arises from this and from the evolutionary theory of intuition: what experiments should we conduct to check which part of our intuition of time and space is inherited, and which one is learned? This way, we could give an experimental answer to a discourse concerning *synthetic* a priori *judgements,* which is over 200 years old.

Another part of the intuitive heritage of humanity is the intuitive understanding of logics related to the quasi-conscious use of language. Note that this understanding is originally acquired through verbal discourses and reinforced by formal learning of logics. Some people have better feeling of logics, such as some of us have a better feeling of time and space. Hoverer, the discussion above clearly indicates that the intuitive heritage of humanity, the intuition of space, time, logics, even ethics, is one of the greatest achievements and, at the same time, a basic fundament of our civilization.

Once we defined the nodes in creative space as ontological elements, we can proceed to the definition and analysis of the diverse transitions between them. A transitions can connect any pair of nodes and there can be various transitions, even in the same direction, between a given pair of nodes. For example, between the nodes of *individual intuition* and *individual rationality* we can observe *internalization* described by Nonaka and Takeuchi, resulting in an increase of individual intuition due to practical application of rational knowledge, but between the same nodes there is also an *expansion*, in the opposite direction, described by Gasson, and in parallel to that, another but perhaps the most important one: *enlightenment* (illumination, aha, eureka, abduction), a fundamental element of all intuitive creative processes. On another level, new facts in the *rational heritage* of humanity might require an *interpretation* using the *intuitive heritage;* and a deficiency in the *intuitive heritage* in the time of a crisis of a knowledge domain might require a *regress* to the *emotional heritage,* as described by Motycka (1998). There are also creative transitions connecting directly the *individual* level with the *intellectual heritage* while possibly bypassing the *group* level: these are well known processes of *publication* of a scientific book or a paper or a *composition* of a work of art.

There is no place here to describe all nodes and transitions in more detail (see Wierzbicki and Nakamori 2006). However, some important general conclusions should be underlined. Firstly, despite the appreciation of the revolutionary importance and character of the *SECI spiral,* it does not constitute a unique and ultimate explanation of creative processes. Nevertheless, we can follow its example and try to present other creative processes in the form of spirals, mostly in the fundamental dimensions of the creative space as represented in Fig. 12.4. Secondly, these fundamental (epistemological and social) dimensions in which the *SECI spiral* was defined do not exhaust all aspects of creative activity. This is stressed particularly by the I^5 *system pentagram* of Nakamori (2000); thus, we should consider also other dimensions of creative space.

12.4 Spirals and Academic Processes of Knowledge Creation

We have already mentioned the theory of creation of fundamental knowledge in the time of a scientific revolution, formulated by Motycka (1998). It can be illustrated by the upper left corner of Fig. 12.4 that is repeated for clarity in Fig. 12.5.

If a group of researchers in a given scientific domain intuitively perceives a crisis of this domain which cannot be resolved by an *abstraction* to the *intuitive heritage,* then a further search might consist of a *regress* to the *emotional heritage,* with its elements of tacit knowledge in the form of *social unconsciousness* of Jung. The results of such search have to influence first the emotions of that group, in a transitions that can be called *mythologization.* In this transition, archetypes of myths and instincts of humanity influence the emotional perceptions of the group. This is not yet equivalent to the impact on *group intuition;* the emotions must be

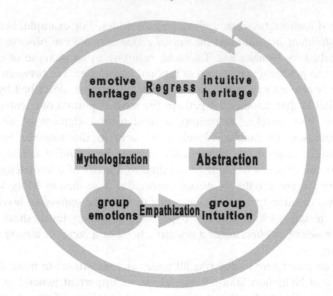

Fig. 12.5 ARME Spiral (Motycka 1998)

subject to a specific discourse with an aim to achieve an empathic intuitive understanding. Thus, the transition from *group emotions* to *group intuition* is called *empathization.*

This is only a very short summary of processes described in more detail in (Motycka 1998), and known in the history of science, e.g. from the discussions on the fundamentals of quantum theory conducted by the group led by Bohr. Here it is only stressed that such a process can be also represented as a spiral, called *ARME spiral*, of following transitions: *Abstraction-Regress-Mythologization-Emphatization.* Such a process can be repeated, and each repetition can contribute to further increase of knowledge. Clearly, similarly to the *SECI spiral,* the *ARME spiral* is actually a metaphor, a model from which many deviations could be observed in practice. There is, however, no doubt that such a metaphor describes a process of knowledge creation that historically led to new, fundamental knowledge.

Once we have a spiral of knowledge creation in the conditions of scientific revolution, we should also have spirals of knowledge creation in the conditions of normal development of science at universities and other research institutions, according to the division between revolutionary and normal development of science proposed by Kuhn (1962). The *SECI spiral* does not describe such a process, it concentrates on creating knowledge according to the interests of a group in a market organization, a process which is accepted and supported by individual members of the group. In the conditions of normal academic knowledge creation at universities, the dominating aspect is the individual interest of a researcher, although it is accepted and supported by the group, since university is a society of researchers supporting each other. According to the classical opinions of Humboldt concerning

Fig. 12.6 EDIS spiral of debate (Wierzbicki and Nakamori 2006)

fundamental academic processes of knowledge creation, we should analyze at least three such processes: a *debate, experiment,* and *hermeneutics,* the latter understood broadly as an interpretation of selected elements of intellectual heritage of humanity, whether they belong to theology, human studies, or strict sciences, and even technical studies, or other fields.

Debate is a process proceeding, approximately, along the same nodes in the creative space as *SECI spiral* or *OPEC spiral,* but with different transitions and interpretation. The process of normal development of knowledge during an academic debate is well known and can be easily recognized in the *EDIS spiral* presented in Fig. 12.6.

An individual researcher, due to her/his intuition, has an idea (smaller or bigger one) and the illumination alone is not enough, the idea must be rationalized, expressed in words or equations; this transition might be called *Enlightenment.* A group can support the researcher by giving her/him a forum to discuss the ideas, the more incisive the discussion, the better the support of the group, and we call this transition *Debate.* We know these two stages well, but here a new stage appears resulting from the rational theory of intuition (Chap. 5), which indicates that a deeper, more incisive discussion will be achieved if we give the group some time for reflection, for ripening of comments, for a transition which is called here *Immersion* of the group rationality in the group intuition. Hence a practical conclusion, which may be called *double debate principle:* discussion of new ideas should be repeated after a week or two, if we want to support the researcher not only with the explicit knowledge of the group, but also with its tacit or intuitive knowledge. After getting the comments of the group, the individual researcher makes a choice, *Selection,* of those comments that should be taken into account in

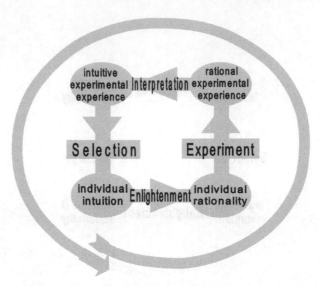

Fig. 12.7 EEIS spiral of experiments (Wierzbicki and Nakamori 2006)

further research. We all know that this choice is made on an intuitive level, that it is not necessarily rational.

One can ask what advantage is in describing well-known creative processes in the form of such abstract, simplified and metaphoric models? The answer is simple: even in a well-known process, by developing such a simple model we derived important practical, supplementary conclusions in the form of the *double debate principle*. We can also analyse what was simplified or omitted in this model. For example, normal processes of knowledge creation usually to a large extent involve rational, emotional or even intuitive heritage of humanity, together with a herme-neutic interpretation of this heritage, and often end in publication; hence the *EDIS* spiral describes only a part of such process that consists of a larger number of transitions between the nodes of creative space. In experimental sciences, beside debates we check our ideas through experiments, which suggests that an experi-mental dimension is missing in the dimensions of creative space in Fig. 12.4, so the number of dimensions of this space is greater. However, before discussing the issue of dimensions of creative space, it is worthwhile to describe the processes of experimental and hermeneutic creation of knowledge with the use of appropriate spirals, originally presented in Wierzbicki and Nakamori (2006, 2007).

I think that it is not necessary to discuss the importance of experimental ver-ification: who does not appreciate it, should not travel by airplanes. The spiral describing experimental processes of knowledge creation is a modification of the *EDIS spiral* but taking into account situations where the verification of new ideas occurs not through debate, but through an experiment. Such modified *EEIS spiral (Enlightenment-Experiment-Interpretation-Selection)* is illustrated in Fig. 12.7.

**Fig. 12.8 EAIR
hermeneutic spiral**
(Wierzbicki and Nakamori
2006)

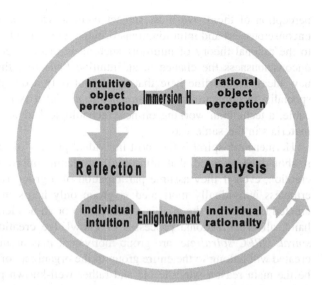

The *Experiment* transition describes experimental verification. However, each experimental researcher knows well that raw experimental data do not tell much, so their *Interpretation* is necessary. In this model, an essential novelty in description of experimental processes is the emphasis that an interpretation of experimental data is actually an immersion of raw data in an experimental intuition of the researcher, intuition based on experience. Similarly as in the *EDIS spiral,* also here a *Selection* of conclusions occurs, this time concerning those aspects of experimental data that have the greatest impact on further development of research ideas. Experiment has often individual character, although larger experiments might be clearly organized and conducted by groups of researchers.

Hermeneutics, originally understood as the art of interpreting Bible, then any text, was usually treated as a typically humanistic activity, even one that distinguishes humanistic sciences from other sciences (see Gadamer 1960). However, as it was already mentioned, some great philosophers of the 20th century treated hermeneutics as a broader concept, encompassing all sciences. Following this idea, I shall understand hermeneutics as the art of interpretation of any element of the intellectual heritage of humanity, thus a part of any research, including academic activities: gathering of materials relevant for a given theme from libraries or from the Internet, their interpretation and reflection on them. Hermeneutic creative activity is described by the *EAIR hermeneutic spiral,* see Fig. 12.8. This spiral is another interpretation of the *hermeneutic circle* with the distinction that its interpretation is naturalistic: it is not closed by transcendence, as postulated by Gadamer, but by the power of natural human intuition.

If we have a new idea coming from individual *abduction* or *illumination, Enlightenment,* we search for related materials in libraries and on the Internet and analyze them rationally, *Analysis.* This is not sufficient to obtain a hermeneutic

234 Creative Space and Micro-Models of Knowledge Creation

perception of them: we must subject them to *Hermeneutic Immersion* into our unconsciousness and intuition, transform them into intuitive perception. According to the rational theory of intuition, such an immersion is possible if we give our unconsciousness the chance of an intuitive *Reflection* that might be a source of new ideas. I underline here that this is not only a description of the work of a specialist in humanities in a *hermeneutic circle;* it is much more broadly applicable, a technician working on new technological ideas reviews her/his research materials in the same way.

Hermeneutic spiral is the most individual process of knowledge creation. Here it should be stressed that almost all academic micro-processes of knowledge creation, even if they assume participation of a group (as in a *debate*), are nevertheless individually motivated: a group only helps in improving individually created knowledge, the goal is a publication or an academic degree. On the other hand, all organizational processes of knowledge creation, brainstorming, *SECI spiral, OPEC spiral* etc, are group-motivated: it is assumed that the knowledge created will belong to the entire group or the organizers of that process. This might be the main reason why the old and rather well-known process of brainstorming did not find a broader appreciation and application in academic conditions. It implies, however, that academic and organizational processes of knowledge creation are essentially different, which might be one of main reasons for the difficulties and delays in transferring knowledge from academia to industry. On the other hand, if we understand these differences and difficulties well, we can try to counteract, e.g. to combine organizational and academic knowledge creation, which will be illustrated below in the *Nanatsudaki septagram.*

12.5 Further Dimensions of Creative Space: Complex Processes of Knowledge Creation

Nakamori (2000) proposed a more extensive model of processes of knowledge creation, expressed by a *pentagram of I^5 system,* see Fig. 12.9.

Two fundamental nodes of this pentagram, *Intelligence* and *Involvement,* can be approximately treated as equivalent to the two basic dimensions of creative space: epistemological and social dimension. However, remaining nodes characterize other aspects of knowledge creation. *Imagination* is related to intuition, but can be also treated as a separate dimension. *Intervention*[4] is in a sense a starting node of the pentagram, it expresses the will to take up the problem. *Integration* is an interdisciplinary synthesis, usually based on systems approach, and represents a final, concluding node.

---

[4] This is not an exact explanation, but consistent with the description and intent of the work (Nakamori 2000).

Fig. 12.9 Pentagram of I⁵ system (Nakamori 2000)

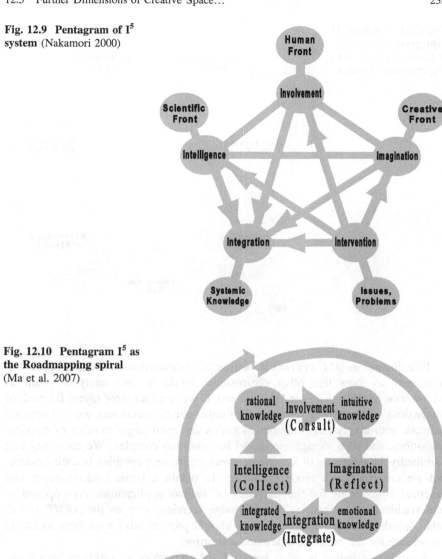

Fig. 12.10 Pentagram I⁵ as the Roadmapping spiral (Ma et al. 2007)

In Ma et al. (2007), a different interpretation of I^5 *system* in the form of a *Roadmapping spiral* is presented (where the concept *Roadmapping* denotes a specific process of forecasting future and planning, see Chap. 14 for a more detailed discussion). This spiral is illustrated in Fig. 12.10, where the initial node is *Intervention,* and the nodes of the pentagram are treated as subsequent transitions in the order of *Intelligence-Involvement-Imagination-Integration.* The nodes between them stand for rational, intuitive, emotional and integrated knowledge.

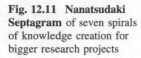

Fig. 12.11 Nanatsudaki
Septagram of seven spirals
of knowledge creation for
bigger research projects

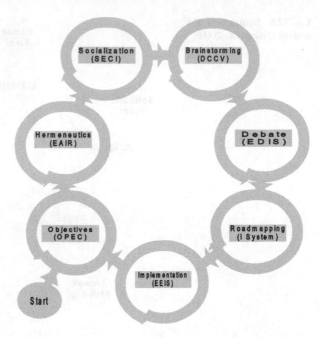

Five dimensions of I^5 *system* can be treated as dimensions of creative space, but
it is easy to show that other dimensions should be also analyzed, such as
Abstraction or *Objectivity* (or *Verification*). If we consider three layers for each of
dimensions of the creative space (as for fundamental dimensions), we will obtain a
network with a very large number of nodes and even larger number of possible
transitions, and such metaphoric model becomes too complex. We could say that
admittedly the processes of knowledge creation are very complex in their essence,
but we construct such models in order to obtain a better understanding and
practical conclusions for the purposes of various applications. As opposed to
philosophical macro-theories of knowledge creation, such as the *ARME spiral,*
other spirals discussed in this chapter should provide advice on how to create
knowledge for the needs of today and tomorrow.

A resulting question is: what approach, or what order of spirals of knowledge
creation described here might be recommended for a larger, group research project
e.g. in the field of technical science? Experience in management of science helped
me to propose the order of such processes in the form of seven spirals, all already
presented above. This is illustrated in Fig. 12.11, presenting the *Nanatsudaki
septagram* of seven spirals of knowledge creation.[5]

The experience mentioned above suggests that we should start from an Anglo-
Saxon approach and discuss goals of the project with the members of the research

---

[5] In Japanese, *Nanatsudaki* means seven waterfalls, a beautiful aspect of a stream flowing in the
forest around the Japan Advanced Institute of Science and Technology in Asahidai near Kanazawa.

group by applying (perhaps partly) the *OPEC spiral*. Afterwards, each member of the research group should start literature studies and interpret obtained ideas by applying the *EAIR hermeneutic spiral*. Next it is good to use the Far East approach and conduct at least *socialization* belonging to the *SECI spiral*, to exchange the ideas informally.

After such hermeneutics and socialization, the research group members are well prepared to take part in a *brainstorming, DCCV spiral*, at least in its divergent part. Its convergent part, and especially crystallization of ideas, can be better supported by a classical *debate*. After crystallization of ideas it is time for a most time consuming, at least in technical sciences, process of experimentation; therefore, it is good to plan it in detail, e.g. using the *roadmapping spiral* before implementing the *EEIS experimental spiral*. To finish the project, it might be good to use the *OPEC spiral* again, perhaps also elements of other spirals.

12.6 Practical and General Conclusions

I underscore here in the first place the practical aspects of conclusions, since the micro-models of knowledge creation concern in a sense recipes of how to develop knowledge for the needs of today and tomorrow. There are many such conclusions, and only several more general ones are discussed below.

Firstly, we should answer the question: what purpose such heuristic models serve? The answer is simple: they have certain cognitive values, but most important are their pragmatic values.

Secondly, in the praxis of knowledge creation it is very important to use various methods to stimulate intuition and creativity, resulting from the rational theory of intuition and other sources. These methods include simple ones, such as the *alarm clock method* (concentration on new ideas just after waking up) described in Chap. 5, but also more complex methods, such as deliberate switching off of the rational part of consciousness, either through Zen meditation and tea ceremony, or through listening to classical music.

Thirdly, a proper stimulation and organization of the fundamental function of a university toward creative individuals: providing a good discussion forum for individual ideas, is especially important for normal development of knowledge at universities. Starting with *Dialogs* of Plato (see e.g. Plato 1993) there exists a long tradition and an extensive theory of debate, hence there are foundations to improve such forum; but today's universities devote not enough attention to it.[6] Additionally, it is possible to use the conclusions from the rational theory of intuition and resulting *principle of double debate*, repeating discussions after a relatively short time in order to *immerse* the subject of discussion in the intuition of

---

[6] For example, while in Japan an informal exchange of opinions during *socialization* works rather well, the abilities of scientific debate are rather poor.

discussants. On the other hand, tools provided by the informational revolution can also be used to support the debate.

Fourthly, also the operation of the *EAIR hermeneutic spiral* can be improved with the use of tools of the informational revolution. An example is the simple computer support for the *Analysis* transition, including the search for useful materials. There are many *search engines* on the Internet, starting with Google with its Google Scholar version, but all of them, in order to function commercially, must concentrate on other goals than best free service for the user. However, it is possible to develop a *human centred* version of a search engine: using ontological engineering and asking the user to define her/his *hermeneutic profile* (a highly personalized version of ontology describing the interests of the user), we can serve the hermeneutic interests of the user (see Ren and Wierzbicki 2007; Granat and Wierzbicki 2009).[7] Similarly, other spirals of knowledge creation can also be computer supported.

Fifthly, after the informational revolution, during the transition towards knowledge based economy and civilization, we should *avoid mental schemas resulting from the tradition of industrial economy and civilization.* This conclusion starts the more general part.

While the 20th century unconsciously succumbed to the instruction of Wittgenstein (1922) that we should not speak about metaphysics, at the end of that century and in the beginning of the 21st century a characteristic change in epistemology occurred: not only the new micro-models of knowledge creation highlighted irrational or a-rational elements traditionally included to metaphysics, such as tacit knowledge, intuition, social unconsciousness, but also philosophy and epistemology proper turned openly towards metaphysics. This chapter has shown the possibility of their integration, a common interpretation as processes occurring in the so-called *creative space,* usually in the form of *creative spirals.* However, it does not express an ambition to document changes in philosophy proper, even if they are visible to an outsider.

One of the elements of the informational revolution is constituted by a change of assumptions that appeared obvious in the industrial civilization. In philosophy, and especially in epistemology, such an assumption was to consider language as an unquestioned tool for not only expression, but also creation of knowledge. New philosophical approaches, similarly as new micro-models of knowledge creation coming from other fields than philosophy, question this assumption and suggest *preverbal sources of creativity; language is only an imperfect code of expressing knowledge about the world surrounding us.*

We should also realize that *many other assumptions should be revised.* Braudel (1979, 1995) defined a *long duration structure,* that is, a kind of civilization epoch, an example of which was the epoch of 1440–1760 considered by him, from Guttenberg to Watt, as a period in which there is a settled way of perceiving the

---

[7] Commercial search engines personalize search and form user profiles, but these profiles serve the commercial interests of the search engine, less the actual interests of the user.

world, thus specific fundamental cognitive assumptions are in force. The epoch of industrial civilization is the period of 1760–1980, until the time of combining personal computers and computer networks that formed the technical foundation of the new era. Thus, we can expect the following conceptual changes, and sometimes we observe them already:

(1) *From a clock to an avalanche.* Industrial civilization understood the world as a great clock, turning with the inevitability of celestial spheres. The complexity of contemporary civilization together with the theory of deterministic chaos suggest that we should see the world as a complex dynamic system with behaviour of avalanche or hurricane type, in which anything can happen. We ceased to believe in the inevitability of historic laws of Karl Marx, and we will soon cease to believe in the inevitability of the invisible hand of market of Adam Smith.

(2) *From industrial economy to knowledge economy.* This is a quantitative change that translates into a qualitative change (in this respect Karl Marx was right): knowledge was always a productive factor, but in knowledge economy it becomes dominant. However, differently than other productive factors, knowledge does not cost more if used repeatedly. Hence the fundamental assumption of the competitive market theory, that with the growth of production volume its marginal costs will grow and thus a competitive equilibrium will be formed, ceased to be applicable to all markets with a large participation of knowledge. Therefore, the prices on such markets do not have much in common with the marginal production costs anymore. The economists, even those correctly stressing the importance of the knowledge-based economy, see e.g. Stehr (1994), Kleer et al. (2009), still do not fully appreciate that fact, hence a basic revision of economic theory is necessary.

(3) *From reductionism to the emergence principle.* The industrial civilization believed in the *reduction principle,* explaining the behaviour of a complex system by reduction to the behaviour of its parts. In complex, nonlinear dynamic systems with feedback we observe in contrast, as explained in the previous chapters, an *emergence,* occurrence of new properties, related not to the behaviour of parts, but to complexity. However, this means an essential change in the way we understand the world during the informational revolution.

The above examples do not exhaust the necessary change of fundamental assumptions. Because of the growing role of science and knowledge, a fundamental conflict of the new epoch will probably concern *property of knowledge.* While understanding its importance in contemporary economy, big corporations try to anticipate the solution of this conflict and to privatize knowledge to the highest degree possible. However, the intellectual heritage of humanity, with its rational, intuitive, emotional parts, was historically treated as a common, public property of humanity, and its importance is tremendous even for creation of small increases of knowledge for market innovations. If we want to have healthy economic principles in the knowledge-based economy, then the taxes for the market use of human intellectual heritage should be high and earmarked for its further development, not for the profit of big corporations. The latter clearly understand this dilemma and exert pressure to fully privatize knowledge, including universities, using neo-liberal doctrine, which, as indicated above, ceased to be valid in the knowledge-based economy. Therefore, it is necessary that people that create knowledge understand this dilemma and take steps to resolve it in the interest of entire humanity. We should return to this problem in Chap. 14.

References

Braudel, F.: Civilisation Matérielle, Économie et Capitalisme. XV-XVIII siècle, Armand Colin, Paris (1979)

Braudel, F.: A History of Civilizations. Penguin Books, London, New York (1995)

Gadamer, H-G.: Warheit und Methode. Grundzüge einer philosophishen Hermeneutik. In: Mohr, J.B.C. (Siebeck), Tübingen (1960)

Gasson, S.: The management of distributed organizational knowledge. In: Sprague, R.J. (ed.) Proceedings of the 37th Hawaii International Conference on Systems Sciences. IEEE C.S. Press (2004)

Granat, J., Wierzbicki, A.P.: Inżynieria wiedzy—nowy obszar badawczy Instytutu Łączności (Knowledge Engineering—a New Research Field of National Institute of Telecommunications). Telekomunikacja i Techniki Informacyjne 3–4, 108–116 (2009)

Jung, C.G.: Psychological Types. Princenton Univeristy Press. Polish translation (2009) Typy 588 psychologiczne. Wydawnictwo KR, Warsaw (1971)

Kant, I.: Kritik der reinen Vernunft. Polish translation (1957) Krytyka czystego rozumu, PWN, Warsaw (1781)

Kleer, J.: Przyszłość zniewolona przez przeszłość (Future Enslaved by Past). In: Kleer, J., Galwas, B., Wierzbicki, A.P. (eds.) Rola nauki w myśleniu o przyszłości (The Role of Science in Thinking about Future), pp. 347–369. Komitet Prognoz "Polska 2000 Plus", Warsaw (2009)

Kuhn, T.S.: The Structure of Scientific Revolutions, 2nd edn., 1970. Chicago University Press, Chicago (1962)

Kunifuji, S., Kato, N., Wierzbicki, A.P.: Creativity Support in Brainstorming. In: Wierzbicki, A.P., Nakamori, Y. (eds.) Creative Environments, op.cit., pp. 93–126 (2007)

Kunifuji, S., Kawaji, T., Onabuta, T., Hirata, T., Sakamoto, R., Kato, N.: Creativity support systems in JAIST. In: Proceedings of JAIST Forum 2004: Technology Creation Based on Knowledge Science, pp. 56–58 (2004)

Ma, T., Yan, J., Nakamori, Y., Wierzbicki, A.P.: Creativity support in roadmapping. In: Wierzbicki, A.P., Nakamori, Y. (eds.) Creative Environments, op.cit., pp. 155–189 (2007)

Motycka, A.: Nauka a Nieświadomość (Science and Unconscious). Leopoldinum, Wrocław (1998)

Nakamori, Y., Sawaragi, Y.: Shinayakana Systems Approach to Modeling and Decision Support. In: Proceedings of MCDM 1992 (10th International Conference on Multiple Criteria Decision Making). Taipei, Taiwan, 2, pp. 77–86 (1992)

Nakamori, Y.: Knowledge Management System Toward Sustainable Society. In: Proceedings of First International Symposium on Knowledge and System Sciences, JAIST, pp. 57–64 (2000)

Nakayama, H., Sawaragi, Y.: Satisficing trade-off method for interactive multiobjective programming methods. In: Grauer, M., Wierzbicki, A.P. (eds.) Interactive Decision Analysis, pp. 113–122. Springer Verlag, Berlin (1984)

Nonaka, I., Takeuchi, H.: The Knowledge-Creating Company. How Japanese Companies Create the Dynamics of Innovation. Oxford University Press, New York (Polish translation: Kreowanie wiedzy w organizacji, Poltext 2000) (1995)

Osborn, A.F.: Applied Imagination. Scribner, New York (1957)

Plato (ca. 380 BC) Dialogi (Dialogs). Polish translation Verum, Warsaw (1993)

Polanyi, M.: The Tacit Dimension. Routledge and Kegan, London (1966)

Ren, H., Wierzbicki, A.: Integrated Support for Scientific Activity. In: A.P. Wierzbicki, Y. Nakamori (eds.) Creative Environments, op. cit. (2007)

Searle, J.R.: The Rediscovery of Mind. MIT Press, Boston (1992)

Stehr, N.: The Texture of Knowledge Societies. In: Stehr, N. (ed.) Knowledge Societies, pp. 222–260. Sage, London (1994)

Vollmer, G.: Mesocosm and Objective Knowledge: On Problems Solved by Evolutionary Epistemology. In: F.M. Wuketits (ed.) Concepts and Approaches in Evolutionary Epistemology. D. Reidel Publishing Co., Dordrecht (1984)

Wierzbicki, A.P.: A mathematical basis for satisficing decision making. Math. Model. **3**, 391–405 (1983)

Wierzbicki, A.P.: On the role of intuition in decision making and some ways of multicriteria aid of intuition. Multi. Criteria Decis. Mak. **6**, 65–78 (1997)

Wierzbicki, A.P.: Konsekwencje popytu na wiedzę: Przestrzeń twórcza i mikro-modele kreowania wiedzy (The Consequences of the Demand for Knowledge: Creative Space and Micro-Models of Creating Knowledge. A lecture for Doctoral Studies at Institute of Systems Studies of P.Ac.Sc., September 2011 (2011)

Wierzbicki, A.P.: Megatrends of Information Society and the Emergence of Knowledge Science. In: Proceedings of the International Conference on Virtual Environments for Advanced Modeling, JAIST, Tatsunokuchi (2000)

Wierzbicki, A.P., Nakamori, Y.: Creative Space: Models of Creative Processes for the Knowledge Civilization Age. Springer Verlag, Berlin-Heidelberg (2006)

Wierzbicki, A.P., Nakamori, Y. (eds.) Creative Environments: Issues of Creativity Support for the Knowledge Civilization Age. Springer Verlag, Berlin-Heidelberg (2007)

Wittgenstein, L.: Tractatus Logico-Philosophicus. Cambridge University Press, Cambridge (1922)

Chapter 13
Philosophy Versus History of Technology

13.1 General Impression of a Technologist

If we select a typical monograph on the philosophy of technology from the turn of the 20th and 21st century, e.g. Scharff and Dusek (2003), then its reading must fill a technologist with a tremendous worry. This anthology of philosophy of technology, edited in Oxford, very comprehensive and thus and appearing to be authoritative, contains 55 papers. The first 14, however, are devoted to the philosophy of strict sciences and the first paper on the philosophy of technology, by Mario Bunge, is based on the assumption that technology is a simple application of strict sciences. After the fundamental text *Die Technik und die Kehre* of Martin Heidegger (in a rather biased translation *The Question Concerning Technology*)[1] there are several papers trying to show that Heidegger has seen in technology an ominous, dark force, without understanding the fact commented in the previous chapters that Heidegger has seen such a force in the fascination of people by technology, while in technology he has seen a creative process of unveiling truth. A technical reader cannot resist an impression that the selection of texts in that monograph simply illustrates an anti-technical attitude of the philosophy of technology, since more neutrally oriented works, discussed below, are significantly omitted in it. From the last seven papers in that monograph only one, *The Social Impact of Technological Change* by Emmanuel G. Mesthene, does not represent an anti-technical attitude, but it is sharply criticized for that by next authors: the paper of John McDermont, immediately following the said text, maintains that technology is an opium for intellectuals. It is also telling that *neither of these 55 texts was written by a representative of technical sciences*. Since the philosophy of mathematics is usually developed by mathematicians, why is it not so with the philosophy of technology?

---

[1] The original title of the work of Heidegger suggests that it is technology that enables social changes; the translators of this paper did not want to admit such possibility.

© Springer International Publishing Switzerland 2015
A.P. Wierzbicki, *Technen: Elements of Recent History of Information Technologies with Epistemological Conclusions*, Intelligent Systems Reference Library 71, DOI 10.1007/978-3-319-09033-7_13

Such anti-technical attitude, prevailing in many papers of that Oxford anthology, is best illustrated by the question of ethics of technology. The discussion of this fundamental question starts the paper of Kristine Schrader-Frechette, who maintains that technology does not appreciate technical risks, hence is unethical. The concept of technology is not defined in that paper, the concept of technological risk is treated in an extremely simplified way, without the necessary distinction of technology proper and its diverse applications. This distinction is necessary, because technological risk does not strongly depend on technology proper, it fundamentally depends on the type of its applications and the manner of applying it. E.g. a technician that constructs hammers is obviously responsible for such a construction so that the head does not fall off when we use the tool, but the user who selects a smith's hammer to fasten small nails is responsible himself for his broken fingers. This does not imply that technology is neutral, only that its applications can have both positive and negative aspects. Most of the negative or even catastrophic results of technology starts with rash or irresponsible applications, or with a social fascination with certain technological products, against which Heidegger warned. Therefore, the arguments of Schrader-Frechette are similar to ascribing the blame to a technician who tries to improve combustion processes that she/he is doing it too slowly, so she/he is responsible for an extensive emission of CO_2 and global warming.

I should stress here again why I perceive the anti-technical attitude of most philosophy of technology as extremely dangerous: because *a wrong diagnosis does not help to cure an illness*. Technicians will reject such a diagnosis as the lack of understanding of technology, while social and humanist sciences have found a scapegoat and do not consider their own responsibility. However, as shown in Chap. 3, both sides should feel responsible. It is true that technology products make it possible nowadays to terminate all life on Earth with them, but this depends on their socio-economic use, on the social system of their utilization. Moreover, if technology corresponds today to a separate cultural sphere, the philosophy of technology cannot be developed without consulting representatives of technical sciences. This results from many reasons, but mostly from common sense. It is too dangerous to allow ourselves not to understand technology that enables not only a fundamental change of our lives, but also, if incompetently or irresponsibly used, threatens to terminate all life on Earth (through an irresponsible use not only of nuclear energy, but also of genetic technologies, even of robots). Postmodern humanists, however, will not understand technology until they master the issue of the objectivity of their own paradigms, or more generally, humanists will not understand technology if they will not be educated in several selected technical subjects. Strict sciences will be still prone to see technology as a straightforward application of their results. All this creates an extremely dangerous situation, and the perception of danger only grows when we analyze contemporary philosophy of technology.

In such a situation it does not seem productive to discuss with the anti-technological attitude of most of the philosophy of technology, but it is necessary to discuss with its hermeneutic perspective, with foundations of its paradigm that result in such an attitude. Therefore, in this chapter I will concentrate on a presentation and discussion of several texts on the philosophy of technology omitted by the Oxford anthology, which are more friendly towards technology, while focusing on the exploration of their horizontal, hermeneutic elements and their relation to the history of technology. The latter aspect concerns an observation that the philosophy of technology, as opposed e.g. to the philosophy of mathematics, seldom analyzes the history of technology, and especially its recent history. As shown in Chaps. 3 and 4, the delays in applications of new technology products make the philosophy of technology irrelevant without recent history of technology.

13.2   Do Scientific Revolutions Happen in Technology? Rachel Laudan: *The Nature of Technological Knowledge*

The Nature of Technological Knowledge: Are Models of Scientific Change Relevant? an excellent monograph edited by Laudan (1984) and not appreciated by other philosophers of technology starts with a clear statement that all contemporary assaults on technology, together with the opposite but equally uncritical admiration of technological achievements, do not contribute to a deeper understanding of technology. The biggest contribution to the understanding of technology comes from its history. And from an analysis of the history of technology comes a clear conclusion that the nature of technology is as of tacit knowledge and thus non-transparent for a historian, sociologist or philosopher. On page 7 of her introduction, Rachel Laudan states that a non-technologist can describe technology products, order them chronologically, describe the biographies of inventors, analyze the impact of inventions on the society, but cannot reconstruct the actions that has led technological practitioners to their inventions. Laudan stresses also the unique and visual character of technology creation. I agree with and share all these opinions of Rachel Laudan and I develop them further in this book; I do not wonder, however, that such opinions are ignored by the mainstream philosophy of technology, since they are against its basic paradigms, discussed in next sections.

Rachel Laudan attacks also, as an imprecise and superficial myth, the opinions that technology is a simple application of the results of strict sciences. She maintains that both science and technology are forms of knowledge. She notes their fundamental differences and connections, but does not recognize the positive feedback relation between these forms of knowledge, does not also stress enough the difference between technological creativity and technological science. The latter is undoubtedly a form of science that supports, but does not determine, technological creativity that can be learned only through practice.

The main goal of the book edited by Rachel Laudan is to analyze a possible connection of qualitative changes in technology with the theory of scientific revolutions and the development of science (by T. Kuhn, I. Lakatos, L. Laudan and others). If technology is not a simple application of scientific results, then do the theories of scientific revolutions apply to technology, or not? To this question, analyzed comprehensively, the book edited by Rachel Laudan gives finally a negative answer.

After a short introduction by Rachel Laudan, the book starts admittedly with a paper of Edward Constant, the author of *Origins of the Turbojet Revolution* (Constant 1980), who concentrates on the breakthrough invention of the turbojet engine, based on the application of the predictions of strict science. Constant sees here a similarity to the scientific revolution in the sense of Kuhn, but stresses the differences between science and technology as two different kinds of knowledge,[2] and resulting differences between a scientific revolution and technological revolution.

However, the next one, very incisive paper *Paradigms, Revolutions and Technology* by Garry Gutting starts with a fundamental critique of the Constant's reasoning by stressing the necessity of using a precise, not a broad interpretation of the Kuhnian concept of paradigm: in original interpretation of Kuhn, a paradigm is an exemplar theory, not a broadly understood meta-theory, a system of meta-premises (which I call in this book hermeneutic horizon or hermeneutic perspective). Further on, Gutting states: "even in the cases where technology is thoroughly informed by theoretical science, particular *inventions* (concrete theoretical achievements) are not just the end-product of technology as applied science but themselves play a major role in stimulating further technological advances".

Gutting sharply criticizes the opinion of Bunge (1966) that technology is a straightforward application of science and is close to the just opposite opinion of (Skolimowski 1966), even if he criticizes also Skolimowski for a too instrumental approach to technology, resulting in an opinion (common to Skolimowski and Bunge) that it is science, not technology, that has the goal of development of knowledge. Gutting, on the other hand, sees in technology a cognitive enterprise, even if different than science.

The book edited by Rachel Laudan contains many other excellent papers, but for me, all of them have three drawbacks. Firstly, all of them do not see clearly technology proper as *the art of creating tools and artefacts*. Admittedly, in the paper *Organizational Aspects of Technological Change* by Norman Hummon the concept of the "art of technology" is used, however without a deeper analysis. Secondly, neither of them takes up the issue of the dynamics of creative processes in technology, nor notes the issue of delays and their importance; as stressed in Chap. 4 of this book, these issues are very important for understanding of

---

[2] Constant explains these differences by more *hierarchical nature of technology* (correct if not quite penetrating, see Chap. 10 of this book on the contribution of technology to the concept of emergence), by more *approximate character of technological solutions* (correct, technology is more pragmatic, hence accepts approximate solutions), and by a larger *influence of socio-economic factors on the selection of technological solutions* (correct, but only in the final selection of applications of technology; original creative selection by a technologist is based more on her/his vision of the future needs, see Chap. 3 of this book).

technological change. Thirdly, while taking the facts of history of technology into account, the authors of that papers concentrate on the older, well described history, and examples given by them concern almost entirely mechanical technology, important for industrial civilization but losing its importance after the informational revolution.

The entire book is concluded by the paper *The Structure of Technological Change* by Peter Weingart stating that "The structure of technological change is so markedly different from that of scientific change that a transfer of approaches from the study of science to the study of technology seems to have very little promise".

13.3   Has a Philosopher of Technology the Right to Decide What is Technology? Carl Mitcham: *Thinking Through Technology*

Mitcham (1994) published an excellent review of the philosophy of technology, worth recommending for all who want to learn about its diverse (and controversial) views. I shall concentrate, however, not on complimenting the Author, but on a critique of his hermeneutic perspective. The Author tries to be friendly to technology but does not note how the lack of a precise definition of technology and the specific paradigm (both in strict and in the broad sense of the word) of humanistic philosophy of technology, which he openly defends, leads him to positions actually unfriendly to technology. Let us illustrate this by an example.

On page 61, when discussing the ideas of Jacques Ellul, Mitcham writes: "... Ellul argues for an ethics of nonpower that would sharply delimit *technical practice* [...] An ethics of nonpower, the root of the affair, is obviously that *human beings* agree not to do everything they are able to do" (words stressed by me). Thus Mitcham implies that *technologists in their practice* make abuses. On the other hand, Ellul expressed a different opinion: the latter part of the quoted opinion of Mitcham is a direct quotation from Ellul who addressed "ethics of nonpower" to human beings, not only to technologists but to *all people using technology products*. All technologists would agree with the opinion of Ellul (not that of Mitcham), but most of them would add that it is the ambition of politicians or the profit lust of technology brokers that is actually responsible for the abuse of possibilities created by technology proper.

Clearly, the error of Mitcham in this example results from the lack of distinction between technological creativity (called by me technology proper) and the socio-economic system of utilisation of technological products, from making the former responsible for the abuses of the latter while including them both to broad "technical practice". Such superficial amalgamation became characteristic for the broad paradigm of philosophy of technology and this is the reason for the ineffectiveness of its opinions: technologists ignore such opinions since they consider them unjust, philosophers found a scapegoat and do not consider their own responsibility. And this responsibility is serious: since philosophers noted abuses that need corrective actions and formulated an erroneous diagnosis, the authors of the wrong diagnosis are responsible for the lack of corrective actions. A correct diagnosis would be that *the socio-economic system of utilisation of technology products has some deficiencies and they should be corrected;* apparently, such a diagnosis was not admitted by the humanist philosophy of technology.

On the other hand, Carl Mitcham is sufficiently objective (as one of few philosophers of technology) in that he openly discusses the issue of differences between the so-called *engineering philosophy of technology* (EPT), developed by technologists, and *humanistic philosophy of technology* (HTP), developed by humanists. Other philosophers of technology usually ignore the problem that philosophers of mathematics are mostly mathematicians while philosophers of technology are usually not technologists. But Mitcham artificially increases the number of engineering philosophers of technology, including to them, beside several actual engineers who started the philosophy of technology already in the 19th century, such as Ernst Kapp or Peter Engelmeier, also philosophers of distinctly humanistic episteme, such as Don Ihde.

Moreover, Mitcham openly defends the right of humanists to develop the philosophy of technology, even if he uses, in my opinion, rather one-sided arguments. On the page 89 of the book of Mitcham there are many such strange arguments, e.g.: "But EPT arose explicitly in the process of rejecting a prior, implicit HPT, even by subversive use of selected aspects of humanities philosophy". There are several insinuations in this single sentence: firstly, since the philosophy of technology was started by engineers, this historical fact should be somehow interpreted in favour of humanists by suggesting the existence of "a prior, implicit HTP" without factual basis. Secondly, engineers should be refused the right to read and interpret philosophy, since they are prone to "subversive use of selected aspects of humanities philosophy" (I beg for pardon, but did Kant develop "humanities philosophy", or simply philosophy? Has an engineer the right to read and interpret Kant, or is this right reserved to philosophers?).

Another strange argument is as follows: "Recognizing that humanities (and philosophy) conceived technology, and not technology that conceived the humanities, grants HPT priority at the level of covert or implicit existence". Firstly, this is an argument of the type "an egg is prior to a chicken". Secondly, who Johann Gutenberg was, a humanist or a technologist? Thirdly, a Renaissance humanist was a polymath, he was also an engineer such as Leonardo da Vinci; a distinction of humanistic sciences and engineering occurred for the first time in the era of industrial civilization, and this era was started by the work of an engineer, James Watt. Therefore, "technology conceived humanities" to an equal of even greater degree than "humanities conceived technology".

Yet another strange argument is: "With regard to functions, making is largely not an end in itself, is not self-justifying". Truly, very few human activities are self-justifying, but in this context this truth insinuates that the function of technology is only production, "making". But if we accept that technology proper is the art of creation of tools and related artefacts, a position that I insistently defend, then its main function is inventing that can be treated as a self-justifying end, or at least sovereign as any other art.

Such strange opinions are repeated on the next pages in Mitcham's book. On the page 91 Mitcham writes: "The primacy of HPT over EPT is thus a primacy of anthropological understanding". He does not note a tremendous inconsistency in this sentence. Today, the fundamental principle of anthropology is that you should not judge another culture without knowing and understanding it well, and the difference of HTP and ETP is indeed the difference of cultural spheres. Thus, a declaration of the primacy of a cultural sphere over another because of primacy of anthropological understanding is an evident oxymoron. This strange opinion Mitcham justifies additionally by quoting an opinion of Socrates on a similar issue, but Socrates had the right not to know the principles of today's anthropology and Mitcham, when using anthropological arguments, has the duty to know and understand them. In other words, the statement about the primacy of HTP over ETP is a simple cultural imperialism.

In further parts of the book Mitcham shows also a lack of understanding of the reasons why a philosopher of technology should diligently study an internal and recent history of technology. He writes (page 116): "Indeed, from the perspective of philosophy, what is needed is what may, for want of a better phrase, be referred to as a history of ideas about technology". Truly, such history of ideas is interesting and useful, I also devoted the second part of this book to the conceptual contributions of information technology, but it should not mean the concentration on the history of what humanists, who usually do not appreciate the conceptual importance of technology, understand by technology. The philosophy of technology will not fulfil its obligations if it will not encompass both the history of ideas about technology and the history of conceptual contributions of technology itself, as well as the future of technological development, and for the latter purpose it is necessary to study recent and internal history of technology in order to note and judge the threats that will soon emerge from already developed but yet not fully implemented technological ideas.

On the other hand, Mitcham excellently describes the issues related to the understanding of the word *techne,* starting with the understanding of this word by ancient Greeks, together with the famous and incisive statement of Socrates that *techne* as *techne* does not work in self-interest, but in the interest of art. My opinion that technology proper, the art of creating tools and artefacts, in its essence remains equivalent to *techne,* even if tools created by it change together with civilization epochs, relates to this statement of Socrates. Hence we can speak, say, about $techne_1$ in the time of Socrates, $techne_2$ in Middle Ages, $techne_3$ in the time of print press, clocks and telescopes, $techne_4$ in the time of industrial civilization, $techne_5$ after the informational revolution, when the most important tools constructed by people are software tools. But in all these cases, in accordance with the opinion of Socrates, *techne* does not act only in the interest of power and money, but in the interest of art; it is only the socio-economic system of technology utilization that acts in the interest of power and money.

In the book of Mitcham, the discussion of more contemporary opinions about *techne,* especially the opinions of Maritain (1962) and Feibleman (1966), is also excellent. The comment of Mitcham to a quotation from Feibleman (page 127) is especially important: "In other words, there is at the heart of technical activity, if not of *techne* itself, an irreducible, nonlogical component. There is an aspekt of *techne* that necessarily cannot be brought into consciousness except through the immediacy of a singular, direct encounter, an encounter that takes place through sensorimotor activity and is properly grounded in one of the various forms of love, *storge, philia, eros, agape.* Only love can encompass or grasp the singular". This is in full accordance with my own opinion, based on over 50 years of experience in technological creativity: *techne is an emotional passion related to intuitive formation of technological ideas, and it can be learned only in direct action.* However, because of this intuitive and emotive character of creation of tools, we cannot be fully sure of their working, from which it results that it is necessary to test them and that the technology proper has a falsificationist nature.

Just after these excellent comments, Mitchel comes to the fundamental controversy of the philosophy of technology: the tension between the understanding of the words *techne* and *technology.* After a long discussion of this controversy (pages 128–136) Mitcham comes to the conclusion that "Such is not to suggest, of course, that this transformation of the term 'technology' took place consciously. This is just the kind of change that takes place, as it were, behind the back of philosophy, and that must be excavated from its sedimentated layers". I should add here that the ambiguity of understanding of the word "technology" is especially characteristic for the English language; other languages, such as Polish or German, are more precise in this respect. Admittedly, the ambiguous use of the word "technology" spreads also in Polish under the influence of English (in correct Polish, "*technologia,* technology" means a recipe for a productive process, while "*technika,* technique" means what is usually expressed by "technology" in English), but in a Polish version of this book I tried to avoid this error.

Further in his book, Mitcham suggests that "perhaps the development of technology, like the development of science, should be viewed as proceeding within the framework of 'paradigms'". This sentence implies that he neither knows nor appreciates the book edited by Rachel Laudan, fundamental for this issue and described above. Moreover, Mitcham does not present any more incisive analysis of the problem that can be treated as fundamental for the philosophy of technology: *how did it happen that the philosophy of technology ceased to distinguish technology proper, techne, from the socio-economic system of technology applications,* who is responsible for this mental shortcut, and what are its consequences? I discuss this problem in more detail in other chapters of this book, but here I will quote the conclusions: responsibility rests with classical writers on the philosophy of technology, such as Ellul or Heidegger, who used the word "technology" ambiguously, changing its meaning according to the theses they wanted to prove. This in turn became a start of a paradigm, an exemplar theory, not in the field of technology (which is not paradigmatic, but rather falsificationist) but only in the field of philosophy of technology: if you use an ambiguous definition of the subject of discourse, you can prove whatever you want about it. This paradigm makes the contemporary philosophy of technology unable to understand technology proper deeply and to draw correct conclusions how to influence its development.

Mitcham understands and stresses the ambiguity of the concept of "technology" and devotes a separate chapter (pages 139–160) to this theme. He even quotes one of the few between many definitions of technology that is acceptable for a technologist (Hannay and McGinn 1980), but soon criticizes it (from the paradigmatic position described above, page 158): "The most obvious weakness of McGinn's descriptive analysis of technology as creative human activity is that it seems to imply a restrictive typology in which both artifacts and their use fail to qualify as primary aspects of technology. An artifact, for instance, is [then] the outcome of technology but not itself technology;...". However, it is McGinn's analysis that is correct, while its critique by Mitcham can be paraphrased as follows: "If we, philosophers of technology, agreed that the word *technology* means also artefacts and their use, it is not correct to characterize technology as a creative human activity". Since I am aware of such an opinion of the majority of philosophers of technology, I call this creative human activity *technology proper*.

This is the essence of this issue: in order to draw correct conclusions how we should influence technological development, the philosophy of technology must correctly differentiate between the many meanings of the word "technology", starting with technology proper, distinguishing it from technological production processes, from technology products or artefacts, from the socio-economic processes of utilization of technology and its products. This was noted by some philosophers of technology, such as Kline (1985), who tried to classify diverse meanings of the word "technology" "as artifacts or hardware, as sociotechnical systems of production, as technique or methodology, and as sociotechnical systems of use", however without a sufficient stress on technology proper, the art of creating tools, or on the fact that today's tools are often virtual software tools.

On page 194 Mitcham, sort of in passing, repeats what I believe is one of the largest mistakes of the philosophy of technology: *the lack of a deeper epistemological analysis of the cognitive horizon or episteme of technology*. Mitcham writes only: "...Various types of technology as knowledge ... are further subject to realist, instrumentalist, pragmatic, and other interpretations, although engineers, like scientists, readily assume the realist stance", which is a tremendous simplification of the complex issue of the distinctiveness of *episteme* and the cognitive horizon of technology.

However, on the same page we have again an excellent comment of Mitcham concerning the importance of the work of Jean Piaget (on the development of intelligence by children) for understanding of cognitive aspects of technology, and also the relations between technology and mythology according to Mircea Eliade. The latter stresses that: "the image, the symbol and the rite anticipate, sometimes even make possible, the practical applications of a discovery.... The hammer, successor to the axe of the Stone Age, becomes the emblem of powerful gods, the gods of the storm". After that, there come (page 196) correct references to the work of Michel Polanyi and a mention of the protest of Schőn (1983) against the "common" (perhaps among humanists, surely not among technologists) understanding of technological rationality as an instrumental approach.

In the further part of his book (pages 199–204), Mitcham gives a very good review of arguments against treating technology as a simple application of strict sciences. Especially important in this context is the opinion of Vincenti (1990), an engineer, stressing the relation of engineering design to analogy and preverbal or visual thinking. Vincenti says that "Outstanding designers are invariably outstanding visual thinkers". However Mitcham appears not to understand the relation of such thinking to tacit knowledge, since he protests against the argument of Vincenti concerning "technology as an autonomous kind of knowledge" saying that the word "knowledge" is used here in too broad a sense (indeed, if we interpret knowledge as explicit only, then it is too broad, but the point is that tacit knowledge is also knowledge).

The lack of understanding of the creative nature of technology returns on page 217 with a contraposition of inventions and design. Mitcham maintains that design implies planning, hence it is not creative, and concludes "Inventors are cowboys, engineers settlers". Here I am forced to make a caustic comment: evidently, Mitcham never invented, or designed, if he did, he would know how much creativity is necessary in any more complex design, also in complex planning. This bias of Mitcham is repeated on page 218, where he writes: "An engineer remains with the familiar, does not venture into the unknown, only orders or reorders the known". I beg for pardon, but this is another example of cultural imperialism: and how a telescope was invented? Who made the first pseudo-random number generator?

Unfortunately, the entire following description (page 219–220) of a process of inventing shows a lack of experience in inventing. Admittedly, Mitcham tries to define stages of this process, makes even a block diagram of it, but the main drawback of that diagram is the lack of feedback or recourse to earlier stages ("linearity" in the sense of social science). As many philosophers, Mitcham seems to have troubles with thinking in terms of feedback.

Repeated examples of criticism of the evidence given by experienced engineers can be also observed in the Mitcham's book. On page 223, he quotes the description of a design process given by Layton (1974), clearly based on personal experience in design. But soon afterwards Mitcham writes: "Layton's mistake here is to call his designing activity primarily a kind of knowledge and to fail to notice that modelling in one form or another goes on not just at the stage of making blueprints …". Again, a very strange objection: a design process is undoubtedly a micro-process of knowledge creation, with participation of tacit knowledge and its transformation into explicit knowledge. Such objections indicate rather a general reluctance to accept the engineering experience.

On page 224, the work of an experienced engineer Ferguson (1992) is quoted. Ferguson stresses the role of visual imagination in the process of engineering design, but Mitcham immediately criticizes it: "they fail sufficiently to see what they themselves describe, that the drawing and modelling central to the engineering designing are inherently miniature makings". Again a very strange opinion: are painters making miniature portraits artists or not? Moreover, Mitcham uses the word "making" in the meaning opposed to "creation" and implies that the process of design is nothing else as repeating known variants and their selection by error and trial. However, visual imagination serves precisely to minimize possible errors; I am not saying that they do not occur in the designing process, but that they are minimized and of secondary importance. Mitcham, paradigmatically, does not want to acknowledge that an experienced engineer might know more about designing than a philosopher.

The paradigmatic position of Mitcham is visible also in the concluding part of the book, where he writes (page 273): "Another approach is that of the humanities. Here the argument is sometimes that certain problems are caused by *the inherent nature of technology itself*, that not more but less technology or alternative technologies are required to deal with the problems of environmental pollution and societal change". I beg for pardon, but what does *technology itself* mean? It is not the *technology proper* that is decisive for environmental pollution and societal changes, but *the socio-economic system of utilization of technology*, people that use technology for power and money. Thus, the humanist philosophy of technology should concentrate on the critique of this system, and if this is awkward, then at least it should realize that putting the blame on *the inherent nature of technology itself* is nothing more than finding a scapegoat.

Despite these all paradigmatic biases of Carl Mitcham, his book *Thinking through Technology* is one of the best monographs of the philosophy of technology in the 20th century. New approaches started to appear only in the 21st century.

13.4 People and Technology Don Ihde *Bodies in Technology* (2002)

There are several books worth recommending in the philosophy of technology, trying to take up new problems after the year 2000. Trying is right word here, because e.g. the book *Philosophy of Technology: 5 Questions* (Olsen and Selinger 2007) concentrates on standard questions, except for the last one: what are the most important philosophical problems for future research? Moreover, the responses of 24 representatives of the philosophy of technology indicate a kind of self-restraint and internal closure of that field, with two exceptions. Joseph Agassi gives a fundamental answer with which I fully concur: "*The most important question we have now is, what we should do to prevent the extinction of human (and other) life on this planet*". Harry Collings refuses to answer, and significantly formulates his position as follows: "The answer to the question how to make sure that important questions *are not identified* is to treat philosophy of technology as a subject that is about the contents of philosophy of technology books rather than about the technological world" (author's underlining).

Two other books discuss the relation of people and technology products (persistently calling these products *technology*). One is a very good book by Peter Paul Verbeek entitled *What Things Do* (2005), focusing on the role of material artefacts, technology products, in contemporary civilization. Another book, describing the relation of people and technology products in the new era is *Bodies in Technology* by Ihde (2002). This book will be presented here in more detail.

This is a very interesting and incisive book. It would represent a significant change in the philosophy of technology, if it did not have a postmodern bias that makes it difficult to understand technology, which is discussed in detail below. Don Ihde is treated by Carl Mitcham as one of "engineering" philosophers of technology (EPT). This opinion, however, is not correct: Don Ihde is not an engineer, he clearly represents a humanistic hermeneutic perspective, he is only more friendly towards technology, at least, more friendly than the majority of philosophers of technology.

That Don Ihde writes from a humanistic and not engineering perspective, becomes clear already on the first page of the introduction to *Bodies in Technology* (page xi), where he defines "embodiment relation" as the relation of perceiving something through the mediation of "artefact or technology". The words "or technology" define artefact as equivalent to technology, while an engineer would rather use a more precise statement "artefact or the product of technology". In further part of introduction, page xiv, Ihde writes "contemporary technohype sometimes wants to extrapolate"; this again indicates a humanistic perspective, since an engineer would limit the concept of "technohype" to unserious media writings, while he would stress that an extrapolation of technology development is the subject of serious studies of *future technology assessment*, conducted in any larger high technology corporation.

Then on pages xv–xvi, where Ihde mentions "science wars", he evidently does not fully understand their reasons, the very differences in hermeneutic perspectives and episteme of strict and natural sciences on the one side and the humanist and social sciences on the other. He maintains that "These worries [about the destruction of objectivity], however, are often expressed without the slightest self-reflection about how late modern (or postmodern?) science produces or constructs its own imagery, particularly in the state-of-the art compound instrumentation made possible by contemporary technologies. Has science become virtual without itself knowing it?" In this way he takes a clear position on the side of humanist sciences in the main subject of "science wars", the dispute about objectivity. Such postmodern position is characteristic for most of humanist and social sciences of the end of the 20th century, but it is difficult to be accepted by a technologist, who naturally knows that he constructs instruments and thus contributes to knowledge development and construction, but he gives no absolute meaning to this simple fact, quite the opposite: he tries to construct these instruments in such a way as to obtain possibly most objective knowledge, not a knowledge manufactured on a production line, as suggested by the postmodern sociology of science.

Another essential question is the understanding of the phenomenon and concept of virtuality. Various interpretations of this concept by postmodernism are evidently exaggerated. It can be observed on several pages (8–11) of the book, where Ihde discusses diverse aspects of the concepts of interactivity and virtuality (neglecting the contributions of informational technology to these concepts, described here in Chaps. 7 and 11 such as interactivity in the contemporary theory and practice of decision support and the concept of *human centred computing*). Ihde devotes much attention to the question "Can VR [Virtual Reality] replace RL [Real Life]?" coming to the correct conclusion that this can happen only if a theatre could replace real life. However, he does not note that the importance given to this pseudo-deep question is a result of an excessive belief in the inflated postulates of postmodernism. The belief that reality is only local results from the belief that all our knowledge about reality is only constructed in a social discourse; this leads to the conviction that all knowledge is virtual. On the other hand, an engineer must exactly distinguish virtual reality (developed originally by technology, see Chap. 7) that he uses commonly in computer simulations of more complex technological systems, from actual reality, since virtual tests can only prepare actual tests and make them less expensive.[3]

Don Ihde perceives the differences between the perspectives or cognitive horizons of social sciences, strict sciences and technology, but does not understand more deeply the cognitive perspective of the latter. On page xvii of the introduction, he gives an account of an interdisciplinary debate between representatives of medial sciences, strict sciences, and information technologists, and stresses that medial sciences represented constructivism, strict sciences, instrumental realism, while information technologists represented, as he maintains, a "hybrid" position: "they knew that the heuristic programs they built... were inventions and could be tinkered with,... they also hoped that in this process they could get close to some kind of reality". Here it becomes clear that such a distinguished philosopher of technology, even suspected of a "technological deviation" by his colleagues, actually does not understand the essence of technological episteme, which is not a "hybrid" of the different epistemai of social sciences and strict sciences, because those are paradigmatic, while technological episteme is falsificationist, see more detailed discussion in Chaps. 2 and 3 of this book. However, Ihde notes at least some important epistemological differences, even if as a philosopher he should look at them more seriously and not impose his perspective onto representatives of other cultural spheres.

The reasons for other philosophers thinking that Don Ihde has a "technological deviation" become clear at the end of the introduction (pages xix–xx) where he declares (as if he felt guilty when opposing the dominating opinions of the philosophy of technology on that issue): "I guess I have to admit that I believe not only that one cannot rationally control technological development .... Rather, one can enter into the situations, and I argue that the entry should be at the research and development stages as well as with the later applied ethics stages, and make nudges

---

[3] Postmodern social scientists dislike this argument, because they dislike the very concept of a test that does not correspond to their concept of a social discourse.

and inclinations". With this principle all engineers would easily agree, but not necessarily all humanistic philosophers of technology, because of the following reason. In order for this principle to be effective, it is necessary to include several technological subjects, e.g. computer hardware and software, robotics, biomedical engineering as mentioned earlier, to the education of philosophers of technology; meanwhile, it was easy to include philosophy into the education of engineers, but reciprocity in that respect seems to be difficult to expect.

Despite such friendly attitude towards technology, Don Ihde remains under the dominating influence of postmodernism, and especially its conviction about a growing union of science and technology, expressed in the slogan of "techno-science", revealing the lack of understanding of the epistemological differences of science and technology. This influence is visible throughout the book. Already in the first chapter, page 3, Ihde writes: "In the cases of human-technology symbi-otics, both mind and muscle transformed our worlds". This is a correct opinion, except one letter: the use of plural in "worlds". The use of plural indicates postmodern conviction about merely local character of knowledge and reality. On the other hand, a technologist agrees that some of our knowledge is local, but believes that for a technical construction, the possibly most universal knowledge should be used. Should different principles of security be used when constructing a bridge in Japan and in the United States? Thus, a technologist would use a singular "world", even if he knows that the instruments constructed by her/him would not necessarily work on Mercury (but she/he hopes that they would). To the adherents of the thesis that all knowledge has merely a local character, I propose the *test of hard wall* described in former chapters: if our knowledge is only local, let us test whether the property of being hard applies also to the wall in our room: close the eyes standing in front of a wall and try to convince yourself that the wall is not hard. If you fail to convince yourself, there are at least some universal aspects of knowledge. But if you succeed, you can always falsify that conviction by stepping ahead with your eyes closed.

The strong attachment of Don Ihde to the postmodern paradigm is shown in many places of his book, e.g. in the last paragraph of his Chap. 3 on visualism. In this paragraph, he repeats his conviction about technical manufacturing of knowledge and describes objectivism in terms of the 19th century, as if Karl Popper in his book (Popper 1972) did not give a much more advanced interpretation what is objective knowledge, valid mostly until today, see also Chaps. 3 and 6 here.

Another example are pages 51–52 of his Chap. 4 *Perceptual Reasoning*, where Don Ihde clearly presents fundamental convictions of postmodern sociology of science: "In its most radical form, social constructionists view science as no different in principle than any other social institution or practice and claim, with admittedly very different degrees of radicality, that the products of science are socially constituted. At least this is to see science as a particular form of social praxis, to understand it as an institution (implicitly as open to and prone to fal-libility and values as any other institution)".

This position is not acceptable for a technologist and requires a radical answer. Technology, as already stressed, is different than science, but requires from science as much objectivism as possible, certainly more than from any other social institution. This results from several reasons, two of which I would like to highlight here: pragmatism and the evolution of civilization.

Pragmatism of technology proper is creative and does not mean that we are satisfied to construct any tool that fulfils its functions here and today. The creative aspect of technology proper means that we want to construct (in a broad sense) the best possible tools, working in possibly broadest conditions (we, technologists, contrary to technology brokers, entrepreneurs who would happily sell products of lower quality). Therefore, a technologist requires the science to give her/him most objective and most universal knowledge (even when understanding that there is no absolutely objective and universal knowledge). If a technologist does not obtain such knowledge from science, she/he will construct tools and artefacts anyway, only with more effort, and she/he will test them in possibly most extreme conditions.

Thinking about evolution of civilization, both a technologist and a representative of strict science (and I hope a representative of social sciences and humanities as well) is motivated by a concern about the well-being of future generations, our children and grandchildren, and wants to leave them possibly best tools and possibly most objective knowledge. A fundamental argument here is the *uncertainty principle* of Rawls (1971): we do not know in what conditions our children will live, what threats they will face. This implies that we should institute such principles of justice that will best serve our children even in worst conditions; moreover (which transcends the arguments of John Rawls) *we should leave them such tools and knowledge that will function in possibly broadest conditions*. Thus, objectivism is similar to justice, they are never fully attainable, but constitute important higher values, fundamental for the development of human civilization. Such fundamental values were many times analyzed by philosophy, e.g. as teleological ideas of Husserl (1973). Moreover, objectivism is a kind of insurance for future generations of humanity against unforeseeable catastrophes (in order to deal with future Fukushimas) and such an insurance is certainly worth a special position of science between other human institutions and at least 0.5 % of GDP, see next chapter, for financing long-term cognitive and civilization aspects of the development of science and technology.

It is obvious that science is a social institution, but it is not true that it does not have special features distinguishing it from other social institutions: in particular, science is obliged to be as objective as possible, and it is in the interest of the civilization development of humanity to give science such privileges that support this goal of objectivism, including the privilege of publishing new results of science for universal access. The deconstruction of objectivity attempted by the postmodern sociology of science is based, as shown in Chap. 6, on an application of an inadequate logics (lack of understanding that an effect often becomes a cause, that the apparent paradox of a vicious circle should be analysed in terms of feedback). However, this deconstruction contains more dangers: by trying to reduce science to power and money, the postmodern sociology of science supports in effect the striving of big corporations for a full privatization of knowledge.

It should be added that the refusal to accept the specific features of strict sciences and technical sciences has a flavour of reluctance towards their deeper understanding, a flavour of cultural imperialism. If, as discussed already in former chapters, there are important differences in the epistemai of the three cultural spheres: technology, strict and natural sciences, and social sciences and humanities, then a judgement concerning the practices of other cultural spheres from the perspective and as part of episteme of only one of them is the grossest transgression against the principles of cultural anthropology.

The said attachment of Don Ihde to the exaggerated postulates of postmodernism is even more strongly accented in the beginning of his Chap. 5 *You Can't Have It Both Ways; Situated or Symmetrical.* In the section *Postmodern Knowledges* he writes clearly: "To use the plural for knowledge, 'knowledges,' initially sounds a bit strange to anglophone ears. But it is more *accurate* today to describe what once was 'Knowledge' as 'knowledges' since one of the features of postmodernity has been *the deconstruction of transcendentals and foundations*, and replacement by local knowledges and particularized knowledge practices. Admittedly, this is not without contestation as the 'science wars' arguments amply illustrate, particularly in North American contexts" (author's underlining).

From the postmodern cognitive perspective, this persuasion is clear and true; however as a technologist, I think that this is precisely an example of a local knowledge, limited to the postmodern hermeneutic horizon, and for many reasons useless for a technologist. Earlier, while commenting the use of plural "worlds", I noted that a technologist prefers the singular "world" since she/he needs knowledge valid in most broad conditions. We, technologists, of course realize that our knowledge is never valid universally, despite all assurances of physics we are not sure that the tools and instruments constructed here, on Earth, even if they were tested in extreme temperatures and pressures, will function as well e.g. on Mercury. We realize also the local and cultural differences, including the differences of the epistemai of three large cultural spheres of technology, strict sciences and social sciences (which is not realized by the postmodern sociology of science). However, this does not imply that we would agree with the exaggeration that *only* local knowledge exists and that the use of plural "knowledges" is more *accurate*.

On the other hand, *"the deconstruction of transcendentals and foundations"* requires an answer that is precisely more fundamental and is based on a deeper understanding of the concept of complexity, presented in Chap. 10 of this book together with the discussion of the emergence principle. Clearly, if we want to reduce everything to power and money, as it seems to be accepted by postmodernism (which supports in that way the privatization of intellectual heritage of humanity), then the discussion ends here, there is no sense in analyzing the concepts on higher levels of complexity, such as justice, objectivity, ethical values, irreducible to lower levels. Many errors of postmodernism consist precisely in its excessive reductionism, in not perceiving that the transcendental, fundamental and basic concepts correspond to attempts to encompass and express the irreducible complexity of the world, perhaps not always successful, but important as stages of the development of intellectual heritage of humanity.

On page 11 of his Chap. 5 Don Ihde expresses also his poststructuralist persuasions "Whatever can be said, can be said meaningfully only within the system of language. And while this move deconstructs, as did phenomenology earlier, the Cartesian spectator consciousness, it now drives whatever vestige of subjectivity there could be in the direction of linguistic-like signifying activity". I am sorry, but apparently Don Ihde never experienced that the body language of his interlocutor contradicted the verbal statement. Body language is not expressed by words; it requires a great intuition, often acquired through many years, to read such language properly. I presented a critique of the poststructuralist reduction of a picture to a sign in Chap. 6 of this book, using the example of comparison of the Japanese 10 yen coin and an actual photograph of Byodoin temple. Hence, the use of audiovisual tools of communication is not only a beautification of signs, as the illumination of letters in medieval texts, but essentially enriches the communication. We know well, who said "the limits of my language mean the limits of my world". But Ludwig Wittgenstein implied that way that we should not discuss metaphysics, while today we need a new rational discussion of metaphysical problems, which are often preverbal, even if we would finally use language for their description.

On page 12 of Chap. 5 Don Ihde adequately recalls Foucault (1972) and his concept of historically and culturally changing episteme. Nevertheless, he does not note that if episteme changes historically and culturally, then it can develop in various directions for diverse cultural spheres, as we observed in the last 50 years in relation to technology, strict and natural science, and social sciences and humanities.

In the section *Slippery Symmetries* Don Ihde correctly stresses the fallibility of reasoning based on the contraposition of only two concepts or two ends of a spectrum. However, as explained in the discussion of *logical pluralism* in Chaps. 6 and 11 of this book, we should go much further and check in all cases if the classical, binary logics is adequate or is too simplified for a given context. Don Ihde writes: "This 'both/and' rather than 'either/or' is a mark of postmodernity". I beg for pardon, but multivalued logic is about 100 years old (Łukasiewicz 1911), the theory of fuzzy sets is around 50 years old (Zadeh 1965), it is broadly used in technology and older than postmodernism, so it is not it's achievement. Instead of binary logics, in several places of this book I proposed to use triple-valued logics and thinking, with the values "true", "false" and "may be" as in the rough set theory (Pawlak 1991), together with the Brouwer (1922) principle of scepticism towards each proof based on reduction to contradiction (since if there is a middle way, a contradiction might be spurious).

The whole book of Don Ihde is devoted to the analysis of the impact of contemporary technology and its products on the instrumentation of scientific research. This impact is tremendous, as correctly stressed by Don Ihde, but it is interpreted from the perspective of his specific hermeneutic horizon, illustrated by the above discussion, without a broader discussion from other horizontal positions.

The first important distinction appears on page 4 of the first chapter of Don Ihde's book. He writes there (and in many other places) about "an embodied and disembodied mode [of parachute jump]". Actually, this distinction concerns all kinds of perception, and also the issue of visualization. Embodied or rather *immanent type of perception* is a personal type of perceiving anything by all senses, "all body"; the whole book of Don Ihde stresses a correct thesis that this type of perception is dominating in subjective personal cognition, so it should be analyzed in a more detailed fashion. We cannot agree, however, with a further conclusion of Don Ihde that the *disembodied perception,* consisting of an external reflection on one's own impressions, is an imperfect cultural habit of modernism or positivism. In his critique of disembodied perception, Don Ihde (under the influence of his postmodern convictions) seems not to note a simple but fundamental fact: such a type of perception is necessary, if we want to convey our impressions to other people, to achieve an intersubjective consensus about them. Thus, it is not an imperfect cultural habit, but an indispensable factor of the development of human civilization, including science and technology.

The lack of understanding of this fundamental fact is evident when Don Ihde criticizes Leonardo da Vinci for the use of drawings in presenting the anatomy of people (page 6 of Chap. 3): "Da Vinci *reduces this anatomy* to a structural and analytical set of drawings…" (author's udnerlining). I beg for pardon, but what da Vinci should do if he wanted to convey his knowledge about anatomy to the reader? Stick only to verbal description? Further discourse of Don Ihde suggests that in such a case he would not accuse da Vinci of reduction, because postmodernism believes in a discursive construction of reality. Apparently, don Ihde never experienced a situation in which we say "this cannot be explained by words alone, I must draw it for you". In our communication, drawings, pictures, body language convey much valuable information.

Entire Chap. 3 of Don Ihde's book analyzes the role of *visualism* in science, that is the trend to present diverse aspects of the world in a visual form. Don Ihde maintains that visualism was a cultural choice, historically reinforcing itself and supported by the development of various visualizing instruments, but not necessary and worse than embodied perception. The description of visualism is rich and interesting; we can agree with the thesis about historical reinforcement of visualism. However, we cannot fully agree with two theses: the main thesis that the development of visualism was not necessary, and the thesis discussed already above, that the disembodied perception, in which visualism is one of essential ways of enriching communication, is imperfect in comparison to embodied perception.

Visualism was necessary because it was an essential aspect of objectivism in science, and if Don Ihde does not notice it, then the possible reason is that as a postmodernist he avoids the concept of objectivism, which only proves that deeper horizontal, hermeneutic convictions can distort the rationality of philosophical analysis. The concept of objectivism appears already in ancient times, it was rediscovered by Roger Bacon in the 13th century, but its broader development was related to two socio-economic needs: the authenticity of banking letters and the objectivity of cartography in the time of geographic discoveries. The first aspect is not necessarily visual (letters must be read, but do not necessarily use pictures), but was essential for the development of the modern banking system with the mutual credibility of banks. The second aspect is fundamentally visual and illustrates the connection of visualism with objectivity of communication: maps of the shores of distant continents are obviously only drawings and are never fully precise, but they should be as objective as possible, not serving the objective to put the competition on unmarked (but known to the map's author) shoals.

Visualism is also a necessary element of scientific and especially technological creation. If Don Ihde had made the effort to ask creators of technology how they achieve their results, create new constructions or tools, he would have obtained diversified answers, but with a common kernel: in their creative work, engineers use visual imagination, they imagine the shape of tools, block-diagrams, architectural forms and plans, etc.; *without visual imagination there would be no new technology.*[4] If even Don Ihde thus far misunderstands technology that he formulates a thesis of unnecessary development of visualism, how can we expect other philosophers of technology to understand it?

Disembodied, communicated perception is obviously less rich than embodied, immanent perception, since we did not develop (yet) good ways of communication of many sensual perceptions, such as taste, smell, full impression of touch. Nevertheless, visual information is much richer than verbal information, which is commonly known in a proverb: "a picture is worth a thousand words". This proverb was corrected to the formulation "a picture is worth at least a ten thousand words" in Chap. 5 of this book in relation to the evolutionary rational theory of intuition and the multimedia principle.

In this respect, Don Ihde uses many ways to show a reciprocal relation, that words, or at least sounds, are more important than pictures (perhaps he perceives intuitively that the admission of the dominance of pictures might result in a crisis of poststructuralism and postmodernism, or at least of their thesis about a discursive construction of reality). One of the ways is to recall phenomenology: "phenomenology holds that I never have simple or isolated visual experience" as if it would prove anything beside the fact that apart from vision we have also other senses that are used in immanent perception. Another way (apparently effective in postmodern discourse) is a hard negation hidden in a seemingly relative statement: "I wish to shift ground a bit to deal with what I take to be a traditional prejudice concerning a presumed superiority of vision over any other human sense" (page 40 in Chap. 3). This statement is both an exaggeration (we should compare vision with language, not with all other senses) and an epithet ("traditional prejudice"). The third way used by Don Ihde is mixing sound with language at an attempt of inexact, physical way to prove that sound carries more information than vision, which was discussed in detail in Chap. 5.

We should add some further arguments. The experiments of Piaget concerning the perception of children, confirmed by anybody who observed the intellectual

---

[4]   One could argue that the development of software requires other imagination than visual, but even in this case we use e.g. visual algorithmic schemes, block-diagrams of software architecture, etc.

development of small children, show that visual perception is decisive in formation of concepts: a child first see a house or its drawing before learning the name "house". Not accidentally men decorated the walls of their caves with pictures of animals. Until today, this is actual also for adults: visual imagination is the foundation of creative intuition, not only for painters and engineers, but also for strict scientists. We should acknowledge, however, that Don Ihde remains self-critical: at the end of the chapter on visualism he admits that "A picture is worth far more than a thousand words these days".

In the section *Perception in Reason* Don Ihde correctly stresses that philosophers of science very often differ in opinions, "how theory-laden or purely given an observation-perception might be"; however, he interprets this question in a way typical for the postmodern sociology of science by suggesting that both hard science and technology treat measurements and perception as "pure data", that they do not realize the relativity of observations and measurements. Such implications appear in the whole book of Don Ihde, starting with already cited "Has science become virtual without itself knowing it?" in the introduction to the book.

From my personal experience of student years 1954–1960, I know well that engineers, taught foundations of philosophy, already then knew the. implications of the work of Heisenberg (1927). Moreover, the fact that a measurement experiment is constructed usually under the influence of a theory is evident for any engineer specializing in measurements; however, knowing this danger we try to counteract it. The results of Heisenberg are significant on the quantum level: the very fact of measurement distorts the results of measurement. However, this also sometimes occurs, usually to a lesser degree, even for macroscopic measurements. Therefore, there is no absolute truth and no "pure" data; nevertheless an engineer needs knowledge and data as objective as possible. Thus technologists searched for a synthesis allowing them to conquer this contradiction and formed an appropriate episteme, which was not noticed by the paradigmatic philosophy of science and technology. I have often encountered accusations from the representatives of postmodern sociology: "Why you, technologists, are positivists? Positivism is a prejudice of the 19th century!". I was forced to patiently explain that we are fully aware of the fallibility of measurements, but since we are motivated by the joy of creation of instruments and tools, we also realize diverse dangers related to that and must submit the created instruments to critical tests. The horizontal attitude of an engineer in this respect is not positivism or modernism, but much rather *postpostmodernism,* while *postmodernism is treated by technologists as an exaggerated intellectual fashion of the end of industrial civilization era.*

In the section *Eyeballs and Instruments* and *A Partial Phenomenology of Scientific Perception* Don Ihde returns to his main idea: the supremacy of embodied, immanent perception over the visual perception. He does not note, however, that he should consider first a more precise classification of different types of perception. The first distinction is between embodied, immanent perception and disembodied, communicative perception. The second distinction is between various types or parts of either immanent or communicative perception, for example between purely verbal perception, with the help of either sound or text, and visual perception, not only with naked eye, but, as correctly stressed by Don Ihde, also by the eye enhanced by diverse instruments. We fully agree that immanent perception is much superior, but not in relation to visual perception only (this superiority is obvious like a superiority of an elephant over its trunk); much more important is the fact (not discussed by Don Ihde) that we still have many difficulties if we try to turn to communicative perception, transfer our immanent observations to another person.

Don Ihde concentrates on the comparison of sound and vision, and presents further strange interpretations, such as (page 54 in Chap. 4 *Perceptual Reasoning*) "But Galileo's visualist science, in which vision was extolled over all other senses, took its position within an overall Renaissance celebration of the visual. Perception as visual [was] correlated with optics as an instrumentarium". On the other hand, these sentences show that Don Ihde is aware of and appreciates the self-reinforcing type of development of science and technology in an evolutionary, positive feedback loop (differently than Bruno Latour, who sees in this fact a "proof" of destruction of objectivity).

Lenses were known already in antiquity, but in the 16–17th century, engineers (such as Dutchmen Jan Lippershey, Zacharias Janssen and Jacov Adrianszoon) constructed telescopes for navigation purposes. Galileo reinvented the telescope, but he used it for diverse astronomic research, which resulted in more interests in optics and its development as a separate field of science, and afterwards in further improvements of telescopes and further ways of using them. Nevertheless, it is an exaggeration that "vision was extolled over all other senses"; it rather occurred that visual perception became an important part of communicative perception: it was not sufficient to speak, it became necessary to show images to listeners.

On page 56 of Chap. 3, Don Ihde writes: "Husserl's critique of the modern early trajectory of Galileo and Descartes contains the observation that such science 'forgets' the plenary perceptual and bodily base of the lifeworld". Read: people perceive immanently, Galileo and Descartes concentrate on pictures (also on words, but Don Idhe does not stress this). This is an important observation, but science does not "forget", it simply did not yet develop means of communicating immanent perception by other senses than hearing and seeing.

While differentiating more precisely between immanent and communicative perception, one can go much further and ask the question: *if we perceive immanently by all senses, with the whole body, and communicate only verbally and visually, what happens with the knowledge acquired by immanent perception and not necessarily communicated?* This partially communicated knowledge contains also much of our visual perception, since, contrary to the theses of Don Ihde, our communication of visual type is also far from ideal. My answer to the above question is that this partially communicated knowledge is accumulated individually, it is the basis of our experience and intuition, a part of our *tacit knowledge,* see Polanyi (1966), Nonaka and Takeuchi (1995), Wierzbicki (1997, 2004), Wierzbicki and Nakamori (2006, 2007). Thus, the phenomenological emphasis on embodied, immanent perception by all senses and whole body confirms the importance of tacit and intuitive knowledge, and the difficulty of conveying this tacit and intuitive knowledge in interpersonal communication as well as in computer-man communication.

It should be added that the phenomenological paradigm, the belief in the power of phenomenological reduction (in short, taking all context outside of parentheses), is nevertheless fallible in describing the issue of creation of knowledge, as described in Chap. 12 of this book. We never create knowledge by taking outside of parentheses all the intellectual heritage of humanity, including its rational, intuitive and emotional parts.

In summarizing Chap. 4, Don Ihde writes: "'Reading' instruments that yield nonisomorphic results, for example, data in the form of numbers, is obviously more hermeneutic in form. Its referentiality is more textlike ...". I must again on the one hand fully agree with the author, but on the other hand again criticize him for the narrow postmodern horizon. Each measurement engineer knows that she/he interprets results of every measurement, independently from its digital, analog, or visual form; in such interpretation we use our intuition in a similar way as a humanist uses her/his intuition while interpreting a text. See the description of the *EEIS Spiral* of experimental creation of knowledge in Chap. 12 of this book and in Wierzbicki and Nakamori (2006). Thus, reading of instruments is obviously related to hermeneutics, but not because instruments give results similar to text (as a postmodernist or rather poststructuralist would prefer), only reversely, because hermeneutics relates to the use of preverbal, tacit and intuitive knowledge.

In Chap. 6 *Failure of the Nonhumans* Don Ihde analyzes the opinions of Bruno Latour on the issue whether arms or rather people kill other people. Latour assumes in this question a strange position that people with arms become quite different people; hence the complex "man at arms" kills other people; Don Ihde has a similar opinion. However, such reasoning would lead us to the conclusion that people with an arbitrary tool or instrument become different (and with each instrument specific?) people, since there are no instruments that cannot be used as arms, such as a needle or a car. Thus, the opinion of Latour obscures instead of clarifying the following essence of this question: *that there are people who are fascinated with arms or instruments (no matter whether these are guns or cars) that kill other people.* This harmful fascination of people is the reason of gross dangers (the number of people killed by car speeding is larger than the number of people killed by guns) and it is to a small degree conditioned by technology, to much larger degree by psychological, ethical, social or even political factors, such as the *pyramid syndrome,* fascination of political leaders by the power of technology that can be used to eternalize their memory.

In Chap. 7 *Prognostic Predicaments* Don Ihde correctly stresses that *technology products* (although he imprecisely calls them *technology*) often have unpredictable consequences in application. This is known to all technologists, therefore, the responsibility for using such products is limited to using them consistently with instructions. But even when writing a user's manual, an engineer must imagine diverse unpredictable behaviour of the user and cannot imagine them all, human innovativeness in finding monkey-like stupid or even malicious ways of behaviour is unlimited. Moreover, instructions of use and laws limiting certain uses of technology products are not kept, particularly if a product, such as a car, becomes the subject of fascination of its users. In this sense, of unpredictable behaviour of the users or their behaviour distorted by fascination, a technologist agrees with the opinion of Don Ihde that *"No technologies are neutral, and all may be expected to have some negative (as well as positive) side effects"*.

When Don Ihde analyzes problems without postmodern bias, as in Chap. 8 *Phil-Tech meets Eco-Phil,* his neutral position towards technology becomes visible, which is actually a friendly position, because the majority of other philosophers of technology cannot be suspected of neutrality or friendliness towards technology. A technologist can fully support the conclusion of this chapter that recent history proves the possibility of solving ecological problems through the creation of modified, different technologies and devices than traditionally used, even if these problems are complex and cannot be solved by abandoning technology, nor by more technology, but only by incisive diagnosis of ecological consequences at the early stages of designing of new devices and technologies (the latter in the narrow sense of recipes of technological production processes).

13.5   Conclusions

Many conclusions might follow from the discussion with several books of the philosophy of technology presented in this chapter; I shall present only selected ones.

In her book, Rachel Laudan enriches the definition of *technology proper:* it is the art of creating tools and artefacts, motivated by practical problems to be solved. From this results the pragmatic character of technology proper, not its instrumentality, because a tool is not only instrumental, it is rather a discovery of truth in solving a practical problem. And since pragmatic character implies lack of paradigmatic character, hence there are no scientific revolutions (such as in strict sciences) in technology; there are technical revolutions, but they are more frequent and do not consist in abandoning of an old paradigm.

Carl Mitcham is sufficiently objective to give a comprehensive review of philosophy of technology before the end of the 20th century, to notice the problem of actual absence of technologists among the philosophers of technology, but he tries to defend the (broadly understood) paradigm of humanistic philosophy of technology without fully recognizing the dangers of such paradigm. To dilemmas discussed by him, an answer of a technologist is simple: philosophers (not only of technology) and humanists must have in their university curricula at least three technical subjects, such as construction and programming of computers, automatics and robotics, and biomedical engineering; without such education, they will not understand current technology and contemporary world.

Don Ihde is a humanist philosopher of technology friendly towards technology, which is a rare case. Despite his postmodern and poststructuralist bias that hinders understanding of technology, he gives many incisive conclusions, from which possibly the most important is: *No technologies are neutral, and all may be expected to have some negative (as well as positive) side effects.*

More general conclusion is that the philosophy of technology should change its (broadly understood) paradigm in order to cope with new challenges. It should start to differentiate technology proper from technology products (artefacts), technology production processes and the socio-economic system of production and utilization of technology; it must analyze the recent history of technology because of inevitable delays in development and social distribution of new technology products. For this, new elements of technical education are necessary, that in line with the principle of reciprocity should be spread between humanists and the representatives of both social and strict sciences.

References

Brouwer, L.E.J.: On the significance of the principle of excluded middle in mathematics, especially in function theory. With two addenda and corrigenda. In: van Heijenoort, J. (ed.) (1967) A Source Book in Mathematical Logic, 1879–1931. Harvard University Press, Cambridge, pp 334–345 (1922)

Bunge, M.: Technology as Applied Science. Technol. Cult. **7**, 329–347 (1966)

Constant, E.: The Origins of the Turbojet Revolution. John Hopkinks Univeristy Press, Baltimore (1980)

Feibleman, J.K.: Technology as Skills. Technol. Cult. **7**(3), 318–328 (1966)

Ferguson, E.S.: Engineering and the Mind's Eye. MIT Press, Cambridge, Mass (1992)

Foucault, M.: The Order of Things: An Archeology of Human Sciences. Routledge, New York (1972)

Hannay, N.B., McGinn, R.E.: The anatomy of modern technology: Prolegomenon to an improved public policy for the social management of technology". Daedalus **109**(1), 25–53 (1980)

Heisenberg, W.: Über den anschaulichen Inhalt der quantentheoretischen Kinematik und Mechanik. Zeitschrift für Physik **43**, 172–198 (1927)

Husserl, E.: Cartesianische Meditationen und Pariser Vorträge. [Cartesian meditations and the Paris lectures.] In: Strasser, S. (ed.) Martinus Nijhoff, The Hague, Netherlands (1973)

Ihde, D.: Bodies in Technology. University of Minnesota Press, Minneapolis (2002)

Kline, S.J.: What is Technology? Bull. Sci. Technol. Soc. **5**(3), 215–218 (1985)

Laudan, R., (ed.).: The nature of technological knowledge. Are models of scientific change relevant? Reidel, Dordrecht (1984)

Layton, E.T.: Technology as knowledge. Technol. Cult. **15**(1), 31–41 (1974)

Łukasiewicz, J.: O wartościach logicznych (On Logical Values). Ruch Filozoficzny **I**, 50–59 (1911)

Maritain, J.: Art and Scholasticism and the Frontiers of Poetry. Scribners, New York (1962)

Mitcham, C.: Thinking Through Technology: the Path between Engineering and Philosophy. The University of Chicago Press, Chicago-London (1994)

Nonaka, I., Takeuchi, H.: The Knowledge-Creating Company. How Japanese Companies Create the Dynamics of Innovation. Oxford University Press, New York (Polish translation: Kreowanie wiedzy w organizacji, Poltext 2000) (1995)

Olsen, J.-K.B., Selinger, E.: Philosophy of Technology: 5 Questions. Automatic Press, UK (2007)

Pawlak, Z.: Rough Sets—Theoretical Aspects of Reasoning About Data. Kluwer, Dordrecht (1991)

Polanyi, M.: The Tacit Dimension. Routledge and Kegan, London (1966)

Popper, K.R.: Objective Knowledge. Oxford University Press, Oxford (1972)

Rawls, J.: A Theory of Justice. Belknap Press, Cambridge, Mass (1971)

Scharff, R.C., Dusek, V. (eds.): Philosophy of Technology: the Technological Condition. Blackwell Publishing, Oxford (2003)

Schőn, D.E.: The Reflective Practitioner: How Professionals Think in Action. Basic Books, New York (1983)

Skolimowski, H.: The Structure of Thinking in Technology. Technology and Culture 7, 371–383 (1966)

Vincenti, W.: What Engineers Know and How They Know It. Johns Hopkins University Press, Baltimore (1990)

Wierzbicki, A.P.: On the role of intuition in decision making and some ways of multicriteria aid of intuition. Multiple Criteria Decis. Making 6, 65–78 (1997)

Wierzbicki, A.P.: Knowledge creation theories and rational theory of intuition. Int. J. Knowl. Syst. Sci. 1, 17–25 (2004)

Wierzbicki, A.P., Nakamori, Y.: Creative Space: Models of Creative Processes for the Knowledge Civilization Age. Springer, Berlin-Heidelberg (2006)

Wierzbicki, A.P., Nakamori, Y. (eds.): Creative Environments: Issues of Creativity Support for the Knowledge Civilization Age. Springer, Berlin-Heidelberg (2007)

Zadeh, L.: Fuzzy Sets. Inf. Control 8, 338–353 (1965)

Chapter 14
Threats and Challenges of the New Era

This chapter is an extended and strongly modified version of my texts (Wierzbicki 2010, 2011). I am addressing first the issue of impossibility and at the same time necessity of forecasting future, then the methodology of analyzing challenges and its relations to known methods of strategic analysis. I suggest that the analysis of challenges should start with the analysis of threats, weaknesses, opportunities and strengths, leading to suggested actions. It is illustrated by a short analysis of a list of global threats, resulting in an identification of four challenges: the challenge of sustainable development, the challenge of new global order, the challenge of informational revolution, and the challenge of biotechnical revolution. A shortened analysis of these challenges is also presented. Finally, challenges related to the development strategy of Poland in next decades are commented as an example.

14.1 Methodological Issues

It is obvious that a precise forecast or prognosis of the future is impossible, but some forecast is necessary, people would not build civilizations if they did not attempt to forecast, with better or worse results. We build a house, forecasting winters and adverse weather, growing family, etc.; but a flood can surprise us. We build robots guided by the vision to replace us in heavy and dangerous work; but the use of robots in an unbalanced socio-economic system oriented only on a short-term profit might lead to excessive unemployment or to utilization of robotics by irresponsible or even criminal social forces.

These are the reasons why forecasting future, even if it was always an inseparable element of civilization development (such as, e.g., forecasting eclipses of the Sun), became at the turn of the 20th and 21st centuries a subject of an assault coming from the neoliberal economic doctrine, see e.g. (Taleb 2007), because if a good forecast was possible, then the state might act more rationally than free

© Springer International Publishing Switzerland 2015

A.P. Wierzbicki, *Techne_n: Elements of Recent History of Information Technologies with Epistemological Conclusions*, Intelligent Systems Reference Library 71, DOI 10.1007/978-3-319-09033-7_14 market that allegedly substitutes forecasting. Allegedly, because market rationality concerns forecasting for at most several, in some subjects maximally 10 or 15 years.[1] Meanwhile, according to the analysis presented in earlier chapters, new technology products might find a broad social application after many decades, and the threats related to them might result not from their technological character, but from a social fascination with the opportunities presented by them.

Therefore forecasting, even if fallible, is nevertheless necessary; this concerns especially forecasting of challenges and threats. It is also obvious that it should not be based on a simple *extrapolation forecast,* or even more complex *prognosis,* but it should involve *constructing future,* a composition of more or less probable, or important, see below, scenarios of possible threats or challenges. We shall call such approaches together *future studies.*

A threat is an inseparable part of a challenge, often most difficult for analysis. The analysis of threats is a fundamental part of future studies. Long experience in future studies indicates that often the greatest social impact resulted from warning prognoses, with a small probability of realization but stressing specific threats.

However, a challenge is more than a threat, it includes also opportunities, and the analysis of a challenge requires also a discussion of strengths and weaknesses, such as in the well-known SWOT strategic analysis. However, in the case of analysis of challenges, the order should be kind of reversed and augmented, *Threats-Weaknesses-Opportunities-Strengths-Actions* (TWOSA): we should first analyse threats, then weaknesses, then opportunities and strengths in using these opportunities, finally actions necessary to rise to the challenge. This last stage of actions actually consists of more detailed stages of setting strategic goals, priorities, instruments and means of achievement of the goals (A = GPIM, *Goals-Priorities-Instruments-Means*), but in this book, this stage will be not analyzed in detail.

The analysis of challenges should be interdisciplinary, similarly as future studies. The most close to it are interdisciplinary systems studies; alike, analysis of threats might be combined with the systemic risk analysis (broadly understood, not only in probabilistic terms, including also rare thus improbable events). As any strategic analysis, the analysis of challenges should have inherently dynamic character, be based on a construction of a dynamic scenario developing in time; therefore, e.g. in the systemic risk analysis we construct scenarios of threats, assess their probability and the scale of threats or damages.

The above comments constitute only general premises; further we concentrate on more specific, even if still general and interdisciplinary analysis: at first of possible threats, then of challenges (together with weaknesses, opportunities, strengths and elements of actions), first on a global scale, then in relation to Europe

---

[1] Defending the short-term market rationality, a classic writer in economy used to say: "In the long term we are all dead". But such attitude, even if typical for economics, is not typical for other disciplines, e.g., technologists often construct tools with a view to future generations. Such an attitude is not sufficient even for economics: the great crisis of the years 2007–2011 has shown clearly that market behaviour even in short perspective can be irrational.

and Poland. I shall provide specific lists of threats and challenges, without trying to make a complete review of them. Such completeness, always only relative, limited, might be achieved only by a group brainstorming, a result of thinking by a large team of people.

Moreover, it is impossible in such short work to analyze deeply even a few challenges, hence the analysis is consciously short, it illustrates methodological issues and the need to determine actions (without analysing them in detail). I shall stress, however, the threats resulting from social fascination by technology and, as an example, the threats and challenges related to the current situation of Poland.

14.2 Examples of Threats

Threats are phenomena or occurrences that we want to avoid, hence we usually hope that they will not happen, and we are ready to attach small importance to them precisely because of such psychological attitude. In this sense, even each larger change of circumstances that endangers our habits and customs might be a threat. All this results in the so-called *Cassandra effect,* well known in the praxis of future studies: the more precisely somebody forecasts larger changes or threats, the less credibility is given to such forecast; it is usually ignored without attaching any larger importance to it. For example, it is not true that the fall of communist system or the development of the Internet were examples of the so-called *black swan,* a phenomenon impossible to forecast (as maintained by Taleb 2007): the fall of communist system was predicted by Toffler and Toffler (1980), and at least one person, Ronald Reagan, believed in this forecast, while many people (including myself) predicted the development of the Internet, the increase of significance of TCP/IP protocol, of computer networks based on this protocol and their further development such as hypertext and WWW. We discussed this process at least for 30 years, even if at the beginning nobody wished to take it seriously until the Internet changed all our lives.

For this reason, *warning prognoses* are very important, even if they are usually exaggerated. Nevertheless, they catch social attention and turn it towards a specific type of threats. The most broadly known, and effective, prognosis of this type was the book *Limits to Growth* (Meadows et al. 1973), presenting a scenario of an exhaustion of natural resources on Earth, actually rather improbable, but supported by a computer simulation which had an impact on popular opinions in the entire world and helped to popularize ecological principles that are today common in education around the globe.

Is *the threat of exhaustion of natural resources of our world (1),*[2] indicated by Meadowses, present also today? Yes, it is, but it is well researched and the

---

[2] In further parts of this chapter, threats are numbered (in parentheses), while capital letters indicate challenges.

scientific and technical research on the entire world tries to counteract this trend. More important is *the challenge of sustainable development (A)*, related to this and other threats, discussed in a further part of this chapter.

Many authors try to repeat the success of Meadowses and describe various spectacular, but not very probable threats. Personally, I ascribe a low probability to *the threat of Third World War (2)*, even if it should be considered and counteracted, mostly when trying to respond to *the challenge of formation of a new world order (B)*, also discussed further on. Similarly, not very probable (at least, on a larger socio-economic scale in the 21st century) is *the threat of radical biotechnical evolution of humans (3)*. Equally low probable, but with a much greater scale of consequences, is *the threat to human civilization related to a space object hitting Earth (4);* counteracting such threat is a strong motivation for responding to *the challenge of formation of a new world order (B)*.

More probable, if with certainly less consequence, is *the threat of anthropogenic changes of global climate (5);* the impossibility of reaching a global agreement on this issue is another example of *Cassandra effect*. More probable, practically certain is also *the threat of losing the dominating position by the group of North American and European countries (6)*, even if I am sceptical about the scale of this threat, about slogans using the metaphor of sinking Titanic. North America and European Union will most probably lose their dominating position in the world soon, but, provided they will act rationally, they might retain important geopolitical role. This loss of dominating position by this group, I believe, will have positive consequences, since it will diminish the scale of another threat, not only probable but certain, that already resulted in the terrorism and the attack on New York on the 11th of September: *the threat of transformation of the growing inequality in the world into anti-American and anti-European attitudes (7)*.

This threat I discuss in more detail, because we should well understand its mechanisms. It is a by-product of the information revolution, that has (at least) three consequences. On the one side, this revolution contributed to globalisation of economy but also to quickly growing economic inequalities, not only inside individual countries (as a result of neoliberal doctrine, see e.g. Wilkinson and Pickett 2009), but also globally (as a result of unlimited transfer of profits of large global corporations to their countries of origin). On the other side, it brought an abolishment of spatial limits to information and knowledge; as a result, the part of global population that lives in poverty can learn without difficulty how the elite of most rich countries lives. Thirdly, it does not help that the globalisation of the Internet and its integration with television resulted in the development of a *new spectacle society,* anywhere on the globe one can participate in this society, in which the domination of advertisement and the neoliberal media of most developed countries propagate the patterns of living of the most rich part of population of the globe. Thus, the new spectacle society threatens the global order: we cannot wonder that it results in jealousy and resentment, we should rather wonder that it resulted in terrorist attitudes only in a younger part of one culture, where the specific religion promises paradise for a martyr's death.

Therefore, we should think how to limit the escalation of anti-American and anti-European attitudes. There are at least two directions of possible action. One is to influence the new spectacle society (if not by ethical values, then by self-interest arguments) to take into account the fact that also the poorest people on Earth will participate in the spectacle. Another, possibly more effective one, is to use the Internet to help in the education of the poorest parts of the world and thus to help in reducing the economic inequality. In the United States, such goals are adopted by the *Open Access* initiative; in Poland, there is a lack of understanding of this problem, media defend rather the so-called intellectual property rights (see e.g. Niezgódka 2007) or promote the neoliberal slogan "everybody should help himself".

This is related also to another, rather probable *threat of Euro-centric or Western-centric cultural imperialism (8);* the scale and negative impact of this threat is usually not perceived. Such imperialism consists in un-appreciation or even contempt for cultural values of distant cultures, together with over-appreciation of culture considered as one's own: Polish, European, North-American. An effort to better understand different cultures, starting from Japanese, Chinese, Indian, etc., is a necessary condition of rationality of political actions in the conditions in which North America and European Union lose their economic domination and try to preserve their geopolitical role.

The informational revolution brings also many other threats. Almost certain, even if in some sense related to other threats, is the conflict about intellectual property rights, or more precisely, *the threat of a conflict between corporate ownership of knowledge and open access to the intellectual heritage of humanity (9),* discussed in more detail in Kamoka and Wierzbicki (2005). Similarly advanced is *the threat of domination of robots, computers and networks over humans (10),* not in the sense of ascribing some bad or malicious features to them, but in the sense of unlimited lack of imagination combined with the desire for power of some people, that can result (and in many cases already resulted) in using the tools of information technology for an excessive social control or an excessive automation of activities that should be reserved for humans.

A socio-economic threat also resulting from the informational revolution is *the threat of anti-democratic and fascist-prone social movements (11).* Some authors maintain (see e.g. Bard and Söderqvist 2006) that the informational revolution results in *netocracy,* an allegedly inevitable social divide as a consequence of that revolution. There are also political slogans about alleged obsolescence of democracy; I believe that such theses are exaggerated and serve the interests of specific social groups. However, what is actually dangerous is a tendency to use new media and tools of information technology to propagate populist fascist (or fascist-prone) slogans; we must remember that it was fascism that used radio as the main tool of propaganda. We observe this today in many countries, not only in Poland, also in Hungary or Italy. The probability of political actualization of such threat is not very large (we can hope that people remember lessons of history), but the consequences might be serious even in a global scale.

An actually certain threat of purely economic character, even if it also results from the informational revolution, is *the threat of high and knowledge-based economy becoming oligopolistic (12)* that will be discussed here in some more detail. The neoliberal economists use the slogan that technical progress results in a destruction of natural monopoly and leads to free market, which is allegedly visible on the market of telecommunications and ICT services. However, precisely on this market a quite different phenomenon occurs: technical progress and the development of knowledge-based economy decreases marginal costs of production to such extent that the classic relation of the price to marginal cost on high technology markets is destroyed. And since this relation is the foundation of the arguments about the efficiency of free market, *the high technology markets are not an example of free market.* The prices actually observed on these markets can be explained only with the situation of *natural oligopoly,* while it is an open question whether these prices result from an unhampered oligopolistic competition, or from a tacit or explicit collusion and cartel price fixing.[3] Therefore arguments that high technology markets should be left to free competition that will solve everything are in fact an expression of the interests of large corporations acting on these markets.

Another threat related to the informational revolution is *the threat of virtualization of economy and systemic crises (13).* By virtualization of economy I understand the fact that the informational revolution made it possible to increase speculative capital in banking to a level several hundred times larger than productive investment capital. Such a situation is obviously prone to systemic crises. In a recent report of the Club of Rome entitled *Money and Sustainability* (Lietaer et al. 2013) it is recalled that too much uniformity (that developed in the finance system because of the informational revolution) makes the system susceptible to crises, the related question is what methods of diversification would be applicable and effective in the financial system.

A consequential, more specific danger of virtualization is the *threat of market corruption due to informational asymmetry.* The great financial crisis of the years 2007–2011 was the result of both the virtualization and of its specific part, informational asymmetry that led to market corruption. There is no doubt that information technology, computer networks and the easiness and speed of complex computing, resulted in the globalization of financial services, enabled to speculate in these services and to inflate the speculative investment bubble at the beginning of the 21st century. However, a more specific example is the story of David X. Lee,[4] an American mathematician of Chinese origin, who developed and sold the so-

---

[3] Since to control such phenomena, classic anti-monopolistic and anti-cartel offices are not sufficient, many countries start to form specialized bureaus to control selected parts of high technology markets. For example, in Poland there is a bureau for control of electronic and telecommunication service markets, while other fields of high technology, such as pharmaceutical industry, are not as effectively controlled.

[4] See Salmon (2009) under a telling title: *Recipe for Disaster: The Formula That Killed Wall Street.*

called *cupola formula* to Wall Street. The formula enabled fast computation of correlation coefficients. Due to this formula, an allegedly "absolutely safe" portfolio of financial investments, derivatives of derivatives, derivatives of housing investments, was created by a combination of uncorrelated investments. The advertisement of such allegedly absolutely safe investments blown up the investment bubble, and average investors believed in the safety of such products, due to information asymmetry. Only experts could know that uncorrelated investments are safe as long as the underlying processes are stationary, while during a crisis that stationary character is lost and uncorrelated investments become fully correlated. We can conclude that new possibilities of information technology, combined with greed and advertisement, resulted in market corruption: they blown up the investment bubble. The bubble could be pierced by anything and the explanations of the neoliberal economists that the crisis was a result of unreasonable decisions of US government are just excuses defending a lost cause. The great crisis of 2007–2011 had at least two important consequences.

Firstly, the crisis marked *the beginning of the end of the era of postmodernism*. Postmodernism was, and still remains, a dominating intellectual trend in social sciences in the end of the epoch of industrial civilization and the beginnings of the informational revolution. Moreover, postmodernism contributed several important concepts, such as the concept of historically changing *episteme* of Foucault (1972) or the concept of *long duration social structure* of Braudel (1979). However, the main paradigm (or, more precisely, a horizontal hermeneutic conviction[5] underlying the paradigm) of postmodernism is that *knowledge has only a local character, it is a result of local social discourse and it is reducible to power and money*. This conviction was obviously negated by the global reach of the crisis, as well by other events, such as the explosion of a volcano in Iceland in 2010 that brought consequences extending far beyond Iceland. As discussed in Chap. 13, postmodernism was never fully accepted by technologists, including information technologists. They construct tools that should be most universal in applications, thus they need possibly universal knowledge. After the crisis it is clear that some statements of postmodernism, e.g. that *each region, because of its unique, local character, should be left to itself to develop locally*, see e.g. (Jackson 2000), are rather exaggerated. Clearly, we should respect the uniqueness and local character of regions, but there are some universal features and means, e.g. the fact that access to the Internet supports regional development and it can contribute to convincing people that they can stay in remote rural areas because the web gives them new possibilities. On the other hand, the postmodern convictions are very popular in social sciences and it will take at least a generation until they peter out.

Secondly, the crisis marked also *an end of neoliberalism*. Under the concept of *neoliberalism* I do not understand a branch of *liberalism*, the noble conviction that *individual freedom is a fundamental value*, but a distortion of liberalism in economy leading to the conviction that *markets should be absolutely free*.

---

[5] See, e.g., the discussion with beliefs of Don Ihde in Chap. 13.

Neoliberalism started to dominate in economy after the fall of communist system, and it was strengthened politically by the so-called *Washington consensus*. Horizontal hermeneutic assumption of neoliberalism was the conviction that *there is only one correct way of capitalism development, the Anglo-Saxon capitalism*. This was clearly opposite to the postmodern convictions, even if in other sense post-modernism supported neoliberalism (e.g. by reducing knowledge to power and money). The events during the great financial crisis 2007–2011 contradicted the assumptions of neoliberalism: markets, if left alone, can be corrupted; today, it is clear that there are many ways of development of capitalism (however perhaps less than the regions in the world), see e.g. (Stehr and Adolf 2008). Neoliberal beliefs will not peter out immediately, particularly in Poland. People change their beliefs slowly, particularly if these beliefs are consistent with economic or political interests. However, neoliberal beliefs will slowly peter out even in economy, because the effects of the informational revolution, inclusive of the oligopolistic character of high technology markets and the corruption of market mechanisms due to irresponsible use of information technology will require a fundamental revision of economic theories, see e.g. (Wierzbicki 2009).

The threats of virtualization of financial markets with corruption of market mechanisms as well as oligopoly on high technology markets result in the need of defining *an industrial strategy and policy,* especially in Poland where these issues were neglected through last 20 years because of the naive belief that free market will replace any industrial strategy. Such a strategy must not be classic, but adapted to the conditions after the informational revolution and aimed at *creation of new working places and investments in human capital according to the requirements of new socio-economic conditions,* where the megatrend of dematerialization of work results in the disappearance of classic professions and ways of employment. If this will not be attempted, the alternative might be the prognosis of Roubini (2011) coming true: "The alternative is, as in the thirties, a permanent stagnation, depression, monetary and commercial wars, limitations of capital flows, financial crises, state bankruptcy and universal socio-economic destabilization".[6]

Another threat discussed recently in media is *the threat of conflict between old generation and young generation (14)*. The increase of the share of older people in the society and many resulting problems are very probable (practically certain, see e.g. Kleer 2009), but the extent of resulting conflict is blown up by media seeking sensation. Thus, I classify this conflict as certain but of small scale.

A threat of large scale even if relatively less probable is the last one (on my list; I am not saying that the list is complete), a *threat of self-annihilation of human civilization*[7] *because of a positive feedback between science and technology on the*

---

[6] Quotation after *Gazeta Wyborcza*, 20–21 August 2011.

[7] I fully agree with the opinion of Joseph Agassi, a philosopher of technology quoted in Chap. 13, that *the most important question today is what we should do to prevent an extinction of human (and other) life on Earth;* however, I try to identify the reasons of this threat more precisely.

one side and the socio-economic system of their utilisation on the other. The existence of such threat might be indicated by the phenomenon of *eerie silence* (Davies 2010), the lack of response from cosmos to our 50 years of radio signalization, indicating that there is intelligence on Earth; the silence might suggest that civilizations such as ours might be doomed. There can be many mechanisms of such self-annihilation, some of them related to threats discussed above, such as a next world war using nuclear weapons, depletion of natural resources, anthropogenic climate changes etc.

The problem relates to the fact that such positive feedback is, at least to some extent, inevitable, even if not caused by technology (as suggested by naive philosophy of technology, see Chap. 13), and can have unforeseeable consequences. Therefore, the issue is how to limit and modify this positive feedback, e.g. by shifting the goal of economic development from GDP growth to the improvement of the quality of life. The most important goal of technology and philosophy of technology is to collaborate in order to identify specific future threats related to this feedback. Examples of such specific threats are the development of robotic weapons to such extent that they might be accessible to fundamentalist religious splinter groups, or an irresponsible development of genetic technology that is left to the market (without taking into account long term threats), etc.

Table 14.1 presents a summary of the threats discussed above, together with subjective assessments of their probability, scale of impact and importance; the latter is understood as a subjective fuzzy logical product of the probability and scale of impact, to illustrate the general principle that even if an event has a small probability it might be of great importance if the scale of impact is very large.

14.3 Challenges

While the number of threats presented here was rather high in order to achieve a possibly broad (even if certainly not complete) presentation, the number of challenges discussed below is consciously limited, to enable presentation of the methodology of their analysis: *a challenge,* related *threats, weaknesses* (of the world, Europe, Poland) in addressing this challenge, *opportunities* (as above, including the ways of benefiting from them), *strengths* (as above) leading to possible *actions.* Thus, I shall address only four challenges: *the challenge of sustainable development (A), the challenge of creation of a new world order (B), the challenge of informational revolution (C) and the challenge of biotechnical revolution (D).*

(A) The Challenge of Sustainable Development

Recall that *sustainable development* (Bruntland 1987) means leaving to our children approximately the same chances and environmental conditions that we have

Table 14.1 A juxtaposition of examples of threats, their probabilities and scales, together with a (subjective) assessment of their importance

No.	Name of threat	Probability	Scale of impact	Assessment of importance
(1)	The threat of exhaustion of natural resources of our world	Small	Large	Middle
(2)	The threat of Third World War	Very small	Large	Small
(3)	The threat of radical biotechnical evolution of people	Very small	Large	Small
(4)	The threat to human civilization related to a space object hitting Earth	Very small	Very large	Large
(5)	The threat of anthropogenic changes of global climate	Large	Middle	Middle
(6)	The threat of loosing dominating position of the group of North American and European countries	Very large (certain)	Small	Small
(7)	The threat of transformation of the growing inequality in the world in anti-American and anti-European attitudes	Very large (certain)	That depends	Middle
(8)	The threat Euro-centric or Western-centric cultural imperialism	Very large (certain)	That depends	Middle
(9)	The threat of a conflict between corporate ownership of knowledge and open access to the intellectual heritage of humanity	Very large (certain)	Large	Large
(10)	The threat of domination of robots, computers and networks over humans	Large	Large	Large
(11)	The threat of anti-democratic and fascist-prone social movements	Large	Large	Large
(12)	The threat of high and knowledge-based economy becoming oligopolistic	Very large (certain)	Rather large	Large
(13)	The threat of virtualization of economy and systemic crises	Very large (certain)	Rather large	Large
(14)	The threat of conflict between old generation and young generation	Very large (certain)	Small	Small
(15)	The threat of self-annihilation of human civilization because of a positive feedback between science and technology on the one side and the socio-economic system of their utilisation on the other	Small	Very large	Large

ourselves.[8] This does not mean that the development will be *balanced,*[9] since any development implies lack of balance; every development will perturb the balance of the Earth's biosphere, but we should try to avoid excessive perturbations. Such undue perturbations are usually not the result of the needs of development, but of the unbridled actions of free market that treats the natural environment as a common good that should be exploited (or first privatized then exploited) at a negligible cost. Therefore, the concept of sustainable development is clearly contradictory to neoliberalism. However, the challenge of sustainable development is related not only to *the threat of exhaustion of natural resources of our world (1)* or to *the threat of anthropogenic changes of global climate (5),* but also e.g. to *the threat of a conflict between corporate ownership of knowledge and open access to the intellectual heritage of humanity (9)* and many other threats, up to *the threat of self-annihilation of human civilization because of a positive feedback between science and technology on the one side and the socio-economic system of their utilisation on the other (15).*

A good answer to the challenge of sustainable development requires long-term rationality, not available in contemporary market economy characterized by short-term rationality.[10] Therefore, the challenge of sustainable development is actually related to *the excessively short rationality horizon of both democratic political systems (4–8 years) and market systems (1–12 years).* This is the basic *weakness* of these systems, all over the world, also in Europe and particularly in Poland, and this is the reason why I cannot agree with neoliberal slogans that we should leave all strategy to markets and everything will be solved. Unfortunately, it is also clear that governments might be even more faulty, while they are the basic force that can counteract the short-term rationality of markets. Thus, the opposition *more market or more government* debated by economists is false; we must find a third way, a third power to resolve this dilemma. A solution might be popularization of

---

[8] This is in a sense similar to the argument of John Rawls about leaving to our children the most just social conditions, or to the extension of his argument presented in Chap. 6 about leaving to our children the best and most objective knowledge and tools. Clearly, all these goals are not fully attainable, but certainly worth striving for as higher values.

[9] Such a meaning was implied by the imprecise Polish translation of *sustainable development, rozwój zrównoważony (equilibrated or balanced development,* since there is no direct equivalent of the word *sustainable* in Polish); after long discussions, a better translation of the term, *rozwój trwały (lasting development),* is slowly accepted in Poland.

[10] This issue was discussed earlier in this chapter; here it should be added that it concerns in particular the present global oligopoly economy on markets of high technology (called by Drucker (1993), post-capitalist economy, since the management of big corporations does not consist of owners of this corporations and is not motivated by a good opinion about the company in a long run) where the time horizon of rationality is at most a dozen years, some even say it is much shorter, like 3 months, and certainly does not take into account the interests of future generations. The neoliberal economy tried to respond to this objection by promoting the so-called *theory of rational expectations* (with the main thesis that each market player forms rational long-term expectations of the future and takes them into account in her/his decisions); however, economic practice together with recent financial crises have shown that market players behave irrationally in a long-term sense of the word. See also (Soros 2006).

the *fundamental values of sustainable development* in the society that would insist on electing governments which are sufficiently motivated by these values. A great achievement of the authors of *Limits to Growth* (Meadows et al. 1973) was the promotion of *ecological values* in educational systems of the entire world; we can therefore draw the conclusion that it is now necessary to promote *values of sustainable development* in these systems. These values can be listed as follows:

(1) The value of leaving to our children reasonable good ecology and natural resources on our globe;
(2) The value of leaving to our children an open (not privatized) access to the intellectual heritage of humanity including most objective knowledge possible and best and well-tested tools possible;
(3) The value of leaving to our children the most just and well-tested social and economic institutions, including methods of prevention of excessive socio-economic disparity.

The certainty of rising to the challenge of sustainable development we shall only obtain when all politicians, entrepreneurs, but also ordinary people and consumers will be educated to respect such fundamental values since childhood.

What are the *chances* of attaining truly sustainable development? A chance is in the development of science and technology, but not only oriented towards a short-term profit, but also towards long-term goals of sustainable development, such as limitation of harmful emissions, moderation of climate changes, etc. This is fully possible with the use of technologies developed today, but requires time and money, and thus strategic determination, and it will not be achieved spontaneously, since free market promotes technologies that bring short-term profit. Therefore, we will face a slow but inevitable process of limitation of the forces of free market by the standards and requirements imposed by "green" consumers. Already today, diverse technologies are developed for the purposes of e.g. more ecologically friendly ways of propulsion of cars, limiting emissions of power plants, ironworks, chemical plants, or providing alternative energy sources. However, such development is typically slowed down because it is not the priority of big corporations.

Another chance of sustainable development is the change of social expectations concerning the goals and measures of development. Purely economic approach is focused on economic growth, measured by GDP increase. However, there are many indications that people live better in countries that concentrate not on economic growth, but on quality of life, such as Sweden or Japan. It is difficult to measure quality of life, but we can include in it the values of sustainable development.

Therefore, the problem how, at what speed and in what proportions we should support development caring for its sustainability, for the resilience of natural environment and the quality of social environment in the interests of our children, will remain a fundamental problem for the global society and will influence the solutions of partially related problems (energy provision, transport, details of environmental protection, life style and socio-economic institutions etc.). To solve these problems we need further scientific and technological development, but the social consciousness, particularly of consumers, will be decisive. Recently, Nico Stehr proposed a theory of *moral or ethical advancement of markets* (Stehr and Adolf 2008): there are moral or ethical values of consumers that decide what products they buy; if large corporations will face the threat of boycott by consumers, they will be forced to accept the values of sustainable development. Thus, ethical advancement of markets is the *strength,* both global and European, that can oppose the fundamental *weakness* of the short-term rationality of politics and markets. Unfortunately, this strength is also manipulated by advertisements and popularization of the values of sustainable development should also use the possibilities provided by the Internet and the *new spectacle society.*

(B) The Challenge of Creation of a New World Order

Many threats discussed above suggest there is a need to create a new world order, even together with elements of world government. Here we can include *the threat of Third World War (2), the threat to human civilization related to a space object hitting Earth (4), the threat of anthropogenic changes of global climate (5), the threat of loosing dominating position of the group of North American and European countries (6), the threat of transformation of the growing inequality in the world in anti-American and anti-European attitudes (7), the threat of Euro-centric or Western-centric cultural imperialism (8), the threat of a conflict between corporate ownership of knowledge and open access to the intellectual heritage of humanity (9), the threat of anti-democratic and fascist-prone social movements (11), the threat of high and knowledge-based economy becoming oligopolistic (12), the threat of virtualization of economy and systemic crises (13),* finally *the threat of self-annihilation of human civilization because of a positive feedback between science and technology on the one side and the socio-economic system of their utilisation on the other (15).* It is telling that several of these threats, say, (4), (9), (11), (12), (13), (15), are classified above as very important. For example, it is difficult to imagine that we could counteract large-scale cosmic catastrophes (4) that would radically change global environment, without some form of global government or at least global consensus.[11]

If there are that many reasons for creating a new world order, why there are so few suggestions to do so? This results from a concatenation of diverse reasons. Firstly, there are geopolitical interests of the USA and other countries dominating in the world. As long as this domination lasts, they are not interested in any change of the existing situation. Secondly, there are the interests of large international corporations: the neoliberal globalization was very advantageous to them, while

---

[11] People might conclude that we are sufficiently rich and have sufficient knowledge to effectively counteract the possibilities of such future catastrophes (until now, we are very poorly prepared for them, which was shown by the recent explosion of the volcano on Iceland or by the catastrophe at Fukushima). Preliminary research on this theme is conducted, but its intensification would require e.g. establishment of an international basis on the Moon in order to better observe incoming space objects and to appropriately react (enforce a change of their trajectory).

global government would constrain them, hence such corporations support the ideology maintaining that governments at all, and in particular global governments, are useless. Therefore, the main *weakness* related to this challenge is the strength of opposite interests.

The *chances* of overcoming these weaknesses result from the current developments in global situation. The decline of the domination of the USA is inevitable in the face of the gradual increase of economic importance of most populous countries of the world and the increasing international debt of the USA. Given the growth of economic importance of China, India, Brasilia, and also Russia, these countries will demand a bigger political role in the world. And the last great financial and economic crisis has proven that the thesis about uselessness of governments is false.

In order to utilise these chances, we must find *strengths* supporting them. This is difficult. The UN is dominated by the diversity of interests of many weak countries, and other international organizations are usually subject to the interests of the USA. However, there is a strength that might support the formation of a new world order; it is the *fear* that sustainable development might be perturbed by the short-term strategies of large corporations, that the recent crisis is systemic and will repeat, etc. The informational technology and biotechnology will drive the changes of the world, as discussed below; but if we leave the exploitation of such demand to large corporations (not to free market, because it does not exist in its pure form for high technology, as discussed above), we should expect another inevitable big crisis, another large bubble of artificially created demand motivated by short-term profit, not by the problems to be solved that torment the world.

Therefore, the vision of the world governed by big corporations is not acceptable, because it would lead to instability. There are many ways of creation of a new world order, but most probable are two. One is a renewed strengthening of the role and competences of the United Nations. Should this organization not undertake new tasks, including both duties and prerogatives, then another international organization (between many existing ones) must fill the vacuum and undertake the task of creation of a new world order. The goals of such organization must be partly political, related to global security (limitation of armaments, elimination of armed conflicts, pacification of regional conflicts etc.), partly economic and regulatory, such as control of cartel behaviour, oligopoly and the pursuit of monopoly by large global corporations, regulation of international banking, etc., and partly developmental, such as overseeing common global projects, on Antarctic, the Moon, other planets of the Solar system.

This is a tremendous challenge, bigger than e.g. the formation of the European Union, but the humanity must manage it in order to look forward with confidence.

(C) The Challenge of Informational Revolution

While accepting the informational revolution as an accomplished fact (or at least happening now) we must admit that it brings both advantages and opportunities, but also serious threats. This is the *essence of the challenge of informational revolution: it is partly accomplished but develops further and we should well understand the related threats,* analyze them as well as related weaknesses, and to define, against that background, the opportunities and strengths enabling to better harness these opportunities.

The threats related to the informational revolution are multiple and diverse. Between threats discussed above I shall list: *the threat of a conflict between corporate ownership of knowledge and open access to the intellectual heritage of humanity (9), the threat of domination of computers, robots and networks over humans (10), the threat of anti-democratic and fascist-prone social movements (11), the threat of high and knowledge-based economy becoming oligopolistic (12), the threat of virtualization of economy and systemic crises (13),* and also *the threat of conflict between younger generation and older generation (14),* even if the last one only to some degree (the dematerialization of work, characteristic of the informational revolution, increases the conflict related to the access to work that can be also generational). These are not the all threats related to the informational revolution; to some extent, also *the threat of self-annihilation of human civilization because of a positive feedback between science and technology on the one side and the socio-economic system of their utilisation on the other (15)* belongs here.

The threat (9) results from the fact that during the all 30 past years of the informational revolution, great corporations tried to maximally privatize the common knowledge of humanity and used the neoliberal ideology of *intellectual property rights* for that purpose, see (Cellary 2011) versus (Lessig 2005; Boyle 2008). However, *knowledge is not a degradable resource,* it usually grows and not decreases when it is intensively, commonly used, thus the classical economic substantiation of privatization of common goods (the so-called tragedy of commons) does not apply to knowledge. Contrariwise, it is in the interests of entire society or even entire humanity to preserve the maximal part of the intellectual heritage of humanity as a common good.

The threat (10) is clearly not a conspiracy of computers and robots, but a weakness of human nature: according to Martin Heidegger (Heidegger 1954) the development of technology results in an effect that "man exalts himself to the posture of the lord of the Earth". We can observe this in many examples, such as the irresponsibility of many information technologists that write software leading to a domination of a computer over people,[12] as well the attempts of politicians to use information technology to attain a full control over society, a realization of Orwellian utopia.

The threat of oligopolization of the economy (12) is a clear outcome of the informational revolution: it is the dematerialization of work and knowledge-based economy that decrease so much the marginal production costs that the free competition prices of free market cease to be used, which is possible only in the

---

[12] E.g. an automatic correction of a text without asking the user about her/his opinion is a clear example of such domination, but there are many other examples.

situation of an oligopoly.[13] The threat of virtualization of economy and systemic crises (14) is another clear outcome of the informational revolution: the banking system was one of pioneers in using information technology for globalization and uniformity of the global financial system, while its virtualization, the domination of speculative capital, was a joint result of human greed and the possibilities of new technologies, another example of the Heideggerian principle "man exalts himself". The threat of conflict between younger generation and older generation (14) is also a partial result of the informational revolution, since it leads to automation and robotization of many human activities (production but also services and bureaucracy) and loss of work, which is particularly painful for the young generation.

Finally, *the threat of self-annihilation (15),* even if it does not result directly from the informational revolution, it is nevertheless exacerbated by it. Because of the revolution, the speed of socio-economic changes is increased to such an extent that it might be difficult to counteract unpredicted and catastrophic consequences of a social fascination with some technology products or applications of scientific results. By its very nature, market economy counteracts the excesses of demand, but after its modifications brought by the informational revolution it might turn out to be an insufficient mechanism for preventing the excesses of social fascination.

Thus, the informational revolution brings tremendous threats, enforced by human or disciplinary *weaknesses:* paradigmatic attitudes of some traditional fields of science, counteracting all changes and thus misunderstanding the importance and impacts of inevitable changes related to the informational revolution; the weakness of economic science, unprepared to analyze knowledge-based economy; the weakness of human nature using new technology either to strengthen the power of authority or for short-term profits; the conflicts related to the access to work; etc.

Despite the above, the *opportunities* resulting from the informational revolution are even more numerous and important than the threats, see (Wierzbicki 2009), and I will not analyze them in detail here. As examples, I list below only some opportunities that appear to be the most important:

- *Annihilation of spatial limitations in terms of access to information and knowledge and in terms of inter-human communication,* brought by network multimedia access and communication. This aspect is perhaps even more important than Gutenberg revolution that popularized the access to information and knowledge through books, but was still subject to spatial limitations;
- *The possibility of turning around (in a long perspective, around the year 2050) the trend of urbanization of the world,* the beginnings of actual realization of the idea of *global village* or even *global forest;*
- *The use of multimedia access and network communication to fundamentally change the education systems.*

---

[13] The informational revolution results also in actions that counteract the general trend of oligopolization of high technology economy, such as the development of software based on *Open Source* licenses. The question is which tendency is stronger.

Between the above chances I shall discuss, shortly, only the chance of a fundamental change of educational systems. It is necessary to counteract the growing global economic inequality that results, among other things, from the pursuit of profit of large global corporations concentrating these profits and knowledge in the most developed countries. The said necessity results at least from the *threat of transformation of the growing inequality in the world into anti-American and anti-European attitudes (7)*. A fundamental way to counteract growing inequalities is a *free network access to educational resources of the best universities in the world,* an initiative of the *Open Access* type.[14] However, it is related to a necessity of a fundamental change of education systems, including not only general university education, but also life-long learning based mostly on a remote, electronic access, together with a multimedia nature of the sources of information and knowledge, and with the usage of this multimedia nature for stimulation of creativity, preparation of people to act in a new society. This requires in turn a fundamental change of paradigms and deep reforms of the entire education system, starting with elementary schools, see also (Auleytner 2009).

What are the *strengths* that drive the utilization of such opportunities? Firstly, a social demand, for multimedia access and network communication, for video collections of new films, for electronic distance and life-long education. Secondly, weariness of people living in big urban agglomeration caused by many hours spent in commuting and by growing costs of life, enforced by the ageing of societies. Thirdly, the commitment of the biggest universities in the USA to the initiative of *Open Access* that results, according to Galwas (2009), in an inevitable trend to create a free network access to educational resources. These are global forces and strengths; in Poland, we could additionally count on a relatively high level (despite accumulating neglect) of education of Polish information engineers and scientists.

Beside such opportunities, a positive aspect of the informational revolution is the fact that it contributes to prolonging the average length of life for all people on Earth. For me, it is the fundamental proof of civilization progress; to the critics of the concept of progress I have the question: would you like to live 200 years earlier, when the average length of life was twice shorter? The informational revolution contributes to lengthening of life not only by propagation of information about various threats to health (the issues related to hygiene, life style, etc.); its essential impact is the progress of biomedical engineering resulting from information technology and allowing earlier and better diagnoses of diseases. Therefore, a question whether it would not be better to stop the informational revolution, for me is without sense: even if possible, it would not be in our interest. Instead, we should rather well understand both the advantages and disadvantages of the informational revolution and prepare actions that would maximize the positive aspects and minimize the negative aspects of it.

A full analysis of the challenge of informational revolution would require a deeper analysis of the actions necessary for good utilization of this challenge.

---

[14] See also (Niezgódka 2007; Galwas 2009).

However, I shall not analyze this issue in detail and will suggest only one aspect of them. Because of the coincidence of the increasing pace of socio-economic changes, resulting from the informational revolution and discussed earlier, and the increasing *threat of self-annihilation of human civilization (15)*, a question arises: what to do, *what actions are necessary to counteract this threat?* Apart from trying to change the goals of development from GDP increase to the growth of quality of life or to develop a better global order, it is necessary to promote a better social imagination as regards possible results of various new technologies in order to counteract their possible negative consequences. This in turn will not be achieved unless an overall social education in new technologies is improved. It is an additional argument for the *introduction of at least three curricular technical subjects, e.g., informatics, robotics and biomedical engineering, to university studies of all fields, especially social and humanistic ones*, postulated earlier.

(D) The Challenge of Biotechnical Revolution

Elementary biotechnologies, such as genetic modification of cultivated plants, etc., already have a significant impact on global economy, but we are still far away from an actual biotechnical revolution. The speculations about *radical evolution,* a vision of a cyber-man as a new, mostly artificial product of biotechnical revolution, are frequent today (see, e.g., Garreau 2008). At the same time, social resistance arises against the excessive or irresponsible automation of diverse human skills, a domination of computers or networks over humans. Will these experiences support the radical biotechnical evolution of people which shall include microprocessors implanted into human organisms? We should rather expect a serious socio-economic resistance that would significantly delay such a radical evolution.

This in a sense intuitive social resistance might result from an unconscious association of the biotechnical revolution with the *threat of self-annihilation of human civilization (15).* That only a very small and the most rich part of society will be able to buy biotechnical implants is quite a possible scenario, but at the same time the information about them will be distributed globally by advertisements in information networks. This will lead to jealousy and sharp conflicts, to a new revolution against the rich. Enrichment as a motive of social development is effective only if it is reasonably universal; otherwise, if it leads to an excessive stratification (and a marginal access to biotechnical implants will be perceived as an excessive stratification), then it results either in mafia-related phenomena, or in revolutionary ideas. Thus, a radical evolution could result in a social revolution which, in the conditions of general access to science and information, could easily result in turn in self-annihilation of human civilization. Therefore, we can only hope that the intuitive social resistance against too speedy biotechnical revolution will be sufficient.

Therefore we can hope that the beginnings of biotechnical revolution and radical evolution of humanity will occur rather in areas of *considerable social demand for biomedical technologies;* it concerns in particular *the healthcare of old people.* An implantation of a microprocessor only to stimulate heartbeat, or artificially cultured bone cells used in order to rejuvenate the bones of old people will not mount much psychological resistance, hence such technologies will meet high economic demand which will help in their refinement and in decreasing their costs, making them commonly available. Together with an intelligent living environment for the purposes of care of old people, monitoring their state of health, or even with mobile robots as companions of the elderly, the biotechnical revolution might become a natural supplement and continuation of the informational revolution.

Thus, this challenge can help to diffuse two threats: *the threat of domination of computers, robots and networks over humans (10)* and *the threat of a conflict between old generation and young generation (14)*. The applications of information technology and biomedical engineering to the healthcare of older people will not meet socio-psychological resistance, but if older people will have only small financial resources, who will finance a universal development of such applications? This divergence between the resources of the elderly in conditions of retirement pensions reforms and the needs of development of biomedical technologies is the fundamental *weakness* of the challenge of biotechnical revolution that will probably lengthen the period of waiting for a social prevalence of results of this revolution.

Nevertheless I believe that the threat of generational conflict between older and younger people is rather exaggerated in the media and there is a serious chance for the beginning of only a *mild* (as opposed to *radical*) biotechnical revolutioninitiated by the older people around 2050. For example, young people could help in buying humanoid robots, at that time commonly accessible, as a companions for their parents, at least in order not to worry about the obligations to provide care for them. The growth of share of the elderly in the society might turn out to be the *strength* of a mild biotechnical revolution.

14.4 Challenges for Poland

Against this general background, it might be interesting to analyze threats and challenges specific for a given region or country. I will present here threats and challenges which are best known to me, concerning Poland. This issue was already discussed in a paper (Wierzbicki 2011), where as the most specific threats *the threat of neglecting the deteriorating civilization distance* of Poland with respect to the most developed European countries and *the challenge to minimize this distance* were indicated.

In the last 20 years, the goal of minimization of the civilization distance was not a priority for Polish governments; the priorities rather were concentrated on economic results and results of next elections, while the economic results were treated as an average that improved significantly but at the cost of an essential deterioration of social stratification and disproportions. The governments often relied on the opinions of international consulting agencies and corporations, without perceiving an explicit conflict of interests in such opinions: consulting

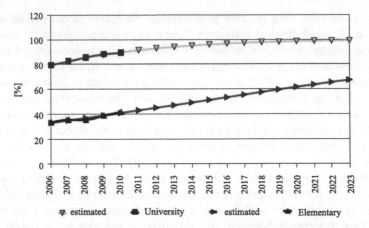

Fig. 14.1 A prognosis (data from the 5 years and estimated logistic curves) for the percentage of people regularly using computers in Poland, broken down into people with elementary and university education (Pietrasz 2011)

corporations are by no means interested in the civilisation development of Poland. As a result, we observed a significant increase of the phenomena of social exclusion (see e.g. Jarosz 2011) that are combined together in loops of positive feedback, self-support. Such phenomena lead to *social division* of many dimensions. The most significant of them is the magnitude of *digital divide* in Polish society. While the European Union warns against *digital divide* and suggests ways to decrease it, Poland did not pay much attention to this issue, which resulted in an increase of *digital divide* to an extent belonging to the biggest in Europe. It is illustrated in Fig. 14.1. (quoted after Pietrasz 2011).

This Figure implies that until 2020 we should attain in Poland a relative saturation of using digital technology (as shown in Pietrasz 2011, also in terms of access to Internet and mobile telephony), but only for people with university education. However, the use of computers by people with elementary education is now in Poland more than twice lower, and a relative saturation can be expected (because of a low rate of increase, about 2 % per year) first around 2040–2050.

Admittedly, we can expect that the participation of young people in university studies will further grow (see Fig. 14.2, quoted after Grzegorek 2011), but this participation influences the percentage of people with tertiary education with a significant delay, because of a natural phenomenon of accumulation of educated people in the society. Moreover, the prognoses presented in Fig. 14.1 are optimistic (as admitted by Pietrasz 2011). E.g. the prognosis for people with elementary education results from an assumption that the percentage of people regularly using computers will eventually grow to 100 %, while the data indicates rather that this percentage for people with elementary education might stabilize at around 50 %.

Although digital divide concerns a relatively small part of society with only elementary education (18.7 % in Poland in the year 2009, see Grzegorek 2011),

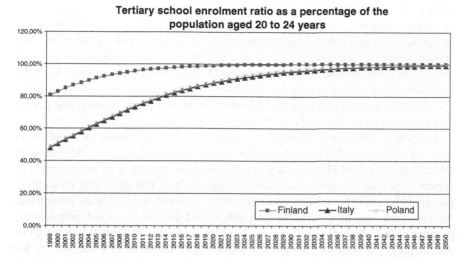

Fig. 14.2 Prognosis of the process of development of universal university education (logistic curves based on data from the last dozen years), participation of young people aged 20–24 in university studies (Grzegorek 2011; the curve for Poland virutally coincides with the curve for Italy)

the mechanisms of positive feedback or self-support result in a self-reproduction of this group: the children of people that do not regularly use computers have much less opportunities to finish secondary schools and to start university education. This is precisely the reason of the *social civilization division:* the group of people systematically excluded petrifies, has much smaller chances of education and employment, and is additionally much more vulnerable to ideological manipulation.

The prognoses presented in Fig. 14.2 are also too optimistic; it is rather improbable that the participation of young people in university education would amount to 100 %, there will be always exceptions. However, we can expect with a large probability that around 2050, this participation will be almost universal, exceeding 95 %. It will be not a result of government actions, but the outcome of consciousness of the Polish society that there will be no chances of good employment, be it in Poland or abroad, without university education, and the financial efforts of families aimed at providing children with good education. Today, Poland takes the fifth place in the world in terms of the percentage of students in the society (Banach 2011). Unfortunately, with the growing number of students, the quality of university education has substantially decreased, see further comments.

We would not have achieved the current index of tertiary education in Poland (around 70.2 % in 2008) without a network of private universities, which are more numerous nowadays than state universities. The quality of education in these networks, both private and state-owned, is diverse, but everywhere the large numbers of students result in an unavoidable decrease of the level of education quality. It is not clear how to counteract this tendency, but it belongs to the challenges related to the process of making the university education universal. The prognoses of further development of university education in Poland and in Italy are almost identical, but also Italy belongs to the European countries which are delayed in terms of education development. Polish civilization distance from the most developed countries may be evidenced by the fact that we shall need probably 25 years to achieve the percentage of university students among young people that today characterizes Finland (close to 95 %).

Much more difficult is to make a prognosis of the percentage of people with university education, because of different interpretations of this term. According to OECD, the so-called HRST index actually includes people with unfinished university education[15]; more exact data for Poland show that the index concerning people with completed tertiary education might be much lower. On the other hand, we have quite long data series for such indices and thus we can make quite reliable prognoses. In Poland, the values of such indices have grown about 9 % during last decade 2000–2009 (from 25.3 to 34.1 % for HRST index, from 9.1 to 18.1 % for the actual percentage of people with completed tertiary education); this indicates a great educational effort of Polish society. But even if such a big educational effort would be maintained in the next decades, we can expect a growth of the percentage of people with completed tertiary education around the year 2050 to 60–70 % at most.

Thus, the educational effort of Polish society is significant when it comes to quantitative data. On the other hand, not only quantitative data, but much more qualitative aspects of education decide about the competitiveness of societies in times of the knowledge-based economy; in this aspect we face rather disturbing signals indicating deterioration of civilization division in Polish society in qualitative terms.

The quality of university education depends mostly on the participation of academic teachers, and also students of later years of study, in current research work. However, for Polish universities, employment at several working places is typical (because of low salaries). Together with large numbers of students this results in a decrease of the intensity of research. Generally, such situation is also caused by a low level of financing of research and development by the Polish state; it decreased to a level characteristic of the least developed countries in Europe, more typical for African countries. In a time when in its subsequent strategies the European Union assumes an increase of R&D financing to an average level above 2 % of GDP, which requires an increase of R&D financing by the state at least to above 1 %, see Fig. 14.3, the financing of R&D by the Polish state as a percentage of GDP during the last 20 years significantly decreased, see Table 14.2.

---

[15] Not only persons "having successfully completed education at the third level", but also people that "are employed in an occupation where such an education is normally required", Canberra Manual, OECD 1995.

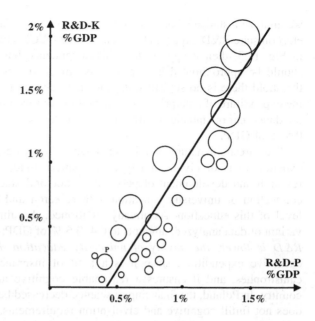

Fig. 14.3 Statistical dependence (linear regression weighted by the GDP level) between public R&D expenses (R&D-P, in % GDP) and commercial R&D expenses (R&D-K, in % GDP) in diverse OECD countries, data from 2003 (Wierzbicki 2008)

Table 14.2 Historical data on R&D financing in Poland as a percentage of GDP (R&D-P, public financing by the state, R&D-K, an estimate, usually approximate, of commercial financing of R&D by enterprises and other sources, see Wierzbicki 2008)

Year	1990 (%)	1992 (%)	1994 (%)	1996 (%)	1998 (%)	2000 (%)	2002 (%)	2004 (%)	2007 (%)
R&D-P	1.2	0.64	0.55	0.44	0.43	0.36	0.32	0.32	0.33
R&D-K	0.7	0.36	0.28	0.23	0.22	0.20	0.19	0.18	0.17

The opinions of international consulting corporations concerning the issue of science financing in Poland can be summarized with the statement that Polish government should not increase state financing of science, because Polish science is inefficient; an increase of R&D financing by market enterprises should be achieved in the first place. One can suspect, however, that such opinions express only a conflict of interests (if Polish science were better financed, there would be no need to employ international consulting corporations). What is worse, such opinions are repeated by governmental publications; e.g. the governmental report *Poland 2030: Development Challenges* presents "a proposition of a civilization project for Poland", but in relation to the Polish science it maintains that it is inefficient and its financing should not be increased.

Such diagnosis is erroneous, which might be proven in several ways. Firstly, Poland is about the 70th place in the world when counting the level of financing of science, in a place slightly better than the 40th when assessing the level of scientific results measured e.g. with publications and their quotations. Secondly, if we analyze a statistical dependence of R&D commercial expenses (by enterprises etc.) on state R&D expenses between diverse OECD countries, which is presented in Fig. 14.3, then it appears that this dependence has a *threshold character:* it should be approximated by a piece-wise linear regression line, below a certain threshold there is no statistical dependence between the expenditure of state and the expenditure of enterprises. The magnitude of this threshold depends slightly on the data (year of statistics used), but amounts to 0.4–0.6 % of GDP, approximately 0.5 % of GDP.

The threshold nature of this dependence has a deep meaning: part of R&D expenditure is of civilization type, it finances fundamental research, humanity research and development of personality, doctoral studies, and finally ensures a connection of university education with research and thus warrants a sufficient level of this education. As already mentioned, this threshold, depending on the variant of data analyzed, amounts to 0.4–0.6 % of GDP; *if the state expenditure on R&D is lower, the quality of university education decreases.* This is not an excessive expenditure, since it is a kind of insurance against possible future catastrophes, and it ensures a reasonable cognitive and civilization level of a country. In Poland, the financing of science decreased below this threshold level, it does not fulfill cognitive and civilization requirements, and this is one of fundamental reasons of civilization divide.

This is best indicated by the index concerning the number of doctoral students (in various S&T fields) in a given age cohort, that in 2008 in Poland amounted to 0.175 %, in Italy to 0.25 %, and in Finland to 1.36 %. When counting in line with this index, Poland is ranked between the three last countries in the European Union, similarly as in the case of the state financing of R&D. However, even if the financing level is very low, we have some practical results of science, implemented in economy, approximated by the expenditure of enterprises and other private sources on R&D in Poland. In the data presented in Fig. 14.3, data for Poland is placed *above* the piece-wise linear regression line, which signifies a better efficiency of Polish science than for countries *below* this line. *Therefore, the opinions about low efficiency of Polish science are erroneous, they express only a conflict of interests of consulting agencies.*

A similar situation occurs in financing of education. The Polish society is to a large extent ready to bear the cost of education of children, which results in in a strong development of private schools, but in last 20 years, the level of elementary education in the poorest rural municipalities, decisive for the civilization level of a country, decreased substantially. Some of the poorest municipalities in Poland are delayed more than 10 years when compared to the Polish average level of equipment in computer laboratories (Grzegorek and Wierzbicki 2009). As stated by Galwas (2011) "a long list of problems opens in education that raises the deepest concern" which substantiates an opinion that "the education system should become the highest priority in Poland".

Because of such indicators of a civilization distance, we can be rather pessimistic about a real advancement of Poland, the realization of a vision of minimization of the civilization distance of Poland to most developed European countries. However, such a vision is essential, we must know what challenges to face and what threats we must overcome in order to realize it. In Poland, the awareness of that is insufficient. Media discourse achieves at most the diagnosis that after 20 years of "civilization breakthrough" it appears that our successes are limited to economic and political aspects. Most of the documents discussing such problems do not note the developing civilization divide, do not stress the fact that a continuation of the current underfunding of science and education in Poland will inevitably lead to a crisis in that respect.

The mechanism of this growing crisis is clear: we introduce (correctly) a legislation limiting the simultaneous employment at many universities, but the financing of universities will not grow sufficiently, and thus the salaries for university teachers will not grow either. In result, the young, most talented alumnae will not start working at the universities and research institutions (we observe this trend already, but it will intensify). This will finally lead to a further, substantial decrease of the quality of both tertiary education and scientific research in Poland.

Nevertheless, in the hope that we can counteract the scenario of deepening civilization divide outlined above, we must have a vision of how to minimize the civilization distance of Poland to most developed European countries. Such vision must be based on a fundamental change of strategic priorities. The civilization divide of Polish society resulted from the lack of sufficient priority given to civilization development, culture, science, education, in the development strategy of Poland during the last 20 years. If we do not counteract this trend, we should lose our competitive opportunities in coming decades. Other countries will try to face up to the challenges of informational revolution and knowledge based economy, but Poland will decline to the position of the least developed countries in the European Union. Therefore, Poland needs a *strategic reorientation: while development priorities during the last 20 years concerned market economy and democratic society, in the next 20 years they should concentrate on civilization development, culture, science, education.* These priorities should be complemented by *a new industrial strategy, aimed at creation of new working places in response to changes resulting from globalization and informational revolution.*

14.5 Conclusions

This chapter discussed fifteen main threats and four selected challenges in a global scale. The list of threats does not pretend to be complete. The goals of the discussion are methodological: it stresses the strategic and interdisciplinary nature of threats and challenges. Moreover, it was shown here how to use an inverted and complemented SWOT method for an analysis of challenges, in the form of *Threats, Weaknesses, Opportunities, Strengths, Actions* (TWOSA). From the analysis of threats and challenges, some important conclusions result.

The last threat of *a self-annihilation of human civilisation* (15), even if not very probable, is of a large importance. This does not mean that science or technology contain such a threat in their "inner nature"; it means that their socio-economic applications are driven by the profit greed of market mechanism. And again it is not an "inner nature" of market mechanism that is responsible for this threat; market itself is a very good, robust tool, even if fallible in some of its aspects, but at least working in broadly changing conditions. *The threat of self-annihilation results from the fact of the existence of a positive feedback loop between science and technology on the one side and a market on the other* (which for some philosophers is a paradox of vicious circle, self-support, hence they refuse to notice the existence of this phenomenon and to analyse it). The danger of positive feedback loop is the danger of avalanche-like development, which, if not stabilized by reflection, ethical values and regulation by state institutions, will accelerate from its "inner nature" and can easily lead to a catastrophe and self-annihilation. We need, therefore, a broad social reflection on such dangers, and such reflection requires a better understanding of science and technology. Such understanding can be achieved, but under certain conditions; one of them is curricular education in the three technical disciplines mentioned above within all faculties of tertiary studies.

The said threat is related to all challenges discussed here: *sustainable development, new world order, informational revolution* and *biotechnical revolution.* Without concern for the sustainability of development with a view to future generations, without creating some form of global government, there are unfortunately considerable odds that negative aspects of the informational revolution, and particularly the biotechnical revolution will dominate over the positive ones. Moreover, the informational revolution has already to a large extent occurred and the subsequent biotechnical revolution is also almost inevitable.

As for Poland, we can observe a serious threat of *civilisation divide,* a deepening of digital divide in the society accompanied by an increasing civilization gap separating Poland from the most developed European countries. Challenges related to this situation can be coped with only by means of a fundamental reorientation of strategic *developmental priorities of the country: while in the last 20 years they concerned market economy and democratic society, in the next decades they should concentrate on civilisation development, culture, science, education, as well as on creation of new working places in response to changes resulting from globalisation and informational revolution.*

References

Auleytner, J.: Uczelnie przyszłości. Czy w Polsce? (Future Universities; in Poland?) Przyszłość:
    Świat. Europa, Polska **2**, 48–61 (2009)
Banach, Cz.: Strategia i kierunki reformy szkolnictwa wyższego w Polsce. (Strategies and
    directions of reforms of tertiary education in Poland). Theses of a paper at the conference
    Poland 2050 organized by the Committee of Future Studies "Poland 2000 Plus" at the
    Presidium of P.Ac.Sc., Jabłonna, January 2010 (2011)

Bard, A., Söderqvist, J.: Netocracy — The New Power Elite and Life After Capitalism, Reuters/ Pearsall UK. Polish translation (2006) Netokracja. Nowa elita władzy i życie po kapitalizmie. Wydawnictwa Akademickie i Profesjonalne, Warsaw (2002)

Boyle, J.: The Public Domain. Yale University Press, New Haven (2008)

Braudel, F.: Civilisation Matérielle, Economie et Capitalisme. XV-XVIII siècle, Armand Colin, Paris (1979)

Bruntland, G. (ed.): Our Common Future. The World Commission on Environment and Development, Oxford University Press, Oxford (1987)

Cellary, W.: Zasoby wiedzy dobrem ekonomicznym w społeczeństwie wiedzy (Knowledge resources as economic good in knowledge society). Przyszłość: Świat, Europa, Polska, nr. 1/ 2011 (2011)

Davies, P.: The Eerie Silence: Renewing Our Search for Alien Intelligence. Harcourt, Houghton Mifflin (2010)

Drucker, P.F.: Post-Capitalist Society. Butterworth Heinemann, Oxford UK (1993)

Foucault, M.: The Order of Things: An Archeology of Human Sciences. Routledge, New York (1972)

Galwas, B.: Edukacja w przyszłości i przyszłość edukacji (Education in Future and Future of Education). In: Kleer, J., Wierzbicki, A.P., Galwas, B., Kuźnicki, L. (eds.) Wyzwania przyszłości – szanse i zagrożenia (Challenges and Threats of Future). Committee of Future Studies "Poland 2000 Plus" at the Presidium of P.Ac.Sc., Warsaw (2009)

Galwas, B.: System edukacji najwyższym priorytetem Polski (System of Education as the Highest Priority for Poland). Theses of the paper for the conference Poland 2050. Committee of Future Studies "Poland 2000 Plus" at the Presidium of P.Ac.Sc., Jabłonna, January 2010 (2011)

Garreau, J.: Radical Evolution Random House. Polish translation (2008) Radykalna ewolucja. Pruszyński i Ska, Warszawa (2005)

Grzegorek, J.: Dane i projekcje statystyczne o rozwoju cywilizacyjnym Polski (Data and Statistical Projections for the Civilisation Development of Poland). Ekspertyzy (Expertises). Committee of Future Studies "Poland 2000 Plus" at the Presidium of P.Ac.Sc., Warsaw (2011)

Grzegorek, J., Wierzbicki, A.P.: New Statistical Approaches in the Systemic Analysis of Regional, Intra-regional and Cross-Regional Factors of Information Society and Economic Development. Mazowsze, Studia Regionalne 3:117–128 (2009)

Heidegger, M.: Die Technik und die Kehre. In: Heidegger, M. (ed.) Vorträge und Aufsätze. Günther Neske Verlag, Pfullingen (1954)

Jackson, M.C.: Systems Approaches to Management. Kluwer Academic – Plenum Publishers, New York (2000)

Jarosz, M.: Wykluczenie w polskim społeczeństwie (Exclusion in Polish Society). Theses of the paper for the conference Poland 2050. Committee of Future Studies "Poland 2000 Plus" at the Presidium of P.Ac.Sc., Jabłonna, January 2010 (2011)

Kameoka, A., Wierzbicki, A.P.: A Vision of New Era of Knowledge Civilization. Ith World Congress of IFSR, Kobe (2005)

Kleer, J.: Przyszłość zniewolona przez przeszłość (Future Enslaved by Past). In: Kleer, J., Galwas, B., Wierzbicki, A.P. (eds.) Rola nauki w myśleniu o przyszłości (The Role of Science in Thinking about Future), pp. 347–369. Komitet Prognoz "Polska 2000 Plus", Warsaw (2009)

Lessig, L.: Free Culture: the Nature and Future of Creativity. Penguin Books, London. Polish translation (2005) Wolna kultura, Wydawnictwa Szkolne i Pedagogiczne, Warsaw (2004)

Lietaer, B., Arnspenger, C., Goerner, S., Brunnhuber, S.: Money and Sustainability: The Missing Link. A report from the Club of Rome – EU Chapter, Finance Watcgh and the World Business Academy, Triarchy Press (2013)

Meadows, D.H., Meadows, D.L., Randers, J., Behrens, W.W.: The Limits to Growth. Earth Island, London. Polish translation (1973) Granice wzrostu, PWE, Warsaw (1972)

Niezgódka, M.: Projekt DIR jako przykład praktycznej realizacji idei Open Access (Project DIR as an Example of Practical Implementation of the Open Access Idea). IV Ogólnopolska Konferencja EBIB Internet w bibliotekach – Open Access (IV Polish EBIB Conference on Internet in Libraries – Open Access). Toruń, 7–8 December 2007 (2007)

Pietrasz, J.: Zaawansowane Technologie Komunikacyjne, W Tym Internetu, Wpływ Wykształcenia Na Ich Stosowanie (Advanced Telecommunication Technologies, Including Internet: Impact of Education on Their Use). Świat, Europa, Polska, Przyszłość (2011)

Roubini, N.: Is Capitalism Doomed? EconoMonitor, August 15 (2011)

Salmon, F.: Recipe for Disaster: The Formula That Killed Wall Street. Wired Magazine 17.03.2009, Tech Biz: IT (2009)

Soros, G.: The Age of Fallibility: Consequences of the War on Terror. PublicAffairs, New York. Polish translation (2006) Nowy, okropny świat: era omylności, Świat Książki, Warsaw (2006)

Stehr, N., Adolf, M.: Konsum zwischen Markt und Moral: Eine soziokulturelle Betrachtung moralisierter Märkte. In: Jansen, S., Schroeter, E., Stehr, N. (red., 2008) Mehrwertiger Kapitalismus. Multidisziplinäre Beiträge zu Formen des Kapitalismus und seiner Kapitalien. VS Verlag für Sozialwissenschaften, Wiesbaden (2008)

Taleb, N.N.: The Black Swan: The Impact of the Highly Improbable. Random House, New York (2007)

Toffler, A., Toffler, H.: The Third Wave. William Morrow, New York (1980)

Wierzbicki, A.P.: Delays in technology development: their impact on the issues of determinism, autonomy and controllability of technology. J. Telecommun. Inform. Technol. 4, 1–12 (2008)

Wierzbicki, A.P.: Wizja roku 2050 a dynamika zmian kapitalizmu (A Vision of the Year 2050 and the Dynamics of Change of Capitalism). In: Kleer, J., Mączyńska, E., Wierzbicki, A.P. (eds.) Co ekonomiści myślą o przyszłości (What Economists Think About Future). Committee of Future Studies "Poland 2000 Plus" at the Presidium of P.Ac.Sc. and Polish Economic Society, Warsaw (2009)

Wierzbicki, A.P.: Konsekwencje popytu na wiedzę: Przestrzeń twórcza i mikro-modele kreowania wiedzy (The Consequences of the Demand for Knowledge: Creative Space and Micro-Models of Creating Knowledge. A lecture for Doctoral Studies at Institute of Systems Studies of P.Ac.Sc., September 2011 (2011)

Wierzbicki, A.P.: Wyzwania i zagrożenia przyszłości; aspekty metodologiczne i interdyscyplinarne (Challenges and Threats of Future; Methodological and Interdisciplinary Aspects). In: Kleer, J., Wierzbicki, A.P., Galwas, B., Kuźnicki, L. (eds.) Wyzwania przyszłości—szanse i zagrożenia (Future Challenges – Chances and Threats). Committee of Future Studies "Poland 2000 Plus" at the Presidium of P.Ac.Sc., Warsaw (2010)

Wilkinson, R., Pickett, K.: The Spirit Level: Why More Equal Societies Almost Allways Do Better. Allen Lane, London (2009)

Part IV
Closing

Chapter 15
Final Conclusions

The essential title of this book, *Techne$_n$*, refers to the conviction of the author that technology develops in a punctuated, evolutionary manner in subsequent civilization eras, as well as to the definition of *technology proper* as *techne, the art of creating tools* characteristic of a given civilization era. *Technology proper* should be distinguished from *technical science,* also from the tools or artefacts, *technology products* it creates, and from the *socio-economic processes and systems of producing and utilizing technology products.* The classical philosophy of technology, starting with the fundamental work of Martin Heidegger *Die Technik und die Kehre,* does not differentiate these meanings to a sufficiently precise degree (this might result also from the ambivalent nature of the English word "technology" which comprises all these meanings) and usually neglects technology proper from which all technological ideas originate. When using the word "technology" in such a broad meaning, the philosophy of technology can prove whatever it wants; but in order to draw correct conclusions about socio-economic policies towards science and technology, it is necessary to differentiate meanings of this word more precisely.

The book has four parts. The first part contains some basic epistemological observations: after a general introduction, the book starts with the question what technology is, then it describes the delays and dynamics of the development processes of technology, then a rational and partly technical but evolutionary theory of intuition, further on the problem area of objectivity and truth, including an emergent, new episteme (in the sense given to this concept by Michel Foucault, but not treated ex post, rather ex ante) based on fundamental naturalism.

The second part describes selected elements of the recent history of information technologies, starting with the history of telecommunications, with a selection of its most important elements, but with a stress on their social and, especially, conceptual importance. The history of automatic control, robotics and analog computers is treated in a similar manner, and later the history of digital computers and transistors together with integrated circuits, then the history of systems theory and technology. Finally the history of the informational revolution.

© Springer International Publishing Switzerland 2015   299
A.P. Wierzbicki, *Techne$_n$: Elements of Recent History of Information Technologies with Epistemological Conclusions,* Intelligent Systems Reference Library 71,
DOI 10.1007/978-3-319-09033-7_15

   The third part contains epistemological and general conclusions, starting with
micro-models of knowledge creation processes and the concept of *creative space,*
then the relation of history and philosophy of technology, the issue of assessment
of the future technology development, challenges and threats brought by a new
civilization era, finally summarizing conclusions.
   The fourth part are these final conclusions.
   The book is written from a cognitive perspective of technology, with a conscious acknowledgement, and discussion, of its differences in relation to the
cognitive perspective of (hard) science or the perspective of social sciences and
humanities. The historical process of fragmentation of the cognitive perspective of
modernism or rather the episteme of the industrial civilization era, is also discussed here. One of main theses of the book is that the informational revolution
will most probably result in formation of a new episteme (essentially different
from the postmodern perspective that was only a nihilistic fashion of the end of a
civilization era), and that the recent history of informational technologies contributed to formation of many new concepts that constitute a foundation, a kind of
conceptual platform, for this new episteme.
   We are living in times of informational revolution. Diverse problems of the end
of a civilization era, the speed of change, the destruction and disintegration of the
old episteme in three different directions (cultural spheres of hard sciences,
technology and technical sciences, and social sciences with humanities) result in a
conceptual confusion, difficulties in reciprocal understanding of the representatives
of various cultural spheres, or even in understanding of transformations of our
world. Thus, a new episteme is necessary.
   In this book, I tried to support a better understanding of these changes,
assuming that an important contribution to them stems from informational technologies. This contribution is only seemingly limited to instrumental aspects;
actually, to a large degree it was a conceptual contribution. Therefore, I started the
book with epistemic observations about some basic concepts related both to philosophy of science and to philosophy of technology. To such observations,
described in **the first part of the book**, the following observations belong:
   A discussion of understanding of the concept of technology and techne or
technology proper. Contrary to the tradition of the philosophy of technology
which understands the word "technology" ambivalently, the meaning of this word
should be differentiated: techne, technology proper, is different from artefacts or
products of technology, different from the socio-economic system of production of
artefacts, different from the socio-economic system of selling and using these
artefacts. Technology proper is the art of creation of tools, techne, which changes
together with a civilization epoch, but this is caused by the change of the nature of
tools typical for a given civilization epoch.
   As it was noted already by Heidegger (1954), *techne means* creative solving of
practical problems, *a process of extracting truth from multiple possibilities offered*
by nature. Techne is not a direct application of theories offered by hard sciences or
even technical sciences; one can learn techne only through immediate experience
in creating tools. This does not mean that techne does not use the results of

scientific research, but that praxis and theory (of technical or hard sciences) act together in a feedback loop: techne uses scientific results, if they are available for solving a given practical problem; if they are not, techne solves practical problems alone, leading to essentially new inventions that in turn stimulate the development of science. As already noted, there were many historical examples when praxis was ahead of the theory.

The issue of delays in technology development with conclusions concerning the fallibility of the paradigm of philosophy of technology. There are significant delays in the recent history of technology (called here civilization delays) between a technical invention and the beginning of its socio-economic penetration, even greater when counting until its universal socio-economic penetration. Television is a significant example here: first ideas in the end of the 19th century, practical inventions of television camera and receiver in 1923–1928, first media transmission in 1936, but until 1948 (for color TV until 1960) the penetration of TV in households of the USA did not exceed 1 % and has grown to 90 % until 1980 (for color TV until 1990). Thus, we observe a *pure delay* of about 23–35 years and then an inertial delay of about 30–35 years. Such phenomena occur for all more significant inventions in information technologies.

The theory of control of processes with delay says that in such cases *it is necessary to anticipate the future development of such process while knowing its precise trajectory at least on the interval corresponding to the time of pure delay.* This principle is contradictory to the critique of *historicism* by Karl Popper (e.g. Popper 1957): it is true that there are no absolute laws of historical or social processes, but it is possible to create models of such processes (e.g. the phenomenon of delays belongs to such models) and it is necessary to utilize a maximally objective knowledge of history, at least concerning the interval corresponding to the delay time.

This principle is also contradictory to the paradigm of philosophy of technology which, since Heidegger (1954) and Ellul (1964), has held that it is sufficient to observe holistic effects of socio-economic applications of technology from outside, and later to comment them critically. *However, if we wait until a new technical solution passes the stage of market penetration and becomes widely used, and then will be evaluated by the philosophy of technology, then obviously this evaluation will come too late.* Thus, it is thus necessary to abandon the external, holistic approach: in order to influence technology development we should concentrate on new ideas that did not found yet a broad socio-economic utilization. *Either the philosophy of technology will stick to its tradition of not becoming involved in the specialized details of technology development, and thus it will continue to see in technology a dark, uncontrollable force, or it will decide to cooperate in controlling technology, but then it must get an insight into details of new technical ideas, not yet transferred to mass production.*

An evolutionary and rational, technical justification of the strength and fallibility of intuition, with resulting multimedia principle. Intuition has fascinated philosophy for a long time, but it was treated as a transcendental phenomenon, in a sense super-natural, and as an infallible cognitive tool; but all examples of ostensible infallibility of intuition turned out to be fallible. A rational, evolutionary and technically substantiated theory of intuition (substantiated by the theory and praxis of information transmission and by computational complexity of its processing) proves that intuition can be treated as a fully natural ability of people, possibly common with animals, distinct from rational cognition by its much greater power, but not always leading to infallible conclusions.

Intuitive cognitive abilities were used already by anthropoid apes and separated in humans as a byproduct of evolutionary development of speech. Speech is a powerful evolutionary shortcut since it simplified reasoning at least ten thousand times as compared to immanent perception and intuitive reasoning; thus, speech uses correspondingly less than 0.01 % neurons in human brain. This explains also a fact known in biology that great apes have approximately the same number of neurons in their brains as humans: after the development of speech human brain became excessive, biological evolution was replaced by civilization evolution, and the excess of neurons in human brains became a tool of creative imagination, as well as of diverse metaphysical and transcendental speculations. To a large degree, technical creativity is also based on intuition, e.g. on visual reasoning.

From this theory of intuition we can infer the *multimedia principle: words are only a simplified code serving to describe much more complex reality, while visual and generally preverbal information is much more powerful,* is related to imma-nent perception, intuitive reasoning and knowledge; future records of intellectual heritage of humanity will have multimedia character that stimulates creativity.

A technological perspective towards the problems of metaphysics, truth, objectivity, evolutionary knowledge creation. Because of the power of intuitive reasoning, metaphysics, understood here as intuitive interdisciplinary reflection, is a necessary complement to the results of all specific disciplinary sciences, including disciplinary philosophy. Because of the fallibility of intuitive reasoning, this does not mean at all that metaphysical reasoning is infallible: for example, in the classical issue of existence and being, mathematics of the 20th century has shown (Banach 1932) that the existence of an entity depends on the domain (space) in which the entity is analyzed. Thus the ideas and forms of Plato of course exist if we only define the domain of their existence as books and minds of philosophers.

Metaphysics can be also understood from the perspective of evolutionism, or *fundamental naturalism* that is consistent with evolutionism and assumes that *humans are not lords of nature, that they are only its parts,* and other parts of nature can be treated as subjects of cognition equally well as humans.

The power of intuition suggests also that all processes of knowledge creation, even if considered in their social aspects, might have a circular, spiral-like char-acter of positive feedback between transcendence and existence, intuitive cogni-tion and creative imagination versus practical verification of knowledge. This circularity is not a paradox, as it was erroneously suggested by many sceptical philosophers (using in their suggestions ostensible paradoxes of vicious circle, self-supporting phenomena, the so-called "hydra" of infinite regress). In a model of such evolutionary knowledge creation process, presented in Chap. 6, a fundamental role is played by the values of objectivity and truth. These values are evolutionary necessary in order to transfer to next generations of humans knowledge that is as objective as possible and will thus help them in facing the uncertainty of future threats. This argument is similar to one used by Rawls (1971) in his theory of justice. Therefore truth and objectivity, similarly as justice, even if not absolutely attainable, are nevertheless higher level values that are irreducible to the values of the lower level such as power or money.

To such values relates also the need of development of a new episteme, common for technical sciences, hard sciences, social sciences and humanities, although such a development will not be easy after the disintegration of the old episteme of industrial civilization during the second half of the 20th century. Chapter 6 presents an outline of such episteme, proposed from the perspective of technology.

In the short review of recent history of informational technologies contained in **the second part of the book**, I concentrated on the importance of new concepts contributed by these technologies. These concepts include:

The concept of feedback. This concept originated in various fields of technology: the first bigger "application" in mechanics is attributed to James Watt around 1760, the definition and name of this concept arose in telecommunications, given by Harry Black 1928, the related theory developed in telecommunications and automatic control soon after Black, but later than seeming "applications", these were actually inventions. Today, this concept is universally used also in computers (positive feedback, units of memory) as well as in robotics and automatic control (negative feedback, e.g. the control of movements of robotic arms). It has a revolutionary importance for understanding of dynamic cause and effect relations, since it proves that the philosophical paradoxes of vicious circle, self-reference and the so-called hydra of infinite regress are only ostensible paradoxes, and not actual ones. The sociology of science contributes this concept erroneously to Wiener (1948) or Forrester (1961).

The concept of network and network relation. This concept originated in telecommunications, in praxis from the first telegraphic and telephone networks in the 19th century (1837–1878), and in theory much later, as from Carson (1926) and Bode (1945). Much later, towards the end of the 20th century, the concept was appropriated by sociology while maintaining that Durkheim (e.g., 1895) and Ferdynand Tönies used it; but these authors analyzed the *structure* of social relations without using the concept of a *network*. For the understanding of contemporary world and society, called *networked* by Castells (2000), this concept is obviously essential.

The concepts of informational and computational complexity and cognitive limitations resulting from complexity. These concepts originated in telecommunications (information theory by Shannon (1948)) and in computer science, both theoretical (theory of computational complexity) and applied (experiences of mathematical modeling and applied computations). Along with the related experiences, they are especially important for understanding of the limitations of our cognitive abilities: even if we knew absolutely precise laws of nature or social development, we could not use them, because our computers and minds cannot solve (in a reasonable time span) more complex problems in which computational complexity grows exponentially together with the dimension of the problem (or amount of information processed). Therefore, *the use of simplified models is necessary not because we do not know more precise models, but in order to be able to rationally draw conclusions*. This is an essential limitation of our cognitive tools, either our minds, or computers and other instruments. Thus, the cognitive abilities of humans are limited not only through their subjective features, but also (and primarily) through the imperfection of tools, including most advanced computers and their software, used in cognitive processes. It seems, however, that humans as cognitive subjects always surpassed (because of huge redundancy of our brains, see comments on intuition) currently available tools, and always tried to improve these tools.

The concepts of deterministic and probabilistic chaos and order emerging out of chaos, resulting in the emergence principle. These concepts originated in mathematics (e.g. Poincare 1890) and its applications, technical (e.g. Van der Pool and van der Mark 1927), meteorological (Lorenz 1963), and financial (Mandelbrot 1963). Again, a decisive factor was the possibility of computerized simulation and analysis of sufficiently complex models. In 1963, as a result of such analysis, Edward Lorenz dared to use the name of *deterministic chaos* to denote observed (simulated, thus virtual) phenomena. Much earlier, however, technical "applications" forestalled the theory: according to the testimony of von Neumann (1951), pseudo-random number generators were used in computers before 1951. Models of nonlinear process dynamics often generate chaotic processes and new order can emerge out of chaos, which was used in technology even before 1963. Later, Ilia Prigogine (e.g. Prigogine and Stengers 1984) showed the possibility of order emerging out of probabilistic chaos. Beside such rational justifications, independent justifications emerged, evolutionary in biology, pragmatic in informational technologies, for a general principle of metaphysical nature, called the *emergence principle* (Wierzbicki and Nakamori 2006): *sufficient complexity of processes or systems can result in emergence of new features of the whole, irreducible to properties of its parts.*

The concept of software independent of hardware as a basic example of emergence. An important civilization phenomenon was the process of spontaneous separation, in practical developments of computer scientists in the years 1950–1970, from the works of Hopper (1952) to the UNIX system (Thompson and Ritchie 1969) of software as independent of hardware. This is a fundamental example of the principle that *an irreducible complexity* (because software is not reducible to hardware, although it cannot function without it) *can spontaneously emerge in the process of evolution*, in this case civilization and technological evolution. We can also find other examples of civilization emergence, but this single example shows not only rationality, but also historical reality of the emergence principle.

The concept of logical pluralism and the resulting question of adequacy of the logics used. The theory of multivalued logics originated in Poland, from Łukasiewicz (1911), but for many years it remained a theoretical curiosity, until the possibility of its practical applications emerged due to computers. Re-discovered by Zadeh (1965) as *fuzzy set theory,* multivalued logics found many technical applications. Zdzisław Pawlak, an engineer and constructor of first Polish computers, developed (1991) *rough set theory* and demonstrated how triple-valued logics (without excluding the middle, with an obvious possibility of a *third way* and thus questioning the *(non)contradiction principle*) results from logical statements on large sets of data when checked in terms of their validity. Engineers started to apply various logics for data processing or in the so-called artificial intelligence, and ascertained in that way that there is no universal logics: *the veracity or rather adequacy of a logics depends on the adequacy of its assumptions for a given application area.* This results in a *logical pluralism,* but it also implies a necessary caution in interpreting all contradictions or paradoxes. It was noted already by Brouwer (1922): if we do not exclude the middle, then the paradoxes of contradictions become comprehensible, and a proof by *reductio ad absurdum* ceases to be valid. A fundamental example of errors that can be made in this way is the attachment of a large part of philosophy to an argumentation based on the paradoxes of vicious circle, self-reference, the "hydra" of infinite regress, while all these paradoxes are ostensible (can be rationally explained with the concept of feedback), similarly as the ancient paradox that Achilles would never overtake a turtle.

Other conclusions. Among other conclusions from the overview of recent history of informational technology it is important to stress *many examples showing that technical praxis was, sometimes significantly, ahead of theory and actually stimulated theoretical developments.* This contradicts the view, popular in the philosophy of science, and sometimes even in the philosophy of technology, that technology is an instrumental application of scientific theories. This does not mean, however, that technical praxis determines the development of scientific theory; there is a third way, namely that *praxis and theory, technology and hard science support each other in a positive feedback loop.* Another important general conclusion already discussed in more detail concerns many historical examples of significant delays between a technical invention and its broad socio-economic application.

The third part of the book is focused on a more detailed epistemic and general conclusions. They include the following:

The issue of creation of knowledge for the current needs and micro-models of knowledge creation. The last twenty years of the 20th century brought an immense increase of the role of knowledge as a fundamental productive resource. This caused in turn a demand for a better understanding of the processes of creation of knowledge for the current needs, in a short-term perspective. In response to this demand, many theories or rather micro-models of knowledge creation were proposed (as opposed to macro-theories of the philosophy of science in a long historical perspective), for various conditions and assumptions. The majority of such micro-models do not originate from philosophy proper and epistemology, but from other disciplines, such as systemic analysis, especially computerized support for decision making, or the management theory, and especially the theory of knowledge management.

It is significant that such theories and models stress the role of a positive feedback between tacit (preverbal, intuitive and emotive) knowledge and explicit knowledge (rational, expressed in words), and also between an individual creating this knowledge and a group supporting this individual. Therefore, such processes are described by *spirals of knowledge creation,* such as, in the case of knowledge created in organizations, SECI spiral (based on socialization), DCCV spiral (brainstorming), OPEC spiral (goal setting), and in the case of academic knowledge creation, EDIS spiral (debate and discourse), EEIS spiral (experiments), EAIR spiral (hermeneutic interpretation), etc. Many conclusions are significant in this context; for example, an analysis of these spirals suggests an essentially different nature of knowledge creation processes between academic environments and industrial or market organizations, which explains the difficulties and delays in transfer of knowledge between universities and industry. Other conclusions are related to the concept of *Creative Space,* a network model integrating such diversified processes and models of knowledge creation. This model distinguishes three types of knowledge, emotive, intuitive, and rational knowledge, and three social levels, individual, group level, and the level of *intellectual heritage of humanity* (which again has emotive, intuitive and rational parts). It is stressed that all processes of knowledge creation use in some sense the intellectual heritage of humanity, thus current trends aimed at enlargement of the so-called intellectual property rights might result in an excessive privatization of the intellectual heritage of humanity.

A discussion of hermeneutic foundations of the paradigms of philosophy of technology. I think the anti-technological attitude of a large part of the philosophy of technology is very dangerous, since *a wrong diagnosis does not help in finding a medical cure.* Technologists would reject such a diagnosis as a lack of understanding of technology, while social science and humanities have found a scapegoat and do not consider their own responsibility. But both sides should feel responsible. If technology constitutes today a distinct cultural sphere, it is impossible to practice the philosophy of technology without consulting engineers or technical scientists. It is simply too dangerous to allow oneself not to understand technology, when it not only contributes to fundamental changes of lifestyle, but also, if incompetently or irresponsibly used, threatens to destroy the life on Earth. However, humanists will not understand it if their university education does not include several courses in selected disciplines of technology. Hard scientists will continue to see technology as an instrumental application of their theories. All this creates and extremely dangerous situation, and the feeling of danger only intensifies when studying contemporary philosophy of technology.

The philosophy of technology, as opposed e.g. to the philosophy of mathematics, very rarely analyzes the history of technology, and especially its recent history. What it postulates is no more than a study of the history of its own ideas about technology, not the history of technology proper. Meanwhile, the delays in applications of new technology products imply that the philosophy of technology becomes groundless without the knowledge of the recent history of technology. Therefore, it is necessary to abandon the assumption that the philosophy of technology has the right to decide what technology is, and it usually understands under this concept a holistic perspective on the socio-economic system of production and utilization of technology products, or sometimes, in newer approaches, the products of technology as such, but it usually ignores technology proper. This state of affairs is related to a visible reluctance to take into account the opinions of experienced engineers and to analyze the manner in which technology is created. The situation is aggravated by the fact that newer works on the philosophy of technology are dominated by a postmodern perspective that is simply useless for its understanding. The postmodern perspective is based on the belief that all knowledge is relative as a result of social discourse, local, while an engineer understands the relativity and locality of knowledge, but nevertheless, in order to ensure that the tools constructed by her/him would work in possibly broadest conditions, needs knowledge that is as universal and objective as possible.

A discussion of threats and challenges of the new era. The analysis of future challenges should start with the analysis of threats, weaknesses, opportunities and strengths, and end with designing of actions. This is illustrated in Chap. 14 by a short analysis of over a dozen global threats, leading to the identification of four challenges: *the challenge of sustainable development, the challenge of global governance, the challenge of informational revolution and the challenge of biotechnological revolution.* Beside these global challenges, Chap. 14 describes also the threat of civilization split (a growing digital divide and other syndromes of stratification) and increasing civilization distance of Poland when compared to the most developed countries in Europe as well as the resulting challenge of a redefinition of strategic priorities.

One of the main conclusions of the book concerns *actions needed for preventing a self-destruction of human civilization in relation to the avalanche effects of the positive feedback loop between science and technology on the one side and the system of their socio-economic, market utilization on the other.* In order to counteract this threat, it is necessary to accelerate the process of social reflection on the possible ill effects of mass utilization of some products of science and technology. In recent history, it usually took many decades (the so-called civilization delay in the development of technology) before an idea of a novel device translated into a broad utilization of the device. However, it is too late to wait with social reflection until a broad utilization is achieved. Such an acceleration of reflection is not possible without a better education of the entire society in important technical disciplines, an *obligatory education in at least three technological courses, e.g. informatics, robotics with automatic control, and biomedical engineering, in all university specializations, including humanities and social sciences.*

References

Banach, S.: Theorie des operationes lineares. Chelsea Publishing Company, New York (1932)
Bode, H.W.: Network Analysis and Feedback Amplifier Design. Van Nostrand, New York (1945)
Brouwer, L.E.J.: On the significance of the principle of excluded middle in mathematics, especially in function theory. With two addenda and corrigenda. In: van Heijenoort, J. (ed.) (1967) A Source Book in Mathematical Logic, 1879–1931. pp. 334–345. Harvard University Press, Cambridge (1922)
Carson, J.R.: Electric Circuit Theory and the Operational Calculus. McGraw-Hill, New York (1926)
Castells, M.: End of Millenium: the Information Age, vol. 1, 2, 3. Blackwell, Oxford (2000)
Durkheim, E. *Les Règles de la méthode sociologique*. Revue Philosophique. Polish translation (1968) *Zasady metody socjologicznej*, Warszawa, Państwowe Wydawnictwo Naukowe (1895)
Ellul, J.: The Technological Society. Knopf, New York (1964)
Forrester, J.W.: Industrial Dynamics. MIT Press, Cambridge (1961)
Heidegger, M.: Die Technik und die Kehre. In: Heidegger, M. (ed.) Vorträge und Aufsätze. Günther Neske Verlag, Pfullingen (1954)
Hopper, G.: The Education of a Computer. In: Proceedings of the Association for Computing Machinery Conference, Pittsburgh, May 1952
Lorenz, E.: Deterministic nonperiodic flow. J. Atmos. Sci. **20**, 130–141 (1963)
Łukasiewicz, J.: O wartościach logicznych (On logical values). Ruch Filozoficzny **I**, 50–59 (1911)
Mandelbrot, B.B.: The variation of certain speculative prices. J. Bus. **36**, 394–419 (1963)
Poincaré, J.H.: Sur le problème des trois corps et les équations de la dynamique. Divergence des séries de M. Lindstedt. Acta Mathematica **13**, 1–270 (1890)
Popper, K.R.: The Poverty of Historicism. Routledge, London (1957). Polish translation (3rd corrected edn.) Nędza historycyzmu. PWN, Warsaw (1999)
Prigogine, I., Stengers, I.: Order Out of Chaos. Bantam, New York (1984)
Rawls, J.: A Theory of Justice. Belknap Press, Cambridge (1971)
Shannon, C.: Mathematical theory of communication. Bell Syst. Tech. J. **27**, 376–405 (1948)
Thompson, K., Ritchie, D.: UNIX System: History and Timeline. http://www.unix.org/what_is_unix/history_timeline.html. (1969)
Van der Pol, B., van der Mark, J.: Frequency demultiplication. Nature **120**, 363–364 (1927)
von Neumann, J.: Various techniques used in connection with random digits. In: Householder, A.S., Forsythe, G.E., Germond, H.H. (eds.) Monte Carlo Method, National Bureau of Standards Applied Mathematics Series, vol. 12, pp. 36–38. U.S. Government Printing Office, Washington (1951)
Wiener, N.: Cybernetics or Control and Communication in the Animal and the Machine. MIT Press, Cambridge (1948)
Wierzbicki, A.P., Nakamori, Y.: Creative Space: Models of Creative Processes for the Knowledge Civilization Age. Springer, Berlin (2006)
Zadeh, L.: Fuzzy Sets. Inf. Control **8**, 338–353 (1965)

Annex
My Experiences and Convictions

Referring to my experiences (as in the Introduction) I should present myself in more detail. I obtained the degree of a master in telecommunication engineering in 1960 at the Faculty of Telecommunications of the Warsaw University of Technology, specialization of automatic control. Already during my studies I worked at various posts, from 1959 as a technician in the Institute of Electrotechnology in Międzylesie near[1] Warsaw; as a part of my duties, I travelled to Polish sugar factories and implemented measurement and control systems in them. After graduation, I went for a yearly stay at the Technische Hochschule Darmstadt in FRG. I made my first invention there, patented both in FRG and later in Poland, concerning the use of a strong nonlinearity in a system with feedback to achieve desired if counter-intuitive properties of a device called three-point stepwise controller. After return to Poland in 1961, I started working as an assistant professor in the Chair of Automatic Control at the Warsaw University of Technology. I specialized further in the analysis of strongly nonlinear dynamic systems with feedback.

In 1964, I defended a doctoral thesis in this field, and afterwards I patented several inventions. After that I slightly changed my research theme, by concentrating on the problems of optimal control and sensitivity of dynamic systems. In this field, I have written and defended my habilitation thesis in 1968. I continued to try to use my results in praxis; I supervised, inter alia, an industrial production of a controller according to my invention. Several other patents concerned the construction of control systems for electric arc steel furnaces that I developed together with my colleagues and implemented in Huta Stalowa Wola.

Then I became interested in the problem of optimal control of dynamic systems with pure delay. Pure delay is an important concept that I discuss in several chapters of this book; but in that time, I have written first of my purely mathematical papers, presenting the proof of a variant of the maximum principle (a

---

[1] Today it is a part of Warsaw. The concept of "technician" I used here in its narrow meaning as a lower-level working post supporting an engineer.

© Springer International Publishing Switzerland 2015
A.P. Wierzbicki, *Techne_n: Elements of Recent History of Information Technologies with Epistemological Conclusions*, Intelligent Systems Reference Library 71,
DOI 10.1007/978-3-319-09033-7

necessary condition of the optimality of control of dynamic systems) for processes with pure delay of control. In connection to these works, I went for a yearly stay at the Faculty of Electrical Engineering in University of Minnesota, Minneapolis, and at the Division of Applied Mathematics in Brown University, Rhode Island, USA. After my return in 1971, the Warsaw University of Technology started a longstanding cooperation with the University of Minnesota. I realized then that in the world of science, the most important are publications, theoretical results, and started to concentrate on them, even if I already had many inventions and patents in my records as well as numerous experiences in industrial implementations.

I started to work intensively on the theory of sensitivity analysis and in 1977, I published my first own monograph (before that I published several co-authored books and chapters in other books) devoted to the sensitivity analysis of optimal control systems. In that book I addressed essentially epistemological problems, concerning social construction of knowledge when designing control systems and the role of objectivity necessary for a technologist despite the appreciation of relativity of knowledge and uncertainty of cognition. Already in that book I came to a conclusion that all human knowledge (including the so-called laws of nature) is composed of more or less precise models, created by people and gradually improved. Moreover, in applications we cannot use entire knowledge, we always select a part of available knowledge appropriately simplified and approximated, because otherwise we could not test resulting conclusions with our computations or computer simulations, since we encounter the phenomenon called in computational techniques the curse of dimensionality (exponential computational complexity). This essentially approximated and uncertain nature of our knowledge does not imply, however, for many reasons, discussed in various chapters of this book, that we should not strive for knowledge as objective as possible (even if absolute objectivity is not attainable).

In the seventies of the 20th century I became first a deputy dean, then the dean of the Faculty of Electronics (former Faculty of Telecommunications) of the Warsaw University of Technology; this was a start of my experiences in management of science. In 1976, I obtained the title of a full professor. The experiences in management caused my interest in the versatility of human motivations, hence in theoretical works I became interested in the multiple criteria optimization and the multiattribute decision making theory.

In 1978, the director of International Institute of Applied Systems Analysis (IIASA) in Laxenburg near Vienna offered me a research position in that Institute. I resigned from trying to get re-election as the dean of the Faculty of Electronics for the next term and for a year I could really work creatively. I originated then the so-called method of reference point in multiple criteria optimization and decision analysis, until today broadly used in IIASA and in the world. This method stresses the sovereignty of a human user working with a computerized decision support system, in the assumption that classical decision theory puts too much emphasis on mathematical modelling of human user's preferences. However, after a year I received a proposal to become the chairman of the theoretical division of IIASA, the Systems and Decision Sciences Program. I worked at this post for almost six years, until 1984, learning much about management of international research programmes.

In 1985, I returned to Poland to the post of a professor at the Warsaw University of Technology. In between, I worked for three months teaching at doctoral studies at the Fernuniverisität Hagen in FRG; I had already quite an experience in teaching in such courses, since I lectured at doctoral studies at the Mining and Steelmaking Academy in Cracow, at the University of Minnesota, at the Technische Hochschule Ilmenau in GDR, at the Faculty of Mathematics in the University of Warsaw. My mathematical colleagues have persuaded me that I could be counted as a mathematician since a sufficient condition for that is a publication of a scientific paper in a mathematical journal, a condition that I fulfilled several times. After return, I introduced the first lecture in Poland (1986) on the art and science of negotiations.

In 1989–1990, upon an invitation from the University of Kyoto, I stayed there a year working mostly in research, with minimal teaching duties, in the Kyoto Institute of Economic Research. I worked with economists before in IIASA and observed the dissimilarity of their episteme. I was interested in the mathematical game theory and succeeded in broadening the theory of multiple criteria games. However, I also became interested in Kyoto in the problem of a rational, technical and naturalistic explanation of the phenomenon of intuition, that can be fallible but is undoubtedly a powerful source of technical and intellectual ideas.

Upon return to Poland in 1991, I was elected a member of the Committee of Scientific Research of Poland (CSR, a governing body of the Ministry of Science) that had the task of reforming the scientific system after the turn towards democracy in 1989. The work on reforming the scientific system gave me many experiences (some of them worth interesting anecdotes[2]). I was also elected the chairman of one of the two Commissions of the Committee, the Commission of Applied Research; I succeeded, among other things, in an acceleration of the development of computer networks for the needs of Polish science, for which I had later obtained an award of Thomas Hofmokl.

In the years 1994–1996, I was free of management duties and I could concentrate on a further development of my approach to the theory of powerful but nevertheless fallible intuition. I retained contacts with CRS, inter alia going to Brussels as an expert of CRS in negotiations concerning Polish preparations to the accession to the European Union. In 1996, scientific management caught me again; I became the director general of the National Institute of Telecommunications in

---

[2] For example, CSR sent me to Brussels with the task of checking at a great exposition of inventions, Eureka, the validity of complains of some Polish inventors that the evaluation of their inventions was not quite objective. Judging inventions is of course a rather subjective process, but I succeeded in solving the problem, I became also a member of the jury of this exposition and the objectivity of evaluations was slightly improved; later, I was awarded for this work an Officer Croix of Belgian Royal Commendation for Inventors. However, for me the experiences gained in this work indicate clearly that inventions are not fabricated, they have rather an accidental, artistic character. There are many other detailed anecdotes related to these experiences.

Miedzeszyn (on the outskirts of Warsaw). I served in this capacity until 2004, when I decided, partly because of health reasons, to retire. At this time, I resigned also from my function as a part-time professor at the Warsaw University of Technology and from many other functions in Polish science (I was also the chairman of a consulting team for international scientific cooperation at the Ministry of Science and the chairman of two scientific councils, of the Industrial Institute of Control and Measurements and of the Scientific Academic Computer Network). I only kept my functions in the Committee of Future Studies "Poland 2000 Plus" at the Presidium of the Polish Academy of Science. I worked in the Committee since 1986 in relation to my interests in future studies and assessment of future technology development, resulting from my work at IIASA. I am still an editor of the journal of the Committee entitled *Future, World, Europe, Poland.* From this work comes also my conviction that epistemology should try to look forward in time, not limiting itself to purely historical studies.

After retirement I went for three years on a purely research position to the School of Knowledge Science in the Japan Advanced Institute of Science and Technology (JAIST) in Nomi, close to Kanazawa. Together with Yoshiteru Nakamori we researched new micro-models of knowledge creation for the current and future needs of knowledge-based economy (as opposed to philosophic macro-theories of creation and substantiation knowledge in a long-term historical perspective, see Chap. 12). We published two books on this theme. I used my experience in management of science and technology; for this work we obtained also an award for the best paper at the Hawaii International Conference on Systems Sciences.

Although in retirement, I still work at the National Institute of Telecommunications as a research professor (the writing of this book was supported also by the Institute). Despite falling health, I try to keep international contacts, for example, I was recently an international auditor of Helsinki universities combined into a new Aalto University. However, when reading books on the philosophy of technology I feel a serious concern (the Heideggerian *Sorge*) about the future of understanding of technology by humanists and social scientists.

Index

A

Abduction (aha, eureka, illumination, enlightenment), 67, 89, 94, 109, 220, 229, 233
Abraham, R. H., 181, 312
Adolf, M., 276, 281, 296
Agamben, G., 24, 312
Agassi, J., 253
Airy, G. B., 144, 149, 312
Allen, P., 165
Ambient intelligence, 20, 51, 133, 165, 193
Analog computer (differential analyzer), 10, 26, 106, 107, 137, 139, 140, 142, 144, 149, 151, 155, 160, 161, 163, 181, 191, 195, 196, 299
Anscombe, G. E. M., 115
Antenna, 122, 129, 130
Armstrong, E. H., 65, 66
Armstrong, M., 66
ARPANET, 164, 189, 193, 194
Aristotle, 89, 176, 186, 213
Arthur, W. B., 31, 49, 63, 73, 158, 194, 198
Asimov, I., 141, 148
Åström, K. J., 146, 155
Asymmetry of information, 274
Auleytner, J., 285, 294
Automatic control, control engineering, 10, 26, 45, 57, 70, 107, 123–125, 129, 137–151, 176, 180, 200, 299, 303, 307
Automation, 19, 34, 37, 52, 67, 125, 132, 141, 142, 144, 146, 147, 149, 158, 165, 190, 205, 207, 208, 273, 284, 286
Autonomy of technology, 44, 57, 58
Autopoiesis, 183

B

Babbage, B.H., , 157
Backus, J., 160

Bacon, R., 260
Bailey, F.N., 146
Balcerowicz, L., 8
Banach, Cz., 289, 302
Banach, S., 91, 107, 302
Bangemann, M., 191
Baran, P., 164, 194
Bard, A., 2, 15, 16, 273
Bardeen, J., 162
Barnes, B., 38
Behrens, W.W., 271, 280
Bell, A.G., 120, 122
Bell, D., 16
Bellert, S., 7
Bellman, R., 145
Benacerraf, P., 79
Bennet, S., 137, 149
Bergson, H., 80, 81
Berkowitz, R.D., 125
Berners-Lee, T., 59, 195
Bertallanfy, L., 140, 177–178
Biomedical engineering, 54, 129–132, 192–193, 219, 256, 265, 285–287, 307
Black, H.S., 138, 139, 149, 303
Blandzi, S., 98, 186
Bloor, D., 38
Bode, H.W., 124, 303
Bogdanov, A., 175, 178
Boltyansky, V.G., 145
Boole, G., 157–158, 212–213
Boone, G., 162
Boulding, K., 177
Boyle, J., 283
Braid, J.L., 121
Braille, L., 119
Brainstorming, 6, 220, 221, 234, 237
Brakefield, T., 89
Brattain, W.H., 162

318   Index

Printed in the United States
By Bookmasters